# GEOGRAPHY, SCIENCE AND NATIONAL IDENTITY

*Scotland since 1520*

Charles Withers' book brings together work on the history of geography and the history of science with extensive archival analysis to explore how geographical knowledge has been used to shape an understanding of the nation. Using Scotland as an exemplar, the author places geographical knowledge in its wider intellectual context to afford insights into perspectives of empire, national identity and the geographies of science. In so doing, he advances a new area of geographical enquiry, the historical geography of geographical knowledge, and demonstrates how and why different forms of geographical knowledge have been used in the past to constitute national identity, and where those forms were constructed and received. This book will make an important contribution to the study of nationhood and empire and will therefore interest historians as well as students of historical geography and historians of science. It is theoretically engaging, empirically rich and beautifully illustrated.

CHARLES W. J. WITHERS is Professor of Historical Geography at the University of Edinburgh. Recent publications include *Geography and Enlightenment* (1999), co-edited with David Livingstone.

*Cambridge Studies in Historical Geography* 33

---

*Series editors:*
ALAN R. H. BAKER, RICHARD DENNIS, DERYCK HOLDSWORTH

---

**Cambridge Studies in Historical Geography** encourages exploration of the philosophies, methodologies and techniques of historical geography and publishes the results of new research within all branches of the subject. It endeavours to secure the marriage of traditional scholarship with innovative approaches to problems and to sources, aiming in this way to provide a focus for the discipline and to contribute towards its development. The series is an international forum for publication in historical geography which also promotes contact with workers in cognate disciplines.

*For a full list of titles in the series, please see end of book.*

# GEOGRAPHY, SCIENCE AND NATIONAL IDENTITY

## Scotland since 1520

CHARLES W. J. WITHERS

*University of Edinburgh*

CAMBRIDGE UNIVERSITY PRESS
Cambridge, New York, Melbourne, Madrid, Cape Town, Singapore, São Paulo

Cambridge University Press
The Edinburgh Building, Cambridge CB2 2RU, UK

Published in the United States of America by Cambridge University Press, New York

www.cambridge.org
Information on this title: www.cambridge.org/9780521642026



First published 2001
This digitally printed first paperback version 2006

*A catalogue record for this publication is available from the British Library*

*Library of Congress Cataloguing in Publication data*

Withers, Charles W. J.
Geography, science and national identity: Scotland since 1520/Charles W. J. Withers.
p. cm. – (Cambridge studies in historical geography; 33)
Includes bibliographical references and index.
ISBN 0 521 64202 7
1. Scotland – Historical geography. 2. National characteristics, Scottish.
3. Nationalism – Scotland – History. 4. Geography – Scotland – History.
5. Science – Scotland – History. 6. Scotland – Intellectual life. 7. Scotland – History.
I. Title. II. Series.
DA757.7.W58 2001
941.1–dc21 2001025556

ISBN-13 978-0-521-64202-6 hardback
ISBN-10 0-521-64202-7 hardback

ISBN-13 978-0-521-02482-2 paperback
ISBN-10 0-521-02482-X paperback

The Earth does not owe gratitude only to those who create books, unite maps to the art of Geography, and fit lands to sky and sky to lands as with a plumbline; but also to those who persuade, urge, correct and increase these works, promote them with money and expense, strengthen the feeble, recover the lost, and give shape and polish to the deformed, so that the illustrious offspring can be born and reborn and appear with beauty into the light and faces of men.

(Johannes Blaeu, *Atlas Novus,* 1654, from
a translation by I. Cunningham)

# Contents

# Figures

# Tables

# Preface

There has been much interest lately and from several quarters in the histories of geography and geographical knowledge, in the situated social nature of science, and in national identity. Amongst historians of geography, questions have been raised about the nature of geographical knowledge in different historical and geographical contexts. Historians of science have considered the geographical nature of science as well as its social construction. Scholars of national identity have addressed the forms taken in such identity, including the representation of geographical space.

This book is an attempt to bring these literatures together in order to understand the geographical 'making' of Scotland as a matter of historical geography. Simply, I am interested in knowing how geographical knowledge, variously understood, was used to give geographical identity and meaning to Scotland in the past. More exactly what is meant by a historical geography of geographical knowledge will become evident in what follows. It is also worth noting what this book is not. I am not here simply or primarily writing a history either of geography or of geographical knowledge *in* Scotland (although the main elements of such a history are present). This is not a book about the historical geography *of* Scotland in any strictly materialist sense, where the concern is with what the human landscapes of Scotland looked like at various moments in the historic present and with knowing their 'processes of becoming'. I am not taking the questions of 'geography', 'science' or, indeed, 'national identity' as simple immutable givens, and I am certainly not proposing either an 'essential' Scottish geographical style, or national identity as a geographical matter, to be simple and shared.

I am, rather, interested in knowing how the ideas and practices of geographical knowledge were used to constitute a sense of what Scotland was geographically. Because geographical ideas and practices were produced by and for people working in particular places in certain ways in the past, there is, we should suppose, a recoverable historical geography of such endeavours. Thus, this book is about the historical geography of a particular form of intellectual and scientific endeavour – geographical knowledge – and with the places of production of such knowledge, the means used to determine, to disseminate and to represent such knowledge, and with knowing who the audiences were for such knowledge.

As will be clear from what follows, more is known about some of these issues than others, and my debts to certain literatures and conceptual questions is greater than to others. Most books are a collective enterprise in one way or another. This one is no exception. Many scholars in different disciplines have discussed my ideas with me. I am grateful to John Bartholomew, Michael Bravo, Alex Broadie, John Cairns, Felix Driver, Roger Emerson, Richard Finlay, Chris Fleet, Michael Heffernan, Michael Hunter, Jean Jones, Colin McArthur, John MacKenzie, Roger Mason, Graeme Morton, Miles Ogborn, Gillian Rose, Jan Rupp, Domhnall Uilleam Stiubhart, Jeffrey Stone, Margaret Wilkes and Heather Winlow. I am grateful to Peter Jones and the staff of the Institute for Advanced Studies in the Humanities in the University of Edinburgh for allowing me, as a Sabbatical Fellow, the time and space to begin writing this book. Conference audiences in Boston, Cambridge, Edinburgh, Fort Worth, Göttingen, Kingston Ontario, London (England and Ontario), Toronto, Vancouver, Victoria and Winnipeg have all helped me refine my thinking.

Three people in particular, however, must be singled out. This book was in part made possible by a Larger Research Grant from the British Academy. Funded from that grant, Andrew Grout was a wonderful research assistant: without his work, this book would have been very different. David Livingstone and Paul Wood, good friends and critical scholars both, have encouraged, criticised and challenged. They may not agree with what I have done, or with how I have done it, but I am nevertheless deeply indebted to them.

In addition to the British Academy, this book was made possible by research support from the Royal Society, the University of Edinburgh, the Carnegie Trust for the Universities of Scotland and the Ratcliff Fund of the National Library of Scotland. I acknowledge this support with thanks. I owe a huge debt to archivists and to librarians in many institutions: to all the staff at the National Library of Scotland and to Christopher Fleet and Diana Webster in the Map Library in particular, but also to staff in the libraries and special collections of the universities of Aberdeen, Cambridge, Edinburgh, Glasgow, Heriot Watt, Oxford, St Andrews, Strathclyde; in local libraries and archives in Aberdeen, Ayr, Dunbar, Inverness, the Mitchell Library, Glasgow, Perth and Stirling; in the Royal College of Physicians, Edinburgh; in the Royal Botanic Gardens, Edinburgh; to Mary Sampson of the Royal Society; to staff in the Royal Library at Windsor; to the librarian of the Society of Antiquaries of Scotland; and to David Munro, Director of the Royal Scottish Geographical Society, for permission to use that Society's archives.

Acknowledgements for illustrations are given for relevant figures but for their help and advice I would particularly like to thank: Rosalind Marshall, Deborah Hunter and James Holloway of the Scottish National Portrait Gallery; Lesley Ferguson at RCAHMS; The Librarian, Pepys Library, Magdalene College, Cambridge; the staff at the UK Hydrographic Office; Mrs Pam Gilchrist of New College Library, University of Edinburgh; Viscountess Melville and Sotheby's in regard to Figure 4.2; Anne Adams of Kimbell Art Museum, Fort Worth, Texas; Jim McGrath at Strathclyde University Library; and Richard Gillanders, Fergus MacTaggart and Robert Crosby of the British Geological Survey. Anona Lyons drew Figures 2.4, 4.1, 4.4, 5.3 and 6.1 from my rough drafts.

Alan Baker as a Series Editor and Marigold Acland of Cambridge University Press have both been very supportive, and I am grateful to Nancy Hynes for her careful copy-editing. My final thanks must go to my family who have had to live with this project for too long: to them, with love, this book is dedicated.

Charles W. J. Withers
Edinburgh

# 1

# Introduction: geography, science and historical geographies of knowledge

This book is an attempt to understand the connections between geography, science and national identity in a particular geographical and historical context, and, in so doing, to write a historical geography of geographical knowledge. The focus is Scotland between the work of late Renaissance humanists concerned to 'situate' their nation historically and the engagement with geography as a form of identity in the work of Patrick Geddes and others in the early twentieth century.

I use the term 'historical geography of geographical knowledge' to signify two central concerns, elaborated upon in what follows. The first concern is with understanding the ways in which geographical knowledge in the past was used to constitute the 'space' that was Scotland and to shape ideas about the nature of Scotland as a geographical entity. 'Geographical knowledge' is understood here as a particular form of intellectual and scientific enquiry encompassing a variety of practices such as, for example, mapping, writing, picturing and natural historical surveying. In this first sense, then, geography itself as one form of intellectual enquiry – however understood by different people at different times in different places – is treated as part of a wider conception of geographical knowledge, part of a range of discursive practices through which ideas about the nation and national identity were realised. My second concern is to recover the sites and the social spaces in which geographical knowledge was undertaken and to plot the connections between the places of geographical knowledge production and its audiences and makers. Taken together, these two concerns inform the historical geography of geographical knowledge as I employ the term as being about *how* and *why* different forms of geographical knowledge were used in the past to constitute national identity, about *where* those different ideas were made and received and *for whom* they had the meanings they did.

These issues reflect wider interests within geography and other disciplines both in the nature of geographical knowledge and in the situated nature of science and other forms of intellectual endeavour. The 'critical turn' within the history of geography, discussed further in this chapter, has been accompanied by a recognition from other disciplines that geography matters. Historians of science, for example, have studied the situated nature and movement of scientific knowledge and science as a social construction. That they and others have considered the place of scientific knowledge in these terms is a conscious rejection of earlier idealist notions of science as a universal practice derived without reference to the spaces of its production. In so doing, the local meanings of science have been brought into focus. Such interests are apparent, too, in the social sciences and in the humanities.

David Livingstone has noted these issues in discussing what he calls 'the historical geography of ideas'. Taking seriously the geography of knowledge or of scientific practice is not simply a matter of site and location. Spaces of and for knowledge are metaphorical as well as material. Place is an ordering term, a relational position for categories of knowing and for the objects of theoretical enquiry as it is also a site of display, for example, or a site either for knowledge production or for the didactic consumption and reception of theories, practices and natural objects. For Livingstone

> Glimmerings of what a geography of scientific knowledge might amount to are thus indeed beginning to be glimpsed as sociologists and historians of science have begun to probe the role of the spatial setting in the production of experimental knowledge, the significance of the uneven distribution of scientific information, the diffusion tracks along which scientific ideas and their associated instrumental gadgetry migrate, the management of laboratory space, the power relations exhibited in the transmission of scientific lore from specialist space to public space, the political geography and social topography of scientific subcultures, and the institutionalization and policing of the sites in which the reproduction of scientific cultures is effected.

As he further notes, a geography of science 'will need to attend to spatial considerations at a variety of scales. Indeed, it will be one of the key methodological issues of such an undertaking to ascertain just what is the appropriate spatial scale at which to conduct any specific historical investigation, and then to determine how the various scales are to be related.'[1]

For Scotland, something of these issues is apparent in a paper to the British Association for the Advancement of Science in September 1885 where H. A. Webster spoke to the title 'What has been done for the geography of Scotland, and what remains to be done'. Of Scotland, a country 'which has been traversed and retraversed in every possible direction by persons devoted to every department of knowledge, in which every district has been mapped and remapped, in which every county and town and parish has its local guide-book, its local antiquary, its local geologist, its local botanist, surely, you say, every geographical fact must have been recorded and made readily accessible to any

[1] D. Livingstone, 'Science and religion: foreword to the historical geography of an encounter', *Journal of Historical Geography*, 20 (1994), 372.

who feel interest therein'. Webster continued in order to note the contrary: 'I hope to show you that, to many questions which the geographer naturally asks, no answer is forthcoming, and that there are whole departments of geographical investigation at which we have only begun to work in a serious and fruitful manner.'[2]

I want to suggest that a historical geography of geographical knowledge can be the subject of 'serious and fruitful' enquiry. I hope to do so for Scotland with reference to questions concerning the different practices of geographical knowledge, the sites and spaces of geography's making, the audiences for such knowledge and the connections with other forms of national knowing. My concern is not only with the national scale. I consider the nature and making of local knowledge, the role of particular institutions and of individuals, even of single texts, as well as the connections between the making of national knowledge in different local places and at different times.

Before considering such questions, however, let me place these intentions in context by first considering recent writings on the history of geographical knowledge and then examining second, work within the history of science on the social and situated nature of scientific knowledge. The final part returns to the question of a historical geography of geographical knowledge through brief consideration of the historical geography of Darwinism and of modernity.

## New histories of geography and of geographical knowledge

The history of geography was not, until recently, a particularly active or prominent field of geographical enquiry. In most cases, writings upon the history of geography were distinguished by uncritical notions of what geography was, by hagiographic portrayals of the subject's 'great men', and by too little attention to the wider social and intellectual contexts in which geography and geographers worked.

In recent years, however, there has been a notable resurgence of interest in the subject.[3] This has been evident in attention to the following: the discursive nature of geographical knowledge; the genealogy of geography's key concepts; the connections between geography and power; and the fact that much work in the history of geographical knowledge has been undertaken by practitioners of other subjects.

[2] H. A. Webster, 'What has been done for the geography of Scotland, and what remains to be done', *Scottish Geographical Magazine*, 1 (1885), 487.

[3] F. Driver, 'New perspectives on the history and philosophy of geography', *Progress in Human Geography*, 18 (1994), 92–100; 'Visualising geography: a journey to the heart of the discipline', *Progress in Human Geography*, 19 (1995), 123–34; 'Histories of the present: the history and philosophy of geography', *Progress in Human Geography*, 20 (1996), 100–9. For the contemporary context, see E. Baigent, 'Recreating our past: geography and the rewriting of the Dictionary of National Biography', *Transactions of the Institute of British Geographers*, 19 (1994), 225–8; J. Sidaway, 'The production of British geography', *Transactions of the Institute of British Geography*, 22 (1997), 488–504.

*On context and the discursive production of geographical knowledge*

For Felix Driver, writing in 1994, the 'progress' evident in work on the history of geography has been apparent in an engagement with the wider academic literature on the history and philosophy of the social sciences, a willingness to consider geographical knowledge as constituted as much by social relations and technical practices as by ideas and individuals, and by critical reflection on the wider purposes of writing about the history and present condition of geographical discourse.[4] These advances have been reflected in and stimulated by Livingstone's *The geographical tradition* (1992). Livingstone's book has been widely and enthusiastically reviewed by geographers and historians of science alike as a key 'moment' in the new critical histories of geographical knowledge.[5] In its attention to geography's context and its defence of 'situated messiness' – the ways in which geographical knowedge was (and is) both discursively complex *and* intellectually shaped in different places by different people at different times – Livingstone's work, argued Driver, 'set a new agenda for the history of geography'.[6] It has done so, too, not just through Livingstone's thorough-going critique of conventional approaches to the history of geography but in his insistence that we must situate geography historically *and* geographically.

Understanding geographical knowledge as a *situated* concern can mean several things. Geographical knowledge, whatever that term means now or meant in the past, cannot be understood as something set apart from the intellectual, social and political milieux of its time. In this sense, recent work by scholars interested in the history and nature of geographical knowledge has been distinguished by attention both to the personal and political connections that underlay geography's emergence as an institutionalised academic subject in Britain from the later 1880s,[7] and to the connections between geography and leading scientific ideas, such as neo-Lamarckianism and Darwinism.[8]

Matters of context are also epistemological. For David Stoddart, the later eighteenth-century encounter between European explorer-navigators and

[4] Driver, 'New perspectives', 92.

[5] Two fuller discussions of the book are A. Werritty and L. Reid, 'Debating the geographical tradition', *Scottish Geographical Magazine,* 111 (1995), 196–8; and 'Conversations in review', *Ecumene,* 3 (1996), 351–60. Good sees one of the book's strengths as its combination of the history of geographical thought with the history of the social and institutional settings of geography's practitioners: G. Good, *Sciences of the earth: an encyclopedia of events, people and phenomena* (New York), Vol. I, xix.

[6] Driver, 'New perspectives', 92.

[7] D. Stoddart *On geography* (Oxford, 1996); Livingstone, *Geographical tradition,* 177–215; T. Unwin, *The place of geography* (London, 1992), 83–5; F. Driver, 'Geography's empire: histories of geographical knowledge', *Environment and Planning D: Society and Space,* 10 (1992), 23–40.

[8] D. Livingstone, 'Natural theology and neo-Lamarckism: the changing context of nineteenth-century geography in the United States and Great Britain', *Annals of the Association of American Geographers,* 74 (1984), 9–28; *Darwin's forgotten defenders: the encounter between evangelical theology and evolutionary thought* (Edinburgh, 1987); R. Peet, 'The social origins of environmental determinism', *Annals of the Association of American Geographers,* 75 (1985), 309–33.

native 'Others' provided the basis for the emergence of 'modern' geography as a whole. Such encounters took place not just in given geographical contexts such as the Pacific Ocean, North America, sub-Saharan Africa or the Indian sub-continent, but in particular 'scientific' ways: realism in description, systematic classification in collection, and comparative method in explanation.[9] Other means of securing geographical knowledge should be noted: trusting native informants; circulating questionnaires; speculative and essentially Baconian fieldwork designed to gather facts about distant places and the unknown near at home. One must also consider the conjoint interests of politicians and natural philosophers whose concerns demanded the institutionalisation of natural knowledge in order to advance it and, of course, the shipping 'home' of new products (and even the people themselves) to become objects of wonder for different audiences within Europe's centres of geographical and scientific calculation.

Matters of context relate also to the politics of doing geography and for whom questions about the recovery of such knowledge have significance. Much recent geographical enquiry in general has been motivated by a desire to give voice to the hitherto marginalised.[10] Such post-colonial perspectives are mirrored in the concerns of some historians of geography and others to understand exactly how geography was implicated in European imperialism and colonialism. At the same time, critical attention to the history of geographical knowledge has resulted from a concern to consider that history in relation to other intellectual or disciplinary histories, such as the history of science.[11] Yet others have discussed the history of geography and of geography's books as part of new perspectives on the history of (geographical) education.[12]

Sensitivity to historical context relates both to the need to consider the making and meaning of geographical knowledge in its own terms and to question the 'idea of geographical traditions' itself.[13] As David Matless notes:

> To raise, as Livingstone does, issues of geography's earthly situation is to question the boundaries of geographical knowledge. . . . Rather than seek a new and all-inclusive definition of geography, we might perhaps recognize that the discipline of geography has been and is now one *genre* of geographical knowledge among many, and that a crucial part of geography's history consists of disciplinary geography's marking out of itself.

[9] Stoddart, *On geography,* 28–40.

[10] D. Gregory, 'Post-colonialism', in R. J. Johnston, D Gregory, G. Pratt and M. Watts (eds.), *The dictionary of human geography* (Oxford, 2000), 612–15.

[11] D. Livingstone, 'The history of science and the history of geography: interpretations and implications', *History of Science,* 22 (1984), 271–302.

[12] T. Ploszajska, *Geographical education, empire and citizenship: geographical teaching and learning in English schools, 1870–1944* (Historical Geography Research Group Publication Series, no. 35), (Cambridge 1999); F. Driver and A. Maddrell, 'Geographical education and citizenship: introduction', *Journal of Historical Geography,* 22 (1996), 371–2.

[13] F. Driver *et al,* 'Geographical traditions: rethinking the history of geography', *Transactions of the Institute of British Geographers,* 20 (1995), 403–4.

Matless has pointed to the implications of such claims for 'a historical geography of the grand categories of geography – region, space, landscape, geography, etc. – and of the role and make-up of the geographer: a genealogy of geography and of the geographical self'.[14]

For Clive Barnett, the history of geography understood as a matter of genealogy and of historical context is not axiomatically useful because such issues have little to say concerning 'the only context that really matters: the contemporary one', and because 'the new contextual and critical histories of geography tend to assume too easily that all geography in the past is the past of today's geography, sweeping any questions about the nature of the historical relation under the cover of expanded notions like 'geographical discourse' or 'geographical knowledge'.[15]

Arguing that the contemporary context is the *only* context that *really* matters smacks of a surrogate presentism, even of the wholesale dismissal of the past. It is one thing to argue that we ought not straightforwardly to see the history of geography as the Whiggish history of today's geography. There I am sympathetic to Barnett. Yet it is another thing to exclude the possibility of a historical investigation of geography's past *in its own terms*, which he seems to suggest. I would want to argue that we must take *more* seriously the attempt to understand geographical knowledge in the past. As with Gillian Rose,[16] my concern is neither to insist upon a genealogy for geographical knowledge nor to privilege the present, but, rather, to recover its historical *and* geographical context as a question of historical geography. As Driver puts it:

> The contextual approach to the history of geography is thus more concerned with mapping the lateral associations and social relations of geographical knowledge than with constructing a vision of the overall evolution of the modern discipline. It demands a far more historically (and geographically) sensitive approach to the production and consumption of knowledge than that provided by more conventional narrative histories.[17]

The idea of geographical knowledge as a discourse does not just refer to that set of intellectual and scientific practices at any given place or moment held to constitute such knowledge. It includes also the languages, the institutions and the different 'modalities' through which we have come to know the world.[18] To argue that geographical knowledge is discursive is to recognise its

[14] D. Matless, 'Effects of history', *Transactions of the Institute of British Geographers*, 20 (1995), 405–6.

[15] C. Barnett, 'Awakening the dead: who needs the history of geography?', *Transactions of the Institute of British Geographers*, 20 (1995), 417–19.

[16] G. Rose, 'Tradition and paternity: same difference', *Transactions of the Institute of British Geographers*, 20 (1995), 414–16; see also D. Livingstone, 'Geographical traditions', *Transactions of the Institute of British Geographers*, 20 (1995), 420–2.

[17] F. Driver, 'Geography's empire', 35.

[18] My use of the term 'modalities' is drawn from Foucault's 'enunciative modalities', where the social subject that produces a statement is understood not as an entity 'beyond' the discourse in question, but is something formed by the statement itself. M. Foucault, *The archaeology of knowledge* (London, 1972), 95–7.

constitutive power; as Driver and Rose note, 'To argue that geographical knowledge is discursively constructed is to insist on the importance of practices and institutions as well as concepts. Discourses always do their work in specific social contexts and with material consequences'.[19]

Considering discourse as specific representations, practices and performances through which meanings are produced, connected into networks and legitimized has been helpful for what Derek Gregory terms the 'revivified history of geography' in revealing the different ways in which geographical knowledge has been made.[20] It is interesting that this revivified history has been paralleled by a more critical history of (the) map(s), for example, and by an interest in 'mapping' as the processes, literal and figurative, of putting things in place.[21] I take such interest to be part of wider concerns with representation in geographical knowledge, apparent in landscape painting and in photography,[22] and with the attention paid to the socially constructed nature of meaning in post-modern human geography.[23]

Certainly, concerns with context and with discourse have shifted the attention of historians of geographical knowledge away from paradigmatic notions of change,[24] and away from conceptions of 'grand theory' and meta-narrative towards the specific, the theoretical and the situated circumstances constituting the conditions of geography's making.

## *Geography as a form of geographical knowledge before* c.*1800*

The 'marking out of itself' of disciplinary geography has been apparent in studies which, while focused upon different time periods and different countries, have collectively challenged the too-often repeated view that academic

[19] F. Driver and G. Rose, 'Introduction: towards new histories of geographical knowledge', in F. Driver and G. Rose (eds.), *Nature and science: essays in the history of geographical knowledge* (Historical Geography Research Group Publication Series, no. 28) (Cheltenham, 1992), 4.

[20] D. Gregory, 'Discourse', in Johnston, Gregory, Pratt and Watts (eds.), *The dictionary of human geography* (Oxford, 2000), 180–1.

[21] D. Burnett, 'The history of cartography and the history of science', *Isis*, 90 (1999), 775–780; J. Brotton, *Trading territories: mapping the early modern world* (London, 1997); D. Cosgrove (ed.), *Mappings* (London, 1999); M. Edney, *Mapping an empire: the geographical construction of British India, 1765–1843* (Chicago, 1997).

[22] S. Daniels, *Fields of vision: landscape imagery and national identity in England and the United States* (Cambridge, 1993); J. Schwartz, '*The Geography Lesson*: photographs and the construction of imaginative geographies', *Journal of Historical Geography*, 22 (1996), 16–45; J. Ryan, *Picturing empire: photography and the visualization of the British empire* (London, 1997).

[23] D. Gregory, *Geographical imaginations* (Oxford, 1993).

[24] P. Haggett and R. Chorley, 'Models, paradigms and the new geography', in P. Haggett and R. Chorley (eds.), *Models in geography* (London, 1967), 19–42; R. J. Johnston, 'Paradigms and revolutions or evolution?: observations on human geography since the Second World War', *Progress in Human Geography*, 2 (1978), 189–206; A. Buttimer, 'On people, paradigms and "progress" in geography', in D. Stoddart (ed.), *Geography, ideology and social concern* (Oxford, 1981), pp. 81–98; D. Stoddart, 'The paradigm concept and the history of geography', in D. Stoddart, *Geography, ideology and social concern* (Oxford, 1981), pp. 70–80; M. Harvey and B. Holly, 'Paradigms, philosophy and geographic thought', in M. Harvey and B. Holly (eds.), *Themes in geographic thought* (Beckenham, 1981), 11–32; A. Mair, 'Thomas Kuhn and understanding geography', *Progress in Human Geography*, 10 (1986), 345–69.

geography in Europe has its 'origins' in the last quarter of the nineteenth century.[25]

Lesley Cormack's *Charting an empire* (1997) examines the nature of geography and geographers in Cambridge, Oxford, and the 'third university' of Gresham College, London from 1580 to 1620. Geography, she claims, was central in inculcating a sense of English national identity that was inward looking in its attachment to local place and country, and outward looking in its attention to the English (and nascent British) empire. She identifies three sorts of geographers and of geography: a first small group focused on mathematical geography and its evident utilitarian connections; a second larger group concerned itself with descriptive geography; and a third group focused upon chorography, understood as regional or local studies. For Cormack, chorography was 'the most wide-ranging of the geographical arts, in that it provided the specific detail to make concrete the other general branches of geography'.[26]

Cormack's work has been criticised for its prosopographical methodology and attention to the ownership of geography books, an approach which too readily divorces geography from its wider intellectual context and the other interests of the individuals concerned.[27] Nevertheless, Cormack's book, and her related work on empire and on geography as a courtly practice,[28] not only extend the chronological period over which geography was part of university education but provide a detailed study of the sites of early modern geography's involvement as a form of state knowledge.[29] Others have shown how geography was, from the later seventeenth century, also part of the rise of experimental science and the 'new' natural philosophy, and that chorography and mapping were practically important both in the emergence of the state and to the idea of national identity in the early modern period.[30]

[25] H. Capel, 'Institutionalization of geography and strategies of change', in Stoddart, *Geography, ideology and social concern*, 37–69; E. Lochhead, 'The emergence of academic geography in Britain in its historical context', (Ph.D. thesis, University of California at Berkeley, 1980); G. Dunbar (ed.), *From traveller to scientist: the professionalization and institutionalization of geography in Europe and North America since 1870* (Dordrecht, forthcoming).

[26] L. Cormack, *Charting an empire: geography at the English universities, 1580–1620* (Chicago, 1997), 163.

[27] R. Mayhew, [Review], *Ecumene*, 5 (1998), 487–8.

[28] L. Cormack, '"Good fences make good neighbors": geography as self-definition in early modern England', *Isis*, 82 (1991), 639–61; 'Twisting the lion's tail: practice and theory at the court of Henry Prince of Wales', in B. Moran (ed.), *Patronage and institutions: science, technology and medicine at the European court* (Woodbridge, 1991), 67–84; 'The fashioning of an empire: geography and the state in Elizabethan England', in A. Godlewska and N. Smith (eds.), *Geography and empire* (Oxford, 1994), 15–30.

[29] Cormack's attention to geography in early modern Oxford is very different from J. Baker, *The history of geography* (Oxford, 1963), 119–29.

[30] D. Livingstone, 'Science, magic and religion: a contextual re-assessment of geography in the sixteenth and seventeenth centuries', *History of Science*, 26 (1988), 269–94; 'Geography, tradition and the scientific revolution: an interpretative essay', *Transactions of the Institute of British Geographers*, 15 (1990), 359–73. J. Black, 'Mapping early modern Europe', *European History Quarterly*, 25 (1995), 431–42; Brotton, *Trading territories*; D. Buisseret (ed.), *Monarchs, ministers and maps: the emergence of cartography as a tool of government in early modern Europe* (Chicago, 1992); Cormack, *Charting an empire;* M. Wintle, 'Renaissance maps and the construction of the idea of Europe', *Journal of Historical Geography*, 25 (1999), 137–65.

Robert Mayhew has advanced our understanding of geography as it was understood in eighteenth-century England in several respects. His work is insistent upon the recovery of geography's textual traditions and its connections with classical education, and, thus, with knowing how it was that geography was defined and used by its practitioners and understood by its audiences. Initial attention to Samuel Johnson's conception of geography as a rational discourse has been developed in further studies.[31] Mayhew's attention to what he has called 'the character of English geography' between *c.* 1660 and 1800 centres upon his analysis of geography books, definitions of their function and audience, their readership and what he calls the 'milieu of book production'.[32] Such a resolutely textual hermeneutic approach shows that geography in England in this period was part both of a commercial and practical tradition, with its emphasis upon practical utility and polite learning, and of a humanistic and scholarly tradition which allied geography with the classics and civil history. Although geography was not 'an independent discipline' in schools and universities in eighteenth-century England, mathematical and descriptive geography were taught to a range of ages and social classes, and grammar schools and Cambridge and Oxford universities taught geography as part of a humanist education. In these ways, geography in eighteenth-century England was understood in particular intellectual contexts and promoted in certain sites as a textual practice designed to enlighten and to politicise civic society.[33]

In France, legislators confirmed the importance of geography for what eighteenth-century commentators understood as the 'Science of Man' by placing it in the Class of Moral and Political Sciences in the new National Institute (in 1795). These initiatives were not continued beyond 1803, however, and human geography was slow to develop in consequence.[34] Even so, scholars such as Turgot placed geography within his progressivist vision for the human sciences. His and others' conception of human progress was fundamentally geographical since the idea of a 'stage-by-stage' development of peoples depended upon global comparisons that were temporal *and* spatial.[35]

[31] Mayhew, *Geography and literature in historical context: Samuel Johnson and eighteenth-century conceptions of geography* (Oxford, School of Geography Research Papers, no. 54, 1997); 'Was William Shakespeare an eighteenth-century geographer?: constructing histories of geographical knowledge', *Transactions of the Institute of British Geographers,* 23 (1998), 21–37; 'Geography in eighteenth-century British education', *Paedagogica Historica,* 34 (1998), 731–69; *idem,* 'William Guthrie's *Geographical grammar,* the Scottish enlightenment and the politics of British geography', *Scottish Geographical Journal,* 115 (1999), 19–34.

[32] R. Mayhew, 'The character of English geography, *c.* 1660–1800: a textual approach', *Journal of Historical Geography,* 24 (1998), 385–412.

[33] R. Mayhew, *Enlightenment Geography: the political languages of British geography, c.1650–1850* (London, 2000).

[34] M. Staum, 'The Enlightenment transformed: the Institute Prize contests', *Eighteenth-Century Studies,* 19 (1985), 153–79; 'Human geography in the French Institute: new discipline or missed opportunity?', *Journal of the History of the Behavioural Sciences,* 23 (1987), 332–40; 'Paris Geographical Society constructs the Other, 1821–1850', *Journal of Historical Geography,* 26 (2000), 222–38.

[35] M. Heffernan, 'On geography and progress: Turgot's *Plan d'un Ouvrage sur la géographie politique* (1751) and the origins of modern progressive thought', *Political Geography,* 13 (1994), 328–43.

Godlewska's *Geography unbound* (1999) examines the several trajectories of geography in eighteenth-century France and emphasises the contemporary search for languages of accurate geographical representation, notably in the mathematical tradition. In that sense, she traces the discursive bases against which France both came to know itself through projects of state mapping and national description and sought to map its overseas territories.[36]

Such work has advanced our knowledge of geography's history and historical geography since, as with Francis Sitwell's summary of what geography books were available before *c.* 1800,[37] the nature of geography and the communities who practised it as an intellectual concern is highlighted for given national contexts and at certain moments. But it is clear, too, that whilst eighteenth-century geography was a textual and institutionalised practice in these terms, geographical knowledge embraced more than either the textual or the disciplinary tradition of geography *sensu stricto*.

Other studies of geographical knowledge in the Enlightenment have shown such knowledge to be altogether more complicated and to have included the classification and display of natural knowledge, the imposition of European ways of thinking on nature's diversity, the visualisation of native 'otherness', and the voyages and travels of explorers.[38] Work on the teaching of geography within universities and in the public sphere also suggests that what was understood as geography varied with context.[39] Alongside an understanding of geography as a textual tradition and of its utility for eighteenth-century scholars in conceiving of conjectural history and the idea of historical change, for example,[40] geographical knowledge in the form of what the Royal Society in the period 1720–79 termed 'natural history' and 'mixed mathematics' (including astronomy, weights and measures, and geometry as well as geography) was an integral part of British commercial and imperial knowledge.[41]

Several things follow from these claims. The first concerns the need to establish the connections between geography and other forms of natural and social knowledge – 'globalising discourses of terrestrial knowledge' as Porter has it[42] – in the eighteenth century and for other times. The second is in showing how

[36] A. Godlewska, *Geography unbound: French geographic science from Cassini to Humboldt* (Chicago, 1999).

[37] O. Sitwell, *Four centuries of special geography: an annotated guide to books that purport to describe all the countries in the world published in English before 1888, with a critical introduction* (Vancouver, 1993).

[38] C. Withers and D. Livingstone, 'Introduction: on geography and enlightenment', in D. Livingstone and C. Withers (eds.), *Geography and enlightenment* (Chicago, 1999), 1–28.

[39] C. Withers, 'Notes towards a historical geography of geography in early modern Scotland', *Scotlands*, 3 (1996), 111–24; 'Towards a history of geography in the public sphere', *History of Science*, 37 (1999), 45–78.

[40] G. Abbatista, 'Establishing "the order of time and place": "rational geography", French erudition and the emplacement of history in Gibbon's mind', in D. Womersley (ed.), *Edward Gibbon: bicentenary essays. Studies on Voltaire and the eighteenth century*, 355 (1997), 45–72.

[41] R. Sorrenson, 'Towards a history of the Royal Society in the eighteenth century', *Notes and Records of the Royal Society of London*, 50 (1996), 29–46; R. Drayton, 'Knowledge and empire', in P. Marshall (ed.), *The Oxford history of the British empire: the eighteenth century* (Oxford, 1998), 231–52; C. Bayley, *Empire and information* (Cambridge, 1996), 300–14.

[42] R. Porter, 'The terraqueous globe', in G. Rousseau and R. Porter (eds.), *The ferment of knowledge* (Cambridge, 1990), 285–324.

the discursive nature of geographical knowledge depended upon the sites of its making. The third has to do with the fact that, as Cormack and Richard Sorrenson's work has shown, geography and geographical knowledge has long been concerned with what David Miller and Peter Reill have termed 'visions of empire'.[43]

### *Geographical knowledge and imperialism*

It is arguable that the enormous range of recent interest in the nature of imperialism, in the post-colonial analysis of culture, and in geography and empire has been prompted by Edward Said's influential *Orientalism* (1978).[44] Amongst historians of geography, Said's attention to the discursive power of 'imaginative geographies' has been drawn upon to re-assess the role of geography in the history of modern imperialism and to explore the enduring ideologies of imperialism in contemporary geographical knowledge.

Many scholars concerned with imperial geographies have considered them, for Britain and the legacy of Britain's empire anyway, in what Eric Hobsbawm termed the 'age of high empire'[45] – *c.* 1870 and 1914. As Driver has shown, the closing decades of the nineteenth century brought into being an altogether different world. It was a world in which exploration shifted from the sea to the land, and the closure of imperial space was accompanied by the popular and racialised representation of the colonialised 'other', not least because the emergence of 'modern' geography as an institutionalised academic discipline was bound up both with the practicalities of empire and with the birth of a certain form of 'modernity' itself.[46]

Two notable essay collections – *Geography and empire* (1994), and *Geography and imperialism* (1995), and Driver's *Geography militant* (2000) – have taken these issues further in directing this 'new agenda for theoretical and historical work on geography and empire'.[47] Significant advances have been made in understanding the connections between the French geographical movement and French imperialism in the nineteenth and early twentieth centuries,[48] and

[43] D. Miller and P. Reill (eds.), *Visions of empire: voyages, botany, and representations of nature* (Cambridge, 1996).

[44] Driver, 'Geography's empire', 30–5; 'New perspectives', 95–6; 'Histories of the present?', 100–2.

[45] E. Hobsbawm, *The age of empire* (London, 1987).

[46] Driver, 'Geography's empire', 26–30.

[47] Driver, 'Histories of the present?', 100.

[48] M. Heffernan, 'The limits of utopia: Henri Duveyrier and the exploration of the Sahara in the nineteenth century', *Geographical Journal,* 155 (1989), 342–52; 'The ambiguity of the Sahara: French images of the North African desert in the nineteenth century', *Maghreb Review,* 17 (1992), 178–88; 'A state scholarship: the political geography of French international science during the nineteenth century', *Transactions of the Institute of British Geographers,* 19 (1994), 21–45; 'The science of empire': the French geographical movement and the forms of French imperialism, *c.* 1870–*c.* 1920', in Godlewska and Smith (eds.), *Geography and empire* (Oxford, 1994), 92–114; 'The spoils of war: the *Société de Géographie de Paris* and the French empire, 1914–1919', in M. Bell, R. A. Butlin and M. Heffernan (eds.), *Geography and imperialism, 1820–1940* (Manchester, 1995), 221–64; M. Heffernan and K. Sutton, 'The landscape of colonialism: the impact of French colonial rule on the Algerian rural settlement pattern, 1830–1987', in C. Dixon and M. Heffernan (eds.), *Colonialism and development in the contemporary world* (London, 1991), 121–51.

between geographical interpretations of Italy and Italian colonialism in the 1920s and 1930s.[49] Advances have also been made in understanding both the Russian imperial geographical imagination[50] and, in Britain, the complex connections between geography's texts, state institutions and the idea of the 'imperial geographical citizen'.[51] Other work has focused on particular figures such as the imperial (self-)propagandiser, Henry Morton Stanley,[52] or upon David Livingstone as imperial missionary,[53] and upon discursive practices such as mapping and photography by which the empire was constituted 'out there' and consumed by domestic audiences.[54] The role of women in the making and representation of the empire, and, thus, the gendering of imperial geographical knowledge, has also been the focus of attention.[55]

In several ways, then, the connections established in this and other work between geographical knowledge and imperialism, notably in the later nineteenth century, have been important to the recent heightened profile of the history of geography and of geographical knowledge. I want here to move away, however, from what Barnett has termed, for the nineteenth century, the 'overwhelming, although not exclusive fascination with geography's historical involvements with empire'.[56] This is borne of a concern to understand the

[49] L. Gambi, 'Geography and imperialism in Italy: from the unity of the nation to the "new" Roman empire', in Godlewska and Smith (eds.), *Geography and empire,* 74–91; D. Atkinson, 'Geopolitics, cartography and geographical knowledge: envisioning Africa from Fascist Italy', in Bell, Butlin and Heffernan (eds.), *Geography and imperialism,* 265–97.

[50] M. Bassin, 'Russia between Europe and Asia: the ideological construction of geographical space', *Slavic Review,* 50 (1991), 1–17; 'Inventing Siberia: visions of the Russian east in the early nineteenth century', *American Historical Review,* 96 (1991), 763–94; *Imperial visions: nationalism and geographical imagination in the Russian far east, 1840–1865* (Cambridge, 1999); J. Pallot, 'Imagining the rational landscape in late imperial Russia', *Journal of Historical Geography,* 26 (2000), 273–91.

[51] R. A. Butlin, 'Historical geographies of the British empire, *c.*1887–1925', in Bell, Butlin and Heffernan (eds.), *Geography and imperialism,* 151–88.

[52] F. Driver, 'Henry Morton Stanley and his critics: geography, exploration and empire', *Past and Present,* 133 (1991), 134–67.

[53] F. Driver, 'David Livingstone and the culture of exploration in mid-Victorian Britain', in *David Livingstone and the Victorian encounter with Africa* (National Portrait Gallery, London, 1996), 109–38; J. M. MacKenzie, 'David Livingstone and the worldly after-life: imperialism and nationalism in Africa', in *David Livingstone and the Victorian encounter with Africa* (National Portrait Gallery, London, 1996), 201–19.

[54] Edney, *Mapping an empire*; Ryan, *Picturing empire.*

[55] M. Domosh, 'Toward a feminist historiography of geography', *Transactions of the Institute of British Geographers,* 16 (1991), 95–104; D. Stoddart, 'Do we need a feminist historiography of geography – and if we do, what should it be?', *Transactions of the Institute of British Geographers,* 16 (1991), 484–7; M. Domosh, 'Beyond the frontiers of geographical knowledge', *Transactions of the Institute of British Geographers,* 16 (1991), 488–90; A. Blunt, *Travel, gender and imperialism: Mary Kingsley and West Africa* (New York, 1994); A. Blunt and G. Rose (eds.), *Writing women and space: colonial and postcolonial geographies* (New York, 1994); D. Gregory, 'Between the book and the lamp: imaginative geographies of Egypt, 1849–1850', *Transactions of the Institute of British Geographers,* 20 (1995), 29–57; G. Kearns, 'The imperial subject: geography and travel in the work of Mary Kingsley and Halford Mackinder', *Transactions of the Institute of British Geographers,* 22 (1997), 450–72; A. Mander, 'Geography, gender and the state: a critical evaluation of the development of geography 1830–1918', (D.Phil. thesis, University of Oxford, 1995); C. McEwan, '"The Mother of all the Peoples": geographical knowledge and the empowering of Mary Slessor', in Bell, Butlin and Heffernan, *Geography and imperialism,* 125–50; 'Cutting power lines within the palace? countering paternity and eurocentrism in the 'geographical tradition', *Transactions of the Institute of British Geographers,* 23 (1998), 371–84; 'Gender, science and physical geography in nineteenth-century Britain', *Area,* 30 (1998), 215–23.

[56] Barnett, 'Awakening the dead', 418.

nature of geographical knowledge in one national context over the *longue durée.* My intentions are lent support by the work of Cormack, Richard Drayton, John Gascoigne, Mayhew and Sorrenson which has extended the historical range of geography's concerns with empire in ways which do not presume to connect the 'age of high empire' to earlier 'moments' for geography and empire but which require, simply, further study in particular geographical contexts. My concern to detail a historical geography of geographical knowledge in one national context has also been stimulated by writings in the history of science.

## The social and situated nature of scientific knowledge

In an editorial published in 1995, Thrift, Driver and Livingstone commented: 'If it were necessary to choose the most vibrant and exciting areas of research in the social sciences and humanities today, then surely the study of science as a social construction would figure large'.[57] Such interest was the result of attention to several things: the theoretical nature of power in society; the reflexivity of knowledge; an understanding of knowledge making as a practical activity; and, notably, an 'emphasis on space'. The study of science as a social construction has, they argued, 'been pursued through a peculiarly spatial imaginary which always attaches insight to the site'. Sites were not, however, simply spatial 'containers': 'The locales in which scientific knowledge is produced are not seen as passive backdrops, but as vital links in the chain of production, validation, and dissemination' [of knowledge]. In considering that the study of science as a social construction has produced its own geographies of scientific knowledge and that such geographies open up possibilities for further research, brief attention was paid by them to three things: to the sites of scientific knowledge; to the 'different networks of people and things which have allowed scientific knowledge to be constructed at a distance'; and to 'the process of constitution of the fields in and through which scientific knowledge can be legitimately gathered'.[58]

As Livingstone has elsewhere noted, attention to space in these ways is apparent in work on the situated nature of rationality, the local nature of culture and the spatial constitution of historical understanding.[59] Against such collective interest in the spatialised nature of knowledge, Livingstone in turn elsewhere sketched out a geography of science by reviewing works in a threefold 'rudimentary taxonomy': studies addressing the regionalisation of scientific style; studies of the political topography of scientific commitment; and those considering what he termed the 'social space of scientific sites'.[60]

[57] N. Thrift, F. Driver and D. Livingstone, 'The geography of truth', *Environment and Planning D: Society and Space,* 13 (1995), 1. [58] Thrift, Driver and Livingstone, 'The geography of truth', 2–3.

[59] D. Livingstone, 'The spaces of knowledge: contributions towards a historical geography of science', *Environment and Planning D: Society and Space,* 13 (1995), 5–34.

[60] Livingstone, 'Spaces of knowledge', 16–29.

Such attention by geographers to the socially constructed and situated nature of scientific knowledge reflects the long standing and more thoroughgoing interests of historians of science in the sociology of scientific knowledge. Prior to Thomas Kuhn's influential *The structure of scientific revolutions* (1962), scientific knowledge was, in general, held to be objective, universal and true; consequently the sites of its making and the conduct of its practitioners were immune from serious scrutiny. Subsequent work by Barry Barnes, David Bloor and others has treated questions of objectivity, truth and the (presumed) universality of knowledge as effects to be explained, however, rather than as direct outcomes of the scientific method. Thus, the sociology of scientific knowledge has increasingly focused on what scientists actually do and, in turn, upon the social and the located nature of knowledge making. As Jan Golinski makes clear, understanding 'scientific knowledge primarily as a human product, made with locally situated cultural and material resources, rather than as simply the revelation of a pre-given order of nature', has initiated a 'remarkably productive period in the understanding of science as a human enterprise'.[61]

### *The 'geographical turn' in the history of science*

Crosbie Smith and Jon Agar review work by historians of science on territorial themes and knowledge making under two general headings: 'Of the Territory' and 'Of Privileged Sites'.[62] Such distinctions have influenced my thinking as the following sections will reveal. But two points are worth making. The first has to do not with sites and intellectual territory as absolute distinctions but rather with the connections *between* them in terms, for example, of the ways in which knowledge travels, in certain discursive forms, from 'there' – a given geographical space *and* intellectual territory – to 'here', a certain site or sites, and *vice versa*. Second, recent expressions of interest in the spatiality of knowledge and, in particular, current localist emphases, are not the only conceptions of the geography of scientific endeavour. Studies of national *styles* of science have considered the different national expressions and origins of, for example, psychology,[63] chemistry,[64] embryology,[65] and ecology[66] as well as what might be held to be a 'national style' in science more generally, either by reference to the cognitive processes employed, to the social

[61] J. Golinski, *Making natural knowledge: constructivism and the history of science* (Cambridge, 1998), 14, 16.

[62] C. Smith and J. Agar, 'Introduction: making space for science', in C. Smith and J. Agar (eds.), *Making space for science: territorial themes in the shaping of knowledge* (London, 1998), 1–24.

[63] K. Danziger, *Constructing the subject: historical origins of psychological research* (Cambridge, 1999).

[64] M. Nye, '"National styles"?: French and English chemistry in the nineteenth and early twentieth centuries', *Osiris,* 8 (1993), 30–52.

[65] J. Maienschein, 'Epistemic styles in German and American embryology', *Science in Context,* 4 (1991), 407–28.

[66] M. Nicholson, 'National styles, divergent classifications: a comparative case study from the history of French and American plant ecology', *Knowledge and Society,* 8 (1989), 139–86; W. Brock, 'Humboldt and the British: a note on the character of British science', *Annals of Science,* 50 (1993), 365–72.

hierarchies of scientific institutions, or to the practical and public consequences of doing science.[67] Others have examined thematic questions within science at the national *scale*. This is less apparent in considering the rise of science across Europe, but is particularly clear, for example, in work on the Enlightenment and on the Scientific Revolution in national context.[68]

This is not to say that scholars working on national styles and on scientific ideas and movements in national context are insensitive to the problematic notion of the 'nation' itself and to the historical construction of ideas of national identity.[69] It is to observe in such work a tendency to assume the nation as *the* frame of reference, *the* spatial scale at which science, however manifest, is to be understood. The same is true, of course, of those who have referred to national 'schools' in the history of geography.[70] I do not want to reify the question of national identity (however understood), by giving it a 'taken-for-granted' status at the outset.

My concern is to reverse the gaze as it were. Rather than start with the nation and with national identity as presumed 'things', in which science (*qua* geographical knowledge) of a certain sort emerges, I want to consider how given forms of geographical knowledge themselves came to constitute the idea of Scotland as a national space. This is what I mean by the 'making' of Scotland through geographical knowledge. Recognising, then, with Livingstone, the need to attend to such issues at a variety of spatial scales, I am further prompted by those who have argued, for example, for an idea of Enlightenment *above* national context in which the exchange of scientific and commercial information can be read as international, transnational *and* between particular individuals and institutions,[71] and by a concern to explore the local nature and sites of scientific knowledge.

## *Local sites of knowledge making*

In discussing the ways in which cartographic knowledge in and of early modern Europe was made in particular locations for particular political

[67] M. Crosland, 'History of science in a national context', *British Journal for the History of Science,* 10 (1977), 95–113; J. Harwood, *Styles of scientific thought* (Chicago, 1993); J. Merz, *A history of European thought in the nineteenth century* (4 volumes, Edinburgh, 1896–1914); A. Crombie, *Styles of scientific thinking in the European tradition* (3 volumes, London, 1994); N. Reingold, 'The peculiarities of the Americans, or are there national styles in the sciences?', *Science in Context,* 4 (1991), 347–66.

[68] D. Goodman and C. Russell (eds.), *The rise of scientific Europe 1500–1800* (Milton Keynes, 1991); W. Clark, J. Golinski and S. Schaffer (eds.), *The sciences in enlightened Europe* (Chicago, 1999). On science in national context, see, for example, R. Porter and M. Teich (eds.), *The enlightenment in national context* (Cambridge, 1981); *The scientific revolution in national context* (Cambridge, 1992).

[69] B. Anderson, *Imagined communities* (London, 1991 edition); E. Said, *Orientalism* (Harmondsworth, 1978); *Culture and imperialism* (London, 1993). In contrast, one recent work on nationhood ignores the role of geographical imagination: A. Hastings, *The construction of nationhood: ethnicity, religion and nationalism* (Cambridge, 1997). Similarly, in the Scottish context, geographical knowledge as a way of constructing the 'shape' of the nation has been ignored: see J. Hearn, *Claiming Scotland: national identity and liberal culture* (Edinburgh, 2000).

[70] Livingstone, *Geographical tradition,* 4–11.

[71] J. Robertson, 'The Enlightenment above national context: political economy in eighteenth-century Scotland and Naples', *The Historical Journal,* 40 (1997), 667–97.

purposes, David Turnbull has commented that 'the picture of science that has emerged from empirical investigations of both contemporary and historical scientists is that all knowledge is constructed at specific sites through the engagements of particular scientists with particular skills, material tools, theories and techniques. . . . Thus a fundamental characteristic of scientific knowledge is its localness'.[72] As Adir Ophir and Steven Shapin point out, this 'influential *localist* genre, marked by attention to national and regional features of an enterprise once regarded as paradigmatically universal', is relatively recent. As they note, localist concerns have close links with the relativist agenda established by sociologists of scientific knowledge: 'relativism can be practically defined through the notion that all knowledge claims and judgments secure their credibility not through absolute standards but through the workings of *local* causes operating in contexts of judgment'.[73] Such claims offer themselves to empirical testing since the making of science (and, thus, of geographical knowledge) may depend upon knowing in *which* sites it was made and what the connections were between them, and upon illustrating the epistemological bases to meaning in and between given sites. Shapin has considered just such issues. We could, he argues, simply take for granted the local nature of science's making:

> That is to say, suppose one regarded it as established beyond doubt that science is indelibly marked by the local and the spatial circumstances of its making; that scientific knowledge is embodied, residing in people and in such material objects as books and instruments, and nowhere else; and, finally, that scientific knowledge is made by and through mundane – and locally varying – modes of social and cultural interaction. If one granted all this, one would be treating the 'localist' or 'geographical' turn in science studies as a great accomplishment – telling us a series of important things about science which previous understandings have systematically ignored or denied.

Admitting such work to be a 'considerable achievement', it is, for Shapin, not enough:

> And yet I also want to say that it is still incomplete and that it is danger of missing something very important about science. The problem here is not that the geographical sensibility has been taken too far but that it has not been taken far enough. We need to understand not only how knowledge is made in specific places but also how transactions occur between places.[74]

In moving, then, to consider the sites of knowledge making and issues to do with 'the territory' – with 'fields of knowledge', with how knowledge 'travels', with how it does so successfully, and with its audiences – my interest is both

[72] D. Turnbull, 'Cartography and science in early modern Europe: mapping the construction of knowledge spaces', *Imago Mundi*, 46 (1996), 5–24.

[73] A. Ophir and S. Shapin, 'The place of knowledge: a methodological survey', *Science in Context*, 4 (1991), 5–6.

[74] S. Shapin, 'Placing the view from nowhere: historical and sociological problems in the location of science', *Transactions of the Institute of British Geographers*, 23 (1998), 6–7.

historical and epistemological. That this is so demands some attention to the nature of science's making as a question of historical geography. It was once generally accepted that 'modern science' was born towards the end of the seventeenth century, a period most contemporaries experienced as one of unprecedented change and crisis in virtually every dimension of European life. The question of *the* 'Scientific Revolution', and, thus, of there being an emergent 'modern science' geographically everywhere the same and shared by all, has been the subject of considerable debate.[75] For Lisa Jardine, the concerns of late seventeenth-century French cartographers and mathematicians to 'break new ground', as she puts it, support the claims of Godlewska and Cormack on early modern geographical knowledge as part of 'revolutions' in conceptions of (e)state measurement.[76] In his *The scientific revolution*, Shapin concentrates upon several issues to do with changes in knowledge about the natural world and changes in the means to securing that knowledge. These, principally, were the mechanization of nature (with reference to the increasing use of mechanical metaphors to construe natural things); the depersonalization of natural knowledge; the attempted mechanization of knowledge making (related to the foregoing in the emphasis placed by contemporaries upon the use of explicitly formulated methodological rules that aimed to discipline the production of *natural* knowledge by managing the effects of *human* intervention); and the intention to use such reformed natural knowledge to achieve given ends.[77] What is also clear, notably in some of Shapin's other work, is that making and disseminating natural knowledge in such ways in the later seventeenth century was always a situated practical activity. One such local site was the laboratory.

Since the seventeenth century, the laboratory has been recognised as 'the preeminent site for making knowledge in the experimental sciences', and is so, notes Golinski, because it 'straddles the realms of private seclusion and public display . . . On the one hand, the laboratory is a place where valuable instruments and materials are sequestered, where skilled personnel seek to work undisturbed, and where intrusion by outsiders is unwelcome. . . . On the other hand, what is produced there is declaredly "public knowledge"; it is supposed to be valid universally and available to all'.[78] Managing the tensions between the private and the public realm was also, however, to make social distinctions concerning who had access to scientific knowledge, who was to be trusted in the production of it, and, in turn, of trusting the forms in which knowledge moved in and out from such sites, as, say, either written reports or verbal accounts. Controlling the venues of knowledge in late seventeenth-century

[75] S. Shapin, *The scientific revolution* (Chicago, 1996); see the theme issue, 'The scientific revolution as narrative', *Configurations*, 6 (1998); J. Henry, *The scientific revolution and the origins of modern science* (London and New York, 1997); P. Dear, *The scientific enterprise in early modern Europe* (Chicago, 1997).

[76] L. Jardine, *Ingenious pursuits: building the scientific revolution* (London, 1999), 177–222.

[77] Shapin, *Scientific revolution*, 65–117.

[78] Golinski, *Making natural knowledge*, 84.

England and establishing the several 'bases of believability' was a matter of bounding and disciplining a community of practitioners, of policing experimental discourse, and of publicly warranting that the knowledge produced in such places was reliable and authentic. Social status mattered here: 'What underwrote assent to knowledge claims was the word of a gentleman, the convention regulating access to a gentleman's house, and the social relations within it'.[79]

Iwan Morus' study of the use of laboratories and other 'spaces of display' among practitioners of electrical science in early nineteenth-century London, for example, notes the distinction between Faraday's use of the laboratory as a private space with no audience participation, in contrast to those 'commercial electricians' for whom experimentation was a matter of public theatre.[80] Graeme Gooday has shown that the nineteenth-century teaching laboratory operated through the 'rigid spatial structuring of laboratory life'.[81] This claim is paralleled in Simon Schaffer's study of the Royal Greenwich Observatory, where demands for accurate measurement necessitated the rigorous bodily control of staff. He has also noted the shifting status of physics laboratories as domestic space and as scientific space within the Victorian country house.[82] In studying T. H. Huxley's working environment – notably, his laboratory and lecture theatre – Sophie Forgan and Graeme Gooday signal towards, as they put it, 'a fully researched historical geography of London science'.[83] Other studies of laboratories have emphasised the spatialised constitution of knowledge making, and the differential social access to such knowledge.[84]

Other work on sites of knowledge, perhaps particularly for the eighteenth century, has considered science's 'audience' and the ways in which a 'public' for science was constituted in certain locales and through performance. Schaffer has shown, for example, how eighteenth-century natural philoso-

[79] S. Shapin, 'The house of experiment in seventeenth-century England', *Isis,* 79 (1988), 404.

[80] I. Morus, 'Currents from the underworld: electricity and the technology of display in early Victorian England', *Isis,* 84 (1993), 50–69.

[81] G. Gooday, 'Precision measurement and the genesis of physics teaching laboratories in late Victorian Britain', *British Journal for the History of Science,* 23 (1990), 25–51; '"Nature in the laboratory": domestication and discipline with the microscope in Victorian life science', *British Journal for the History of Science,* 24 (1991), 307–71; Teaching telegraphy and electrotechnics in the physics laboratory: William Ayrton and the creation of an academic space for electrical engineering in Britain 1873–1884', *History of Technology,* 13 (1991), 73–111; 'Instrumentation and interpretation: managing and representing the working environments of Victorian experimental science', in B. Lightman (ed.), *Victorian science in context* (Chicago, 1997), 409–37; 'The premisses of premises: spatial issues in the historical construction of laboratory credibility', in Smith and Agar (eds.), *Making space for science,* 216–45.

[82] S. Schaffer, 'Late Victorian metrology and its instrumentation: a manufactory of ohms', in R. Bud and S. Cozzens (eds.), *Invisible connections: instruments, institutions and science* (Washington, 1992), 23–56; 'Physics laboratories and the Victorian country house', in Smith and Agar (eds.), *Making space for science,* 149–80.

[83] S. Forgan and G. Gooday, 'Constructing South Kensington: the buildings and politics of T. H. Huxley's working environments', *British Journal for the History of Science,* 29 (1996), 435–68.

[84] This work is reviewed in L. Pyenson and S. Sheets-Pyenson, *Servants of nature: a history of scientific institutions, enterprises and sensibilities* (London, 1999).

phers used their lectures – indeed, their own bodies – to constitute particular moral and political claims about the workings of the natural world.[85] Stewart has added to our understanding of the 'map' of scientific knowledge in eighteenth-century London through attention to the Royal Society and to London's coffee-houses as information exchanges.[86] Other sites of scientific knowledge have been the subject of similar attention: the lecture hall,[87] the library,[88] the museum,[89] botanical and zoological gardens,[90] even the public house.[91] Much further work remains to be done on these and other 'spaces of science'.[92]

Yet the social and spatialised nature of scientific knowledge – and a historical geography of geographical knowledge as I propose it here – cannot ever be just a matter of sites. For one thing, sites, certainly in the institutional sense, are never single places. In his use of the term 'heterotopia', Foucault encapsulated the sense in which several spatial settings with different purposes coexist in given sites. For Ophir and Shapin, 'The development of modern science – both natural and human – is closely linked to the institutionalization of special heterotopic sites. By the mid-seventeenth century one could already point to the chemical laboratory and the mechanical operatory, the observatory, the botanical garden, and the room of curiosities'. As they note and as the above has shown, other sites were established later.[93] Foucault's notion of heterotopic sites extended from his work on the nature of power embodied in institutionalized sites such as prisons and the asylum, and, thus, from a concern to see power relations as *spatial* relations.[94] Such matters of power and of its spatial constitution are always social *and* epistemological, always a matter of warranted authority in terms of who has the power to undertake and to make knowledge in certain ways, and who not. The making of what, at any given time and in any given place, becomes regarded as scientific knowledge is also dependent upon the movement of such knowledge across the boundaries between the site itself and what lies, to cite Golinski, 'beyond the laboratory walls'.[95]

[85] S. Schaffer, 'Natural philosophy and public spectacle in the eighteenth century', *History of Science*, 21 (1983), 1–43.

[86] L. Stewart, 'Public lectures and private patronage in Newtonian England', *Isis*, 77 (1986), 47–58; 'Other centres of calculation, or, where the Royal Society didn't count: commerce, coffee-houses and natural philosophy in early modern London', *British Journal for the History of Science*, 32 (1999), 133–54.

[87] P. Bourdieu, *In other words* (Cambridge, 1992).

[88] R. Chartier, *The order of books* (Cambridge, 1994).

[89] S. Forgan, 'The architecture of display: museums, universities and objects in nineteenth-century Britain', *History of Science*, 32 (1994), 139–62; Pyenson and Sheets-Pyenson, *Servants of nature*, 125–49; C. Yanni, *Nature's museums: Victorian science and the architecture of display* (London, 1999).

[90] A. Cunningham, 'The culture of gardens', in N. Jardine, J. A. Secord and E. C. Spary (eds.), *Cultures of natural history* (Cambridge, 1996), 38–56; Pyenson and Sheets-Pyenson, *Servants of nature*, 150–72.

[91] A. Secord, 'Science in the pub: artisan botanists in early nineteenth-century Lancashire', *History of Science*, 32 (1994), 269–315.

[92] D. Livingstone, *Spaces for Science* (Chicago, forthcoming).

[93] Ophir and Shapin, 'The place of knowledge', 14.

[94] M. Foucault, *The order of things: an archaeology of the human sciences* (London, 1970); 'Of other spaces', *Diacritics*, 16 (1986), 22–7.

[95] Golinski, *Making natural knowledge*, 91–102.

*Of movement: travelling knowledge, 'the field' and audiences*

It may be claimed that the 'modern' science which emerged from the later seventeenth century depended more upon the testimony of nature than upon the testimony of humans, more, that is, upon the personal experience of nature than upon what others might say or ancient authorities propound. For Shapin, 'Here is the root idea of modern *empiricism*, the view that proper knowledge is and ought to be derived from direct sense experience'. It is also, as he acknowledges, a problematic route to reliable knowledge about the world: 'And here too are the foundations of the modern mistrust of the social aspects of knowledge making: if you really want to secure truth about the natural world, forget tradition, ignore authority, be skeptical of what others say, and wander the fields alone with your eyes open'.[96]

Such remarks about knowledge making and the implications arising from them are of considerable importance for questions of geographical knowledge understood historically. For one thing, the emergence of 'modern' science is, profoundly, a matter *of* geography – not alone in terms of the sites of its making, but in relation to being 'in the field', to exploring the world in various embodied and instrumentalised ways. For another, early modern geography and geographical knowledge as discerned by Cormack and others was clearly part of the emergent 'new' natural philosophy. There is a third sense, however, in which being 'in the field' is fundamentally geographical; it concerns the geographical movement of knowledge itself and the displacement of knowledge from one site to another.

Not everyone could or did travel. Voyages of discovery to new empires were expensive, and, at smaller scales, even national or regional travels and surveys could be arduous and costly. Furthermore, even where possible, wandering the fields alone is of little value unless one's measured results can be made sense of by others elsewhere. This presents problems both of how to get reliable knowledge in ways understandable to others, and, if one cannot travel, of how to get such knowledge from distant authorities. Drawing upon the above remarks about trust, this is, as Shapin has stressed, a matter of trust 'inscribed in space'.[97]

Such questions have received attention in work on the importance of travel and travel narratives for the early Royal Society, and the problems of communicating 'at a distance' in the Scientific Revolution.[98] The use of circulated

[96] Shapin, *The scientific revolution*, 69–70.

[97] S. Shapin, *A social history of truth: civility and science in seventeenth-century England* (Chicago, 1994), 245.

[98] D. Carey, 'Compiling nature's history: travellers and travel narratives in the early Royal Society', *Annals of Science*, 54 (1997), 269–92; J. Carrillo, 'From Mt Ventoux to Mt Mayasa: the rise and fall of subjectivity in early modern travel narrative', in J. Elsner and J-P. Rubiés (eds.), *Voyages and visions: towards a cultural history of travel* (London, 1999), 57–73; R. Iliffe, 'Material doubts: Hooke, artisan culture and the exchange of information in 1670s London', *British Journal for the History of Science*, 28 (1995), 285–318; D. Lux and H. Cook, 'Closed circles or open networks?: communicating at a distance during the scientific revolution', *History of Science*, 36 (1998), 179–211; Shapin, *A social history of truth*, 243–309; C. Withers, 'Reporting, mapping, trusting: making geographical knowledge in the late seventeenth century', *Isis*, 90 (1999), 497–521.

questionnaires by the Royal Society, and of guides on how to travel and what to see, were means to produce knowledge in ways which eliminated some of the difficulties arising from deriving strange 'facts' about strange places by people one did not know. As David Lux and Harold Cook note, 'Without the ability to place trust in reports of matters of fact that had not been personally experienced by people like oneself, the new philosophy would have remained fragmented and isolated in local social and geographical spaces'.[99]

The production of (geographical) knowledge in such ways should not, however, simply be seen as a direct 'dislocation' of material, since new facts could effect a different conception either of the problem or of the place in question and, thus, alter geographical understanding. Whilst the early Royal Society adopted a strategy for guiding the activities of travellers, it is clear that geographical discovery, or, more properly the reports arising from it, guided the society: 'Knowledge of nature was not an organized and structured entity but a diverse field, potentially reshaped and altered by the arrival of each new ship'.[100]

My understanding of 'the field' is, then, of a mutual constitution between given intellectual practices (rather than subject areas a priori), the epistemological questions that sustain them and their practitioners. The historical and situated making of geographical knowledge in these ways was, of course, a matter for some people of being in 'the field'. But like Golinski, I do not straightforwardly want to see geographical knowledge as a 'field' of enquiry in the sense of it simply as 'a fieldwork science'.[101] As Bruno Latour has shown, 'fieldwork' is a matter of sites *and* of movement. Latour has considered the movement of the practitioner in 'the field' as a question of 'translation' in terms both of movement over space and of epistemic equivalence; that is, of applying methods which effect a 'representation' of the real world in order to bring it back for further local use. Such a representation is, to use his terms, an 'immutable mobile', a form of knowing which is indicative of the real world object in question. Examples would include specimens of exotic plants or animals which would serve as representatives of distant flora and fauna, and the returns from questionnaires, or maps, or instrumental readings – each functioning as a 'material trace' of the distant geographical 'real'. These 'immutable mobiles' are made sense of by being accumulated in what Latour terms a 'centre of calculation', where knowledge of the features in question and, thus, of the world beyond the laboratory or other site, is accumulated.[102]

These notions have been drawn upon by several scholars, particularly in

[99] Lux and Cook, 'Closed circles or open networks?', 181.

[100] Carey, 'Compiling nature's history', 277.

[101] Golinski, *Making natural knowledge,* 98.

[102] This summary is taken from B. Latour, 'Give me a laboratory and I will raise the world', in K. Knorr-Cetina and M. Mulkay (eds.), *Science observed: perspectives on the social study of science* (Beverly Hills and London, 1983), 141–70; and *Science in action: how to follow scientists and engineers through society* (Cambridge, 1987).

relation to the growth in natural knowledge during the eighteenth century. In considering Latour's work in relation to Joseph Banks' calculated control of natural historical knowledge in and through London, for example, Miller and Gascoigne have stressed the connections between such natural knowledge and political authority over the spaces themselves. 'It would be fair to say', notes Miller, 'that Latour reconceptualizes eighteenth-century voyages of exploration as attempts to re-create at sites in Europe as much information as possible in a form which would extend the imperium of European powers'.[103] Michael Bravo has discussed Latour's notions with reference to Inuit native knowledge and early nineteenth-century Arctic exploration and Lapérouse's voyages of discovery around Sakhalin, *c.* 1785–8.[104] Like Anthony Pagden before him,[105] Bravo problematises Latour's ideas by suggesting that the contact between the European 'scientist' and the distant 'native' should be seen as more a matter of overcoming incommensurability – in ways of reckoning space, in what a map meant and so on – than simply translating knowledge in the form of material traces.[106] And as Ingjerd Höem, Daniel Clayton and others show, we should not treat of 'native knowledge' and its systems of local meaning as absolute and unchanging, or, indeed, without any effect on social and scientific understanding 'back home'.[107] To take a later example, Alfred Russel Wallace's fieldwork in the Amazon and in the East Indies between 1848 and 1862 was a collective enterprise dependent upon native helpers and informants, the British Royal Navy, prompts from distant authorities and trust in his fellow naturalists.[108]

What such work brings into focus is the question of audience, of the different publics receiving geographical and scientific knowledge in the form, for example, of newspaper accounts, books, maps, rumour or lecture. Late seventeenth- and eighteenth-century Britain witnessed the remarkable rise of

[103] D. Miller, 'Joseph Banks, empire, and "centers of calculation" in late Hanoverian London', in Miller and Reill (eds.), *Visions of Empire,* 25; 'The usefulness of natural philosophy: the Royal Society of London and the culture of practical utility in the later eighteenth century', *British Journal for the History of Science,* 32 (1999), 185–202. See also J. Gascoigne, *Science in the service of the empire: Joseph Banks, the British state and the uses of science in the age of revolution* (Cambridge, 1998).

[104] M. Bravo, *The accuracy of ethnoscience: a study of Inuit cartography and cross-cultural commensurability* (Manchester Papers in Social Anthropology, 2, Manchester, 1996); 'Ethnographic navigation and the geographical gift', in Livingstone and Withers, *Geography and enlightenment,* 199–235.

[105] A. Pagden, *European encounters with the new world* (New Haven and London, 1993).

[106] Bravo, 'Ethnographic navigation and the geographical gift', especially 226–33.

[107] D. Clayton, *Islands of truth: the imperial fashioning of Vancouver island* (Vancouver, 1999); A. Duranti, 'Mediated encounters with Pacific cultures: three Samoan dinners', in Miller and Reill (eds.), *Visions of empire,* 326–34; I. Höem, 'The scientific endeavor and the natives', in Miller and Reill (eds.), *Visions of empire,* 305–25. The question of how natives reacted to European explorers has, for the case of Cook in the South Pacific in the eighteenth century, been the subject of a heated debate: for discussions of this, particularly of the differing views of Obeyesekere and Sahlins, see M. Bravo, 'The anti-anthropology of highlanders and islanders', *Studies in the History and Philosophy of Science,* 29 (1998), 369–89.

[108] J. Camerini, 'Wallace in the Field', in H. Kuklick and R. E. Kohler (eds.), *Science in the field, Osiris,* 11 (1996), 44–65; 'Remains of the day: early Victorians in the field', in Lightman, *Victorian science in the field,* 354–77.

science as public discourse.[109] Most studies of the audiences for eighteenth-century science draw in one way or another upon Jurgen Habermas' idea of the 'public sphere' in which civil society was made in and through new forms of polite sociability (chiefly among the urban *bourgeois*).[110] For Thomas Broman, this was because of Habermas' attention to new forms and sites of sociability, such as coffee houses and salons, the phenomenal increase in the eighteenth century of newspapers and periodicals, and the development of a new kind of critical discourse altogether.[111]

Public science in these ways had national, regional and local geographies. The nature of chemistry as a public science differed, for example, between Joseph Black's and William Cullen's work in Enlightenment Scotland and Joseph Priestley's work in the English provinces.[112] Science lecturing in eighteenth-century London drew audiences from different social groups and parts of the city depending upon the subject, although it is not easy to know with what intentions such people went – to be, for example, informed, amazed or mystified by the performance.[113] In several senses, the idea of science as public knowledge is a geographical concern: not just in terms of the local sites of knowledge making, but in terms of eighteenth-century coffee houses (to give one example) being sites for the critical reception and popular discussion of new knowledge, or of bookshops as sites for the purchase of the latest works. And the audiences for scientific knowledge also moved: to and between the sites of knowledge making, to hear papers read, to purchase maps, to view new specimens on public display.

There is much still to know about the question of audiences for knowledge, perhaps particularly so for studies of popular geographical knowledge in historical context. I mean this in just the ways in which others have discussed popular science, for example,[114] and in terms of such questions as 'Who came to hear public geography lectures?', 'What were the patterns of circulation –

[109] J. Golinski, *Science as public culture: chemistry and enlightenment in Britain, 1760–1820* (New York, 1992); L. Stewart, *The rise of public science: rhetoric, technology and natural philosophy in Newtonian Britain, 1660–1750* (New York, 1992); J. Money, 'From Leviathan's air pump to Britannia's voltaic pile: science, public life and the forging of Britain, 1660–1820', *Canadian Journal of History/ Annales Canadiennes d'Histoire*, 28 (1993), 521–44.

[110] J. Habermas, *The structural transformation of the public sphere*, translated by T. Burger with F. Lawrence (Cambridge, Mass., 1989).

[111] T. Broman, 'The Habermasian public sphere and "science *in* the Enlightenment", *History of Science*, 36 (1998), 123–49; see also H. Mah, 'Phantasies of the public sphere: rethinking the Habermas of historians', *Journal of Modern History*, 72 (2000), 153–82.

[112] This is the essential thrust (for Scotland anyway) of Golinski, *Science as public culture.*

[113] Schaffer, 'Natural philosophy and public spectacle in the eighteenth century'; A. Morton (ed.), 'Science lecturing in the eighteenth century', *British Journal for the History of Science*, 28 (1995), 1–100; V. Jankovic, 'The place of nature and the nature of place: the chorographic challenge to the history of British provincial science', *History of Science*, 38 (2000), 79–113; S. Shapin, 'Science and the public', in R. Olby, G. Cantor, J. Christie and M. Hodge (eds.), *Companion to the history of modern science* (London, 1990), 990–1007.

[114] Shapin, 'Science and the public'; R. Cooter and S. Pumfrey, 'Separate spheres and public places: reflections on the history of science popularization and science in popular culture', *History of Science*, 32 (1994), 237–67.

of people and of ideas within and between these audiences?', 'Did they get "the message" of what was told them?' and, in those terms, 'What exactly was the process of "translation"?' As Cooter and Pumfrey note, audience studies in these terms raise difficult questions, perhaps particularly so in a historical context where the evidence for why people attend things, how they read books and how they instruct themselves and others is hard to determine.

## Towards a historical geography of geographical knowledge

The notion of geographical knowledge as having a historical geography in the ways sketched above is apparent in work on the historical geography of Darwinism and in work on the historical geography of modernity.

The relationship between scientific thought and religious belief, as it was apparent in the encounter between Darwinism and Protestantism, is understood to vary by national contexts.[115] As Ronald Numbers has shown for the United States, this relationship is also very much alive today: the different geographies of scientific belief attaching to Darwinism or Creationism differ by state and, within states, by school board, in modern America. In terms of the nature of the engagement of American naturalists with evolutionary ideas after the 1859 publication of Darwin's *Origin of species*, Numbers' work demonstrates a complexity to the encounter evident in the age of the proponents of Darwinism, their disciplinary training and intellectual background and their location, religious commitment and institutional affiliation. In America and elsewhere, the encounter with Darwinism had shifting contours of acceptance, local sites of resistance, and paths of popular public and scientific engagement determined by the availability of the editions concerned, the lecture itineraries of advocates or denouncers, and the circulation patterns of local newspapers carrying the debates.[116] Such national geographies have been complemented by local historical geographies, in terms both of a comparative analysis of the reception of Darwinism in different towns and an analysis of the exact personal and institutional circumstances which shaped the nature of the encounter in given places.[117]

What such work demonstrates is the need to understand 'science', 'religion' (and, thus, geographical knowledge) not as absolute and universal, but as locally negotiated categories of meaning. The question of such categories having discernable geographies, at whatever scale, is one that depends upon such geographies being understood in a *constitutive* sense, where knowledge

[115] T. Glick (ed.), *The comparative reception of Darwinism* (Austin, Tex., 1974).
[116] R. Numbers, *Darwinism comes to America* (Cambridge, Mass., 1998).
[117] D. Livingstone, 'Darwinism and Calvinism: the Belfast-Princeton connection', *Isis,* 83 (1992), 408–28; D. Livingstone, D. Hart and M. Noll (eds.), *Evangelicals and science in historical perspective* (New York, 1999); D. Livingstone, 'A chapter in the historical geography of Darwinism: a Belfast–Edinburgh case study', *Scottish Geographical Magazine,* 113 (1997), 51–7; J. Scowen, 'A study in the historical geography of an idea: Darwinism in Edinburgh 1859–1875', *Scottish Geographical Magazine,* 114 (1998), 148–57.

takes the shape it does because of where and how it is made, and how and why it moves between the places of its production. The sense that ideas have a historical geography is especially clear in Ogborn's discussion of modernity in London in the century after 1680 and in his concern both to examine the nature of modernity in different places and to conceptualise modernity as a matter of the hybrid relationships between those places.[118]

Modernity is not one moment or age, 'but a set of relations that are constantly being made and remade, contested and refigured, that nonetheless produce among their contemporaneous witnesses the conviction of historical *difference*'.[119] Ogborn's work highlights amongst such witnesses the importance of *geographical* difference by examining five sites within London: the Magdalen Hospital, the street (with especial reference to Westminster), Vauxhall Gardens, the workings of the Excise Office and the Universal Register Office. To take just one, Ogborn shows with reference to the Excise Office that modernity was not just constituted in given sites but was also made through the control of space, nationally and globally in the sense of Britain's expanding empire, and conceptually and epistemologically in regard to the state's capacity to know and to govern itself. Geographical knowledge was a state enterprise of measurement and surveillance, of Excise officers (and surveyors and map-makers and circulated questionnaires) being 'out and about': it was, to use a term of Latour's, a matter of metrology, of state measurement and national constitution.[120]

Ogborn's work is a sustained attempt to think through the ways in which geography matters to the making of modernity and to scrutinise the ways in which modernity was 'grounded' in sites and practices. In noting that place matters, our understanding of the historical geographies of modernity in eighteenth-century London might well have been different had Ogborn focused, say, not on the Magdalen Hospital with its morally reformative spaces for women, but on the analytic spaces of the anatomy theatre where, as has been shown for the Netherlands, dissection had moral overtones about the constitution of the 'body politic', and on the nature of audiences whose social standing and capacity to ask questions was closely controlled.[121]

### *Historical geographies of geographical knowledge: Scotland since 1520*

My concern to understand the historical geography of geographical knowledge in Scotland in relation to the above ideas has parallels with Jacques Revel's discussion of the 'making' of the national territory of France.[122] Revel

[118] M. Ogborn, *Spaces of modernity: London's geographies 1680–1780* (New York, 1998).
[119] K. Wilson, 'Citizenship, empire and modernity in the English provinces, *c.*1720–1790', *Eighteenth-Century Studies,* 29 (1995), 71. [120] Golinski, *Making natural knowledge,* 173–7.
[121] J. Rupp, 'The new science and the public sphere in the premodern era', *Science in Context,* 8 (1995), 487–507.
[122] J. Revel, 'Knowledge of the territory', *Science in Context,* 4 (1991), 133–61.

argues that, by about 1300, the idea of France had become associated with a specific geographical area and that, in the following centuries, political and public authorities sought to gather knowledge about that geographical space in order both to control it and to render its representation standard through agreed weights and measures, for example, and through national mapping. Statistical and natural historical surveys in the eighteenth century were undertaken to 'know France' as national space, the first through use of common national standards, the second by recording local difference. His claims that cartography was inseparable from the affirmation of monarchic power and that forms of state knowledge in the wake of the French Revolution replaced interests in regional diversity are echoed in Godlewska's work on the languages of representation within French geographical knowledge.

Like Revel, my treatment of the 'making' of Scotland through geographical knowledge is historical *and* geographical. I do not see the national territory that was Scotland in the past as a geographical 'given', whose identity can be interrogated for given historical periods assuming that that geographical space was itself securely known. I want to suggest that the discursive forms used to constitute Scotland geographically themselves had a historical geography and, thus, to show that questions of national identity are dependent upon certain processes of intellectual production which have sites and contexts of meaning. What I mean by these remarks is, I suggest, symbolised in a remarkable map of Scotland produced by John Cowley in 1734 (Figure 1.1). Cowley was factor to the Duke of Argyll, an able surveyor and map-maker and author in 1742 of a geography text for schools, *A new and easy introduction to the study of geography.* In 1734, Cowley produced a map entitled 'A Display of the Coasting Lines of Six Several Maps of North Britain', which showed several different Scotlands according to the surveys of earlier map-makers: John Adair (against whose outline Cowley places the other Scotlands), Herman Moll (author in 1725 of *A set of thirty six new and correct maps of Scotland*), Robert Gordon of Straloch, John Senex, and the French map-makers Charles Inselin and Nicholas Sanson. Here, in a single image of the nation, is what Cowley called 'the Disagreement among Geographers in their representations of the Extent and Situation of the Country'. Cowley went on to say that he was much discouraged in pursuing his own plans for a new map of Scotland 'when I reflected on the Discord that appears among the works of so many able Geographers, and the impossibility of either reconciling them to each other, or of framing what may justly bear the Title of a correct Map, without the Assistance of a Survey of the Whole Kingdom'. He proceeded, however (although he never completed his project), 'in hopes that my Errors may induce some more capable Person to furnish the World with a more correct map'.[123]

[123] On Cowley and the wider context of his attempt to 'know Scotland', see C. Withers, 'How Scotland came to know itself: geography, national identity and the making of a nation, 1680–1790', *Journal of Historical Geography,* 21 (1995), 371–97.

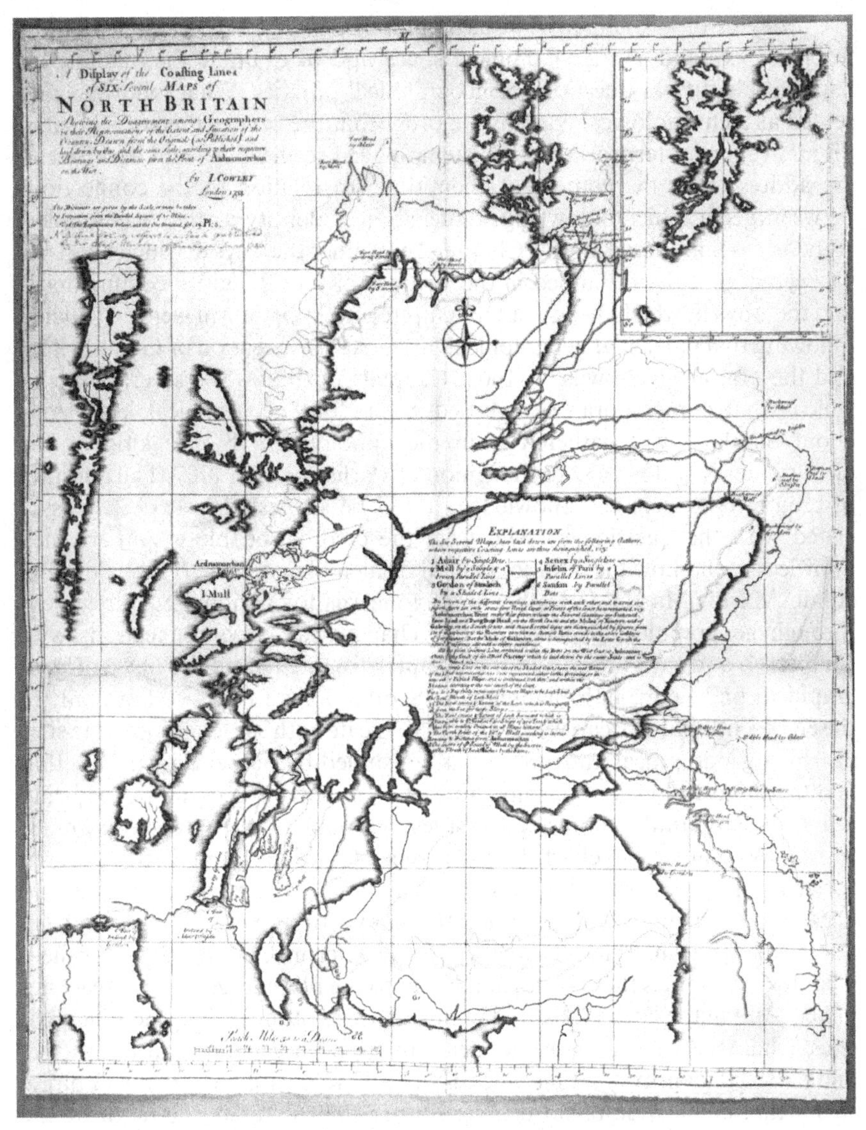


1.1 John Cowley's 'Display of the Coasting Lines of Six Several Maps of North Britain' (1734) illustrates well the sense in which the geographical shape of Scotland was not agreed upon.

Cowley's words and map serve as a point of departure to my own attempts to chart Scotland as a geographical space and to examine the 'making' of national identity as questions of geographical practice. What follows is presented as a chronological narrative in order that the central themes – of sites, discursive forms, territorial and epistemological connections and audiences – are addressed in the context of their time. In relation to the connections between geographical knowledge and national identity as they were understood in early modern Europe, Chapter 2 examines the ways in which Scotland was geographically fashioned in the writings of late Renaissance humanists, and the ways in which debates about empire and imperialism were articulated following that particular geographical 'moment', the creation of Great Britain and the Union of Crowns, in 1603. Chapter 3 explores the several ways in which what contemporaries regarded as credible geographical knowledge about Scotland was gathered from the appointment by the king of the Geographer Royal in 1682 to the Union of Parliaments in 1707. The question of credible geographical knowledge, it will be suggested, was closely associated with the question of who were the credible people to impart such knowledge, and how they did so. Insofar as men like Sir Robert Sibbald, John Adair, Martin Martin and others had concerns for national enlightenment through geographical knowledge, this chapter has close connections with Chapter 4, and my concerns in that chapter to understand the place of geographical knowledge in Enlightenment Scotland. As I hope to show, only in the eighteenth century can we first begin to locate with some certainty the sites in which geographical knowledge was consumed by public audiences within Scotland and to know, too, something of what sort of geography was being taught in Scotland's schools, universities and public sphere. It is also during this period that we can chart the emergence of the idea and practice of territorial geographical survey to know the nation. By the end of the second decade of the nineteenth century, the connections between geographical knowledge and what I have here called 'civic enterprise' were well established. Chapter 5 examines these issues in relation to the place of geography teaching in Scotland in the later 1830s, and in the activities of local civic scientific societies. Chapter 6 considers the connections between geographical knowledge and national identity at a time when Scotland's role within the British Empire was a matter of widespread debate. Rather than reiterate the myth of the 'Imperial Scot', the chapter discusses the ways in which national identity in Scotland was constituted through various practices of scientific survey, in the founding of a national geographical body, and, in rather different ways, in the visionary work of Patrick Geddes and his concerns to see geographical knowledge as a form of citizenship.

This obligation to narrative, as Golinski puts it,[124] is accompanied by an

[124] Golinski, *Making natural knowledge,* 186–206.

equal obligation to connect the central themes of the book *over* time. There are, then, connections to be made between chapters: to take the idea of empire and its different meanings as one illustration, between Chapter 3 (pp. 104–8) and Chapter 6 (pp. 195–207); or the place of geography in Scotland's public sphere from the late seventeenth to the early nineteenth centuries (pp. 97–104, 119–33, 160–72); and the idea of survey (pp. 79–96, 142–53, 210–25). Chapter 8 brings these connections into sharper focus. No doubt, to echo John Cowley, my errors in thus 'mapping' this historical geography of geographical knowledge for Scotland will either induce some more capable person or, perhaps, stimulate similar work in other national contexts.

# 2

# Geography, identity and the making of the nation, 1520–1682

This chapter examines how geographical knowledge was used to give Scotland a sense of national identity in the early modern period. Attention is paid to the ways in which Scotland was geographically imagined and rhetorically 'self-fashioned' from the Renaissance; a process, I suggest, that begins with John Major's 1521 *History of Greater Britain* and Hector Boece's *History of Scotland* of 1527, and which is apparent in the important but neglected geographical introduction to George Buchanan's 1582 *History of Scotland.*

These works, and the historiographical tradition of which they and others were part, drew upon earlier accounts of Scottish identity. That this is so means that we should not see the geographical 'making' of Scotland as having a formal beginning in the first quarter of the sixteenth century. Similarly, 1682 – the year of Sir Robert Sibbald's appointment as Geographer Royal – does not mark any clean epistemological break with earlier ways of national self-knowing since much of his work drew upon earlier geographical enquiry. The dates of this chapter are, however, not entirely ones of convenience. Major's emphasis upon a union between Scotland and England and the creation of a greater Britain distinguishes his work from other discussions of Scotland's historical and geographical identity, even if Boece was largely content to repeat earlier myths. It is also the case that these and other works endured without overmuch scholarly criticism until the later seventeenth century, and that developments then in historical and geographical scholarship were themselves bound up with advances in early modern natural philosophy. Major's work is a point of departure, then, not a definitive beginning, and 1682 marks a new emphasis in geographical studies within Scotland and more widely rather than distinctive differences in either concern or method. In the same way, those other forms of geographical knowledge discussed here – regional 'Descriptions', Timothy Pont's chorographical survey of the 1580s, debates about Scotland's place in the geographical making of a 'Greater Britain' following the Union of 1603, and triumphal processions and theatrical productions designed to present Scotland's national and imperial identities – have both antecedents and longer-term influence. Making sense of such matters depends in the first instance upon understanding the diverse nature of early modern geographical knowledge.

## Geographical knowledge and national identity in early modern Europe

At least some historians and historians of science now consider geography central to, rather than simply consequent upon, the rise of modern science. Hooykaas' claim that Prince Henry the Navigator inaugurated the Scientific Revolution was based upon the observation that it was through Portuguese seafarers that empirical experience of the globe challenged established views. For Parry, the global navigations of the Portuguese and others were the 'catalyst inducing the emergence of modern science in Western Europe'.[1] Historians of geography have only recently begun to acknowledge this central place for geography in the rise of modern science. Several distinguishing features may be noted. This section outlines their principal characteristics in order to put Scotland's geographical making in context.

### *Geography, natural magic and astrology*

Natural magic was, and continues to be, a belief system underpinning that knowledge which sought to control innate natural forces and processes for human advantage. The knowledge of nature that the magi attempted to gain in the Middle Ages and Renaissance was not an end in itself, but, rather, part of a set of procedures, like the astronomical mathematics of the astrologer or the alchemist's crafts, designed to harness nature. It will not do, therefore, to describe either natural magic or astrology as 'non-rational', not least because many of the leading intellectuals in the so-called 'Scientific Revolution' regarded natural magic, numerology and knowledge reliant upon natural lore as rational and legitimate sources of truth about the world. The inventor of logarithms, John Napier, is said to have valued them, for example, because

[1] R. Hooykaas, *Humanism and the voyages of discovery in 16th-century Portuguese science and letters* (Amsterdam, 1979), 16; 'The rise of modern science: when and why', *British Journal for the History of Science*, 20 (1987), 453–73; J. Parry, *The age of reconnaisance: discovery, exploration and settlement, 1450 to 1650* (Berkeley, 1981); D. Banes, 'The Portuguese voyages of discovery and the emergence of modern science', *Journal of the Washington Academy of Sciences*, 78 (1988), 47–58. On the development of science, see, for example, P. Bowler, *The environmental sciences* (London, 1992), 66–139; I. B. Cohen, *Revolution in science* (Cambridge, Mass., 1985); T. Goldstein, *The dawn of modern science* (Boston, 1988); R. Merton, 'Science, technology and society in seventeenth-century England', *Osiris*, 4 (1938), 360–632; M. Hunter, *Science and society in Restoration England* (Cambridge, 1981). For more explicit attention to 'Scientific Revolution', see D. Lindberg and R. Westman (eds.), *Reappraisals of the scientific revolution* (Cambridge, 1990); J. Schuster, 'The scientific revolution', in R. Olby, G. Cantor, J. Christie and M. Hodge (eds.), *Companion to the history of modern sciences* (London, 1990), 217–39; R. Porter and M. Teich (eds.), *The scientific revolution in national context* (Cambridge, 1992); S. Shapin and S. Schaffer, *Leviathan and the air-pump: Hobbes, Boyle and the experimental life* (Princeton, 1985); S. Shapin, *A social history of truth: civility and science in seventeenth-century England* (Chicago, 1994); *The scientific revolution* (Chicago, 1996).

they aided calculations of the significance of the mystic figure 666, the 'number of the beast' in Revelation.[2]

Livingstone has shown how magical knowledges in Renaissance and early modern Europe found expression in geography. William Cuningham's *The cosmographical glasse, conteinying the pleasant principles of cosmographie, géographie, hydrographie or navigation* (1559) was an important utilitarian text with emphases on navigation, commercial enterprise and the skills of triangulation. For Cuningham and for others, the study of geography as knowledge of the terrestrial sphere was grounded in mathematical and astrological study of the celestial spheres. John Dee, late Renaissance map- and instrument-maker and mathematician, understood geography as the science of location and description based on mathematical truths. Dee's involvement with 'political magic' and 'emblematic knowledge' sought to link the celestial, the terrestrial and the intellectual worlds. John Dee intended such conjunctions to be to the political benefit of England both in restoring ancient unities and in promoting an imperialising nation with a global reach. In Leonard Digges' *A prognostication everlastinge* (1576), we find a mathematical-cum-astrological attempt to establish a cosmologically-derived reliable basis to weather forecasting. In early modern 'secrets books' about nature, for example in Thomas Blundeville's *Brief description of universal mappes and cardes* (1589) and his 1594 *Exercises,* a work with emphases on cosmography, astronomy and geography, there is further evidence that geographical knowledge was indissolubly tied, in theoretical speculation and for practical purposes, to natural magic and astrology.[3]

### *Mathematics and political arithmetic*

Eva Taylor and Lesley Cormack have discussed the principal components of 'mathematical geography'.[4] The evolving tradition of mathematical geography has its roots in the Renaissance rediscovery of Ptolemy's *Geographia,* a circumstance which lent mathematical examination renewed emphasis from the late fifteenth century: as a utilitarian practice; as a form of philosophical

[2] C. Hill, 'Science and magic in seventeenth-century England', in R. Samuel and G. Stedman Jones (eds.), *Culture, ideology and politics* (London, 1982), 177. On natural magic and astrology, see M. Feingold, *The mathematicians' apprenticeship: science, universities, and society in England, 1560–1640* (Cambridge, 1984); J. Henry, 'Magic and science in the sixteenth and seventeenth centuries', in Olby, *et.al.* (eds.), *Companion to the history of modern sciences,* 583–96; W. Eamon, *Science and the secrets of nature: books of secrets in medieval and early modern culture* (Princeton, 1994); C. Webster, *From Paracelsus to Newton: magic and the making of modern science* (Cambridge, 1982); P. Curry, *Prophecy and power: astrology in early modern England* (Cambridge, 1989).

[3] D. Livingstone, 'Science, magic and religion: a contextual reassessment of geography in the sixteenth and seventeenth centuries', *History of Science,* 26 (1988), 269–94.

[4] L. Cormack, '"Good fences make good neighbours": Geography as self-definition in early modern England', *Isis,* 82 (1991), 639–61; 'The fashioning of an empire: geography and the state in Elizabethan England', in Anne Godlewska and Neil Smith (eds.), *Geography and empire* (London, 1994), 15–30; *Charting an empire: geography at the English universities, 1580–1620* (Chicago, 1997), 90–128.

enquiry concerned with numbers as symbolic of essential natural qualities; and as a language for describing Nature.

For Cormack, English mathematical geographers had three main areas of study: the determination of longitude at sea; the question of safe navigation in the northern waters given magnetic variation and the inadequacy of maps; and the development of a mathematical map projection that would allow sailing lines to be drawn as straight lines. Such issues were shared by others, notably by the Dutch.[5] Mathematical geography was of profound importance in respect of the 'self knowledge' of European nations and of their emergent territories overseas. Colonial expansion depended upon accurate navigation and the capacity to put the 'new world' to order in the form of maps and cadastral surveys.[6] Likewise, knowing one's own realm depended upon the use of mathematical knowledge in surveying and estate planning. In England, such issues are apparent, for example, in the attempts of the Hartlib circle and others to delineate new agricultural geographies in the newly-drained Fens and in the attempted geodetic survey of England and Wales by John Caswell and John Adams in 1681–4.[7] Similar circumstances were apparent in France and in Germany, and in Russia found perhaps their best expression from the late seventeenth century where geography's practical application in the form of map-making, navigation and survey was an essential element in the making of national space.[8] Simply, geography employed in such ways was a crucial means to state making in the early modern world.[9]

Mathematical geography was also taught as a means to polite learning and world knowledge. As Cormack has shown, the teaching and discussion of mathematical geography between 1580 and 1620 promoted connections to geometry and to astronomy and advanced a common conception of geography as a utilitarian enterprise.[10] John Milton termed geography 'both profitable and delightful': John Locke considered geography 'absolutely necessary'.[11]

[5] Cormack,'"Good fences make good neighbours"', 650.

[6] J. Brotton, *Trading territories: mapping the early modern World* (London, 1996); B. Mundy, *The mapping of New Spain: indigenous cartography and maps of the Relaçiones Geograficas* (Chicago, 1996); M. Wintle, 'Renaissance maps and the construction of the idea of Europe', *Journal of Historical Geography*, 25 (1999), 137–65.

[7] E. Gilbert, *British pioneers in geography* (Newton Abbot, 1972), 55; R. Grove, 'Cressey Dymock and the draining of the Fens: an early agricultural model', *Geographical Journal*, 147 (1981), 27–37.

[8] On France, see N. Broc, *La géographie de la Renaissance* (Paris, 1980), and J. Konvitz, *Cartography in France 1660–1848: science, engineering and statecraft* (Chicago, 1987); for Germany, see G. Strauss, *Sixteenth-century Germany: its topography and topographers* (Madison, Wis., 1959); for Russia, see D. Shaw, 'Geographical practice and its significance in Peter the Great's Russia', *Journal of Historical Geography*, 22 (1996), 160–76. On England, see N. Thrower (ed.), *The compleat platt-maker: essays on chart, map and globe-making in England in the seventeenth and eighteenth centuries* (Berkeley, 1987).

[9] D. Buisseret (ed.), *Monarchs, ministers and maps: the emergence of cartography as a tool of government in early modern Europe* (Chicago, 1992); J. Anderson, *The rise of the modern state* (Brighton, 1986); C. Tilly (ed.), *The formation of nations in western Europe* (Princeton, 1975).

[10] Cormack, *Charting an empire*, 90–128, 203–22.

[11] J. Baker, *The history of geography* (New York, 1963), 21.

The teaching of mathematical geography in seventeenth-century Oxford was paralleled in Cambridge, where Isaac Newton, as part of his responsibilities as second Lucasian Professor in Mathematics and Natural Philosophy, was charged with instruction in geography (although there is no direct evidence he ever taught the subject). Newton's textual revisions and additions to Bernhard Varenius's *Geographia generalis* of 1650 show him concerned with the application of geometry to geographical phenomena. The work in Cambridge of men like James Jurin (who produced an edition of Varenius' *Geographia generalis* with additional mathematical and geographical appendices), Richard Bentley, Roger Cotes and Edmond Halley helped promote, initially anyway, a Cartesian and neo-Platonic discourse in which, to use Varenius' words, geography was 'scientia Mathematica mixta' (a science mixed with mathematics). For Warntz, who has shown that editions of Varenius, Newton and Jurin were studied in the early American colonial colleges, 'It can be concluded that the *Geographia Generalis* commanded significant concern and respect from Newton and the Newtonians. Their attention to it was not aberrant or spasmodic but an integral part of their concerted thrust in science'.[12]

Globe-making and the use of globes was also bound up not only with the mathematical refinement of navigation and with advances in scientific instrumentation, but also with the promotion of geographical knowledge as a form of polite discourse.[13] Geographical and mathematical knowledge also came together in what William Petty called 'political arithmetic', understood as the survey of peoples. Such concerns were not the preserve of any one subject. They were reflections of the increasingly widespread acceptance of mechanist natural philosophy during the seventeenth century and of the utilitarian emphases outlined in Baconian empiricism. Political arithmetic in such terms was an essential part of the making of national identity. Livingstone has extended such claims also to the emphasis on geography as a useful subject in, for example, the Puritan 'social programme' and the work of Joseph Glanvill, amongst others. The emergent philosophy of utility during the seventeenth century was also a technical matter and linguistic matter, concerned with accurate instrumentation and with the development of a precise rhetoric for the description of experimental philosophy and the wider world. Such work is apparent in later seventeenth-century Britain, for example in John Graunt's *Natural and political observations* (1662) and in Gregory King's *Natural and*

[12] W. Warntz, 'Newton, the Newtonians, and the *Geographia Generalis Varenii*', *Annals of the Association of American Geographers*, 79 (1989), 165–91; '*Geographia Generalis* and the earliest development of American geography', in B. Blouet (ed.), *The origins of academic geography in the United States* (Connecticut, 1981), 245–63; J. Baker, 'The geography of Bernhard Varenius', *Transactions, Institute of British Geographers* 21 (1955), 51–60.

[13] H. Wallis, 'Geographie is better than divinitie: maps, globes, and geography in the days of Samuel Pepys', in Thrower (ed.), *Compleat platt-maker* (Chicago, 1986), 1–43; P. Barber, 'Necessary and ornamental: map use in England under the later Stuarts, 1660–1714', *Eighteenth-Century Life* 14 (1990), 1–28; S. Tyacke, *London mapsellers 1660–1720* (Tring, 1978), 14; S. Johnston, 'Mathematical practitioners and instruments in Elizabethan England', *Annals of Science*, 48 (1991), 319–44.

*political observations and conclusions upon the state and conditions of England* (1696). Petty's first practical engagement with social survey in such ways came with his Down Survey in Ireland, undertaken between 1655 and 1659, and published in 1685. The work, also cartographic, was an administrative assessment of colonial economic potential and drew upon a longer tradition of state survey apparent in several earlier *Descriptions* of Ireland. The political and methodological intentions underlying Petty's emphasis in his *Political arithmetick* (1690) are evident in his attention to the measurement of peoples, their produce and their ways of life in order that, as he put it, 'wee shall bee able to Methodize and regulate them to the best advantage of the publiq and of perticular persons'.[14]

### *Of empire and empiricism*

Geography was closely linked with the idea of empire in three senses. The first was in a continuing concern, in Britain and elsewhere, to describe distant places and peoples: to bring to order – by means of map and globe and text – the geography of the world. The second was in the ways in which geography was embraced by Baconian claims to the progress and ordering of such knowledge. In Francis Bacon's relational classification of subjects in his *Instauratio magna*, for example, geography sits between mathematics and natural history.[15] Indeed, the image on the work's engraved title page – the 'ship of learning' laden with new ideas and the facts of geographical discovery sailing back through the Straits of Gibraltar, which traditionally marked the limits of human knowledge of the world – is symbolic of the connections between geographical knowledge, Britain's overseas empire and the authority of empirical methodology (see Figure 2.1). The third was in the development of different theories of *imperium*: not just in terms of the geographical bases to the different theories of the English, the Spanish and the Dutch, but in helping to make a particular empire, Great Britain.[16]

Cormack has addressed these issues with reference to Richard Hakluyt, his literary executor Samuel Purchas, and works on the Virginian colony.[17] In his 1584 *Discourse of western planting*, Hakluyt saw the founding of an English

[14] Petty Papers, Vol. I (IV), in Charles Hull (ed.), *The economic writings of Sir William Petty* (New York, 1963), 25, 90; J. Mykkanen, '"To Methodize and Regulate Them": William Petty's governmental science of statistics', *History of the Human Sciences* 7 (1994), 65–88; T. Aspromourgos, 'Political economy and the social division of labour: the economics of William Petty', *Scottish Journal of Political Economy* 33 (1986), 28–45; P. Buck, 'Seventeenth-century political arithmetic: civil strife and vital statistics', *Isis*, 68 (1977), 67–84; M. Ogborn, 'The capacities of the state: Charles Davenant and the management of the excise', *Journal of Historical Geography*, 24 (1998), 199–224.

[15] B. Vickers, 'Francis Bacon and the progress of knowledge', *Journal of the History of Ideas*, 53 (1992), 495–518; C. Withers, 'Encyclopaedism, modernism and the classification of geographical knowledge', *Transactions of the Institute of British Geographers*, 21 (1996), 363–98.

[16] A. Pagden, *Lords of all the world: ideologies of empire in Spain, Britain and France, c.1500–1800* (Yale, 1995).

[17] Cormack, '"Good fences make good neighbours"'; *Charting an empire*, 135–9.

2.1 The 'ship of learning' and the pillars of the 'Old World', from the frontispiece to Francis Bacon's *Novum Organum* (1620).

colony overseas as a cure for England's unemployment and overpopulation, and a means to establish a greater England. Hakluyt's major work, *The principal voiages and navigations of the English nation* (1589), together with later editions, listed the voyages of discovery of Englishmen in the Americas and Asia. Cormack argues that the work, and that of others by John Brereton and Purchas, helped legitimise the English view of themselves as world leaders, distinct from the rest of the world and helped promote Baconian inductive empiricism for political ends: this work was an apologia for the Old (World) over the New, and for the new 'science' over old beliefs – 'books to build an empire' as Parker has it.[18]

Yet Cormack's attention to the connections between empire and geographical knowledge misses a crucial point. From 1603, England (and Wales) and Scotland had become, in the 'United Kingdom', themselves an empire. Her emphasis on England is unfortunate, since, as I show below, early seventeenth-century contemporaries considering theories of empire were faced with the problem of forging a new, *British,* identity, a problem which involved the reconciliation of different historiographical traditions for Scotland and England.

### *Geography, travel and natural history*

Margarita Bowen has claimed that 'the rise of natural history marked a crucial stage in the encounter between geography and scientific empiricism'.[19] This is sustainable in terms, perhaps, of the writings of the late seventeenth-century physico-theologists like Thomas Burnet, William Derham and John Woodward. It is sustainable, too, in terms of those natural histories of the New World which provided descriptions and pictures of creatures not before seen by Europeans. Through geographical discovery and in concert with the expansion of political empires, the empire of nature was being revealed and, in cabinets and museums, being displayed.[20]

Bowen's claim is misleading, however, unless account is taken of major conceptual shifts within what was held to be natural history over the period covered by this chapter. Of particular importance was the demise from the early 1600s of 'emblematic' natural history, in which understanding was sought by way of similitudes and resemblances between objects and their supposed innate qualities (matters which had close affiliation with interests in natural magic). Such ideas were increasingly replaced by systematic natural

[18] J. Parker, *Books to build an empire* (Amsterdam, 1965).

[19] M. Bowen, *Empiricism and geographical thought: from Francis Bacon to Alexander von Humboldt* (Cambridge, 1981), 107; C. Raven, *English naturalists from Neckham to Ray: a study of the making of the modern world* (Cambridge, 1947); N. Jardine, J. Secord and E. Spary (eds.), *Cultures of natural history* (Cambridge, 1994).

[20] P. Findlen, 'Courting Nature', in Jardine, Secord and Spary (eds.), *Cultures of natural history,* 57–74; *Possessing nature: museums, collecting and scientific culture in early modern Italy* (Berkeley, Calif., 1994); R. French, *Ancient Natural History* (London, 1994), 114–48; V. Dickenson, *Drawn from life: science and art in the portrayal of the new world* (Toronto, 1998).

history, in which the natural world and its products were subject to empirical investigation, to an increasingly precise descriptive language and to 'modern' classification.[21] The move to systematic natural history was related to the advance of 'modern' science, was promoted by bodies such as the Royal Society and the Académie des Sciences, and was further assisted by the concerns of the antiquarian movement with its interests in historical accuracy and chrorographical description. Natural history in these rather more defined terms had close links with geographical enquiry, as we shall see for Sibbald's geographies in particular.

### *Chorography, map-making and national identity*

Strictly speaking, chorography as the description of *parts* of the known world, as opposed to geography's concern with the whole (and cosmography's with the earth's celestial location), dates from the works of Ptolemy, Strabo and Eratosthenes. This difference in principle, if still recognised in the seventeenth century, was always blurred in practice. Some historians of early modern geography have identified a continuing tradition for chorography in regional geography, and others even claim an intellectual genealogy for nineteenth- and twentieth-century regional geography in these early writings.[22] Three principal features of chorography in Renaissance Europe have been noted. First, Renaissance chorography originated in Italy and emphasised the local in both historical and geographical senses: the surveying of estates close to home as matters of useful knowledge, the genealogies of local families of note, local traditions and antiquities. Second, chorography had political significance in that it sought to place local families and local features in ancient roots, to situate them, as it were, historically *and* geographically. Third, Cormack argues, the practice of chorography, more than mathematical and descriptive geography, encouraged the development of an inductive and public spirit in the human sciences: local knowledge, in wider context, was publicly useful. For Britain, these features merged with older chronicle traditions 'to form a chorographical study that was uniquely British and helped to develop pride in country and to breed familiarity with a rapidly developing inductive approach to the natural world'.[23]

The first important English chorography in these terms was William Camden's *Britannia* (1586), which drew together extant chorographical

[21] M. Foucault, *The order of things: an archaeology of the human sciences* (London, 1974); W. Ashworth, 'Natural history and the emblematic world view', in Lindberg and Westman (eds.), *Reappraisals of the scientific revolution*, 303–32; P. Sloan, 'Natural history, 1670–1802', in Olby, *et.al* (eds.), *Companion to the history of modern sciences,* 295–313.

[22] E. Taylor, *Late Tudor and early Stuart geography,1583–1650* (London, 1934), 39–52; F. Emery, 'English regional studies from Aubrey to Defoe', *Geographical Journal,* 124 (1958), 315–25; R. Butlin, 'Regions in England and Wales', in R. Dodgshon and R. Butlin (eds.), *An historical geography of England and Wales* (Cambridge, 1991), 235.

[23] Cormack, '"Good fences make good neighbours"', 657.

traditions in England, went through seven editions by 1607 and influenced several later geographical ventures. Camden's *Britannia* sought to draw past and present together as part of contemporary geographical enquiry. Chorography was allied with map-making and mathematical knowledges in seeking to construct a survey of the present state of the place, by which was meant either the county or the nation. Chorography's goal was a contemporary view of a social stability which, for the elite in society, depended upon the continuing authority of the monarchy and the maintenance of that social order of which they were part. This important point is underplayed in others' studies. Chorography's place in what have been described by Helgerson and Greenblatt as processes of Renaissance 'self-fashioning', of 'writing the nation',[24] extended not just to the geographical object of enquiry, but also to the social authority of the enquirer and of the enquired-into. In these terms, chorographical work was closely connected both with certain forms of mapped representation and to the credibility of the gentleman natural philosopher or other socially appropriate correspondent. The first is evident, for example, in the maps in John Speed's *The theatre of the empire of great britaine* (1611) which have, for each county, a symbolic border of gentry and nobility: the nation as a whole is, if you will, framed by such people. The second is apparent in the late seventeenth century in the reliance placed upon circulated queries as a method for gathering knowledge, queries that were directed at socially credible groups and individuals. These notions of self-identity and socio-political authority depended to a large extent upon establishing the historical longevity of local gentry, the nobility, and, of course, the crown, to see them as 'natural' phenomena in their own right.

Even from such a limited survey, it is clear that geographical knowledge in the early modern period was neither an adjunct to contemporary practices in natural philosophy nor apart from the social and political processes of nation building. There was not an immutable core to what contemporaries understood as geography. As a means to know the world and to give nations identity, geographical knowledge took various representational forms and embraced mathematics, geometry and political arithmetic, natural magic, the taxonomic imperatives of systematic natural history and different political and philosophical notions of utility, and was part of emergent theories of empire.[25]

[24] S. Greenblatt, *Renaissance self-fashioning from More to Shakespeare* (Chicago, 1980); R. Helgerson, 'The land speaks: cartography, chorography and subversion in Renaissance England', *Representations* 16 (1986), 51–85; *Forms of nationhood: the Elizabethan writing of England* (Chicago, 1992).

[25] Lesley Cormack, 'Twisting the lion's tail: practice and theory at the court of Prince Henry of Wales', in B. Moran (ed.), *Patronage and institutions: science, technology, and medicine at the European court* (Woodbridge, 1991), 67–84. See also S. Mendyk, 'Early British chorography', *Sixteenth Century Journal*, 17 (1986), 459–81; *Speculum Britanniae: regional study, antiquarianism, and science in Britain to 1700* (Toronto, 1989). On Ireland, see K. Hoppen, *The common scientist in the seventeenth century: a study of the Dublin Philosophical Society* (London, 1970); F. Emery, 'Irish geography in the seventeenth century', *Irish Geography*, 3 (1958), 259–70; Y. Goblet, *La transformation de la géographie politique de l'Irlande au XVIIe siecle* (Paris, 1930), Vol. I, 147–9.

With these ideas in mind, let me return to those texts of Renaissance humanism in Scotland with which I began this chapter.

## 'Writing Scotland': historical traditions, chorography and national identity

The appearance of a self-sustaining Scottish identity – of Scotland as a country and as a people defined by the kingdom itself – is first evident by the later thirteenth century. The self-conception of the Scots as separate and different, notably from England, was clear by the first quarter of the following century: 'in the three decades between 1290 and 1320 a new sense of Scottish identity and nationhood was refined and articulated'.[26]

This identity, neither inevitable nor innate, is explained by several things: the emergent understanding, by Scots themselves, of Scotland as that geographical area ruled by the king of Scots; the idea of constitutional independence from England, evident in a rhetoric stressing, above all, freedom; the creation of the 'community of the realm', that body of persons within Scotland able to speak for the nation; and, crucially, the role of the crown as the mainstay of the country's independence. These circumstances are apparent, for example, in the Wars of Independence with England culminating in the Battle of Bannockburn in 1314, and in the 1320 Declaration of Arbroath. This declaration, which took the form of a letter to Pope John XXII by the 'community of the realm' in Scotland asking, essentially, that Edward II leave Scotland alone, was one of many similar entreaties in early medieval Europe and had important antecedents within Scotland.[27] The rhetoric of freedom informing the Declaration ofArbroath is important, but its significance rests also upon the fact that it helped establish a constitutional theory of monarchy – simply, that the king of Scots was the rightful guardian of Scotland, but could be replaced if found wanting – and it established a legitimacy for Scotland because of the attention it paid to the historical lineage of Scotland's kings.[28]

From the fourteenth century, drawing upon earlier accounts such as, for example, the geographical description of Scotland north of the Forth in the late twelfth-century manuscript 'De Situ Albanie', these origin myths found expression in John Barbour's epic narrative poem *The Brus,* in Blind Harry's *Wallace* (in the 1480s), and in that chronicle tradition most clearly apparent

[26] E. Cowan, 'Identity, freedom and the Declaration of Arbroath', in D. Broun, R. Finlay and M. Lynch (eds.), *Image and identity: the making and re-making of Scotland through the ages* (Edinburgh, 1998), 38; D. Broun, 'The origin of Scottish identity in its European context', in B. Crawford (ed.), *Scotland in dark age Europe* (St Andrews, 1994), 21–31; , 'Defining Scotland and the Scots before the wars of independence', in Broun, Finlay and Lynch (eds.), *Image and identity,* 4–17; Fiona Watson, 'The enigmatic lion: Scotland, kingship and national identity in the wars of independence', in Broun, Finlay and Lynch (eds.), *Image and identity,* 18–37; E. Cowan, 'Myth and identity in early medieval Scotland', *Scottish Historical Review,* 63 (1984), 116–22.

[27] W. Ferguson, *The identity of the Scottish nation: an historic quest* (Edinburgh, 1998), 36–55.

[28] Cowan, 'Identity, freedom and the Declaration of Arbroath', 51; D. Broun, 'The birth of Scottish history', *Scottish Historical Review,* 76 (1997), 4–22.

in John of Fordun's chronicle of *c.*1380, *Chronica gentis scotorum.*[29] Fordun's work – known usually as the *Scotichronicon* – sought to vindicate Scotland's independent identity by emphasising the legitimate hereditary succession and lineage of Scotland's kings:

> Stripped to its barest essentials, the account of the Scottish past which Fordun inherited from his largely anonymous predecessors combined a myth of racial origins – the Scots' descent from the Greek Prince Gathelus and the eponymous Scota, daughter of Pharoah – with the belief that the Scottish kingdom itself was founded by Fergus I in 330 BC and had had a continuous and independent existence under a line of over one hundred kings ever since that date.[30]

Fordun is also important because he dismisses the Brutus origin myth advanced by Geoffrey of Monmouth, itself used by the English kings to legitimise their claims over Scotland: namely, that a Trojan, Brutus, had conquered the island of Albion and divided it among his three sons. Geoffrey's influential *History of the kings of Britain,* written *c.*1135, promoted the idea of Britain as a single geopolitical entity under English hegemony. Fordun's chronicle sought to establish a Celtic origin for Scotland in opposition to Geoffrey's emphasis upon Brutus and the Britons. Fordun's work, and that of others in Scotland's chronicle tradition, is illustrative of what has been termed that 'dynamic and historically-based *mythomoteur* capable of explaining the community to itself (and others)'.[31] It was for these reasons that it attracted the attention of Renaissance humanists and others in Scotland concerned to give historical legitimacy to Scotland's geographical identity.

### *Renaissance humanism, historical identity and chorography*

Roger Mason has claimed that in his *Scotorum historia* (*History of Scotland),* published in Paris in 1527, Hector Boece – friend to Erasmus and first principal of King's College in Aberdeen – not only established the Scottish *mythomoteur* in 'its most elaborate and exotic form', but that 'Scotland's status among the nations of Europe was asserted more powerfully than ever before'. Following Fordun, Boece charted the history of the Scots from their origins in Greece and Egypt through the foundation by Fergus I in 330 BC of a kingdom that ran uninterrupted through over one hundred kingly descendants. He stressed Scotland's past as one of Europe's most ancient kingdoms, a nation that had resisted Roman imperialism (when England and much of mainland Europe had succumbed) and one, moreover, that had never been

29 Fergusson, *Identity of the Scottish nation,* 43–50; D. Broun, *The Irish identity of the kingdom of Scots* (London, 1989).

30 R. Mason, *Kingship and the commonweal: political thought in Renaissance and Reformation Scotland* (East Linton, 1998), 84.

31 Mason, *Kingship and the commonweal,* 79; R. Mason, 'Chivalry and citizenship: aspects of national identity in Renaissance Scotland', in R. Mason and N. Macdougall (eds.), *People and power in Scotland* (Edinburgh, 1992), 50–73.

subjected to slavery. For Mason, the translation of Boece's *History of Scotland* into Scots by John Bellenden (*c.*1531–6) 'was a resounding declaration, not only of the Scots' belief in their history, but also of their belief in themselves'.[32]

Boece, I want to argue, was historically and geographically 'making Scotland' in his writing: he was giving the nation identity. Boece emphasises Scotland's martial glories, praises Highlanders for their native simplicity and places huge importance upon the ideas of the 'commonweal' and the defence of freedom. In those terms, his work is a historical account of Scotland, an account, albeit, that repeats rather than challenges earlier chronicle traditions. Boece's work also contains geographical descriptions of Scotland in order to allow readers to site historical events and to know the location of mountains and rivers. Of course, his geographical descriptions were no more the result of accurate testing than was his use of Fordun. One later commentator has noted that 'his geographical knowledge appears to have been inconsiderable; and, accordingly, his description of Scotland is inaccurate'.[33] Accuracy, historical or geographical, is not, however, why Boece is important. In drawing upon such earlier chronicle traditions, in considering the history of a nation as influenced by its geography, and in his attention to the sustaining antiquarian beliefs upon which people drew, Boece was working within the traditions of early European chorography. His attention to local geographical circumstances in order to understand the whole – 'Becaus we have discrivit [described] all regionis of Scotland in speciall, we will schaw sum things concerning thaim in general', as he put it[34] – is exactly consistent with that contemporary distinction between special and general geographical description informing works, by definition anyway, of chorography and geography. His *History* is rich in local geographical description: of the 'Est, West, and Middill Bordouris of Scotland', of 'notabill Townis', and of the nature of natural resources in Scotland. In these terms (although some of the headings are the work of later editors), and, for example, in his attention to 'the maist notabill thingis thairof' in his description of the Outer Isles – that is, to unusual events and to local antiquities and beliefs – Boece provides a chorographical description of Scotland.

In suggesting that Boece be read this way, I am not seeking to challenge the treatment of him by modern historians but, rather, to add to it by attempting to see his work as part of wider Renaissance interests in knowing one's nation. Boece is important, moreover, because of his influence upon the work of the humanist George Buchanan, whose own *History* of 1582, as I contend below,

[32] Mason, *Kingship and the commonweal*, 95.

[33] This comment is made in the Biographical Introduction to the 1821 edition of Boece: H. Boece, *The history and chronicles of Scotland: written in Latin by Hector Boece, Canon of Aberdeen; and translated by John Bellenden, Archdean of Moray, and Canon of Ross* (Edinburgh, 1821), Vol. I, xxvii. In the 1541 folio edition of Boece's *History*, the title has the words 'with the Cosmography and Description thairof'.

[34] Boece, *History and chronicles*, (1821 edition), xxiv.

should also be seen in chorographical terms. Boece is of interest, too, because, as Ferguson has shown, it is with criticism of his work and of the chronicle traditions, of which his *Scotorum historia* is the exemplar, that a critical historiography of Scotland's historical identity emerges in the later seventeenth century.[35] Finally, Boece merits our attention because his work was more influential than that other contemporary work exploring Scotland's national identity, John Major's *History of greater Britain as well England as Scotland (*hereafter, simply, *Greater Britain*), begun in 1518 and published in 1521.

Major or Mair (I use the Latinised 'Major' in order to distinguish him from a later geographical John Mair) was born at Gleghorn fermtoun near North Berwick, and educated at Cambridge and at Paris, and was a European scholar of great distinction. The style and content of his *Greater Britain* is very different from Boece. Major dismisses the Trojan myth of the English and the Gathelus-Scota myth of the Scots, criticises Scotland's nobility as having too great an influence over the king, and, of importance here, proposes a geographical union of the English and Scottish kingdoms.

Major's historical reasoning that Scotland's interests be subsumed beneath a singular British identity should be seen as an expression of a geographical, even an imperial, consciousness. His dismissal of origin myths did not mean he denied Scotland's historical claims. Rather, such claims were to be accorded equal place in a union 'of equal partners each voluntarily recognising the liquidation of their separate sovereign status and the simultaneous creation of a new all-encompassing sovereignty'.[36] In creating that new geopolitical entity, a greater Britain, the very term 'Scotland' would itself be history, even though the term 'Britain' suggested an authority to the Brutus myth over other origin stories. In this regard, Major's argument foreshadows the Union of 1603 as a geographical question. Although he is much neglected outside of modern historical scholarship, his work echoes in late twentieth-century British political discourse. As Mason has shown, Major's work has importance for any understanding of the history of British imperialism and of Anglo-Scottish political relations.[37] Others have seen Major as a figure of 'pivotal significance' in the evolution of political ideology in Britain and as a leading European logician.[38]

[35] Ferguson, *Identity of the Scottish nation*, 120–43. [36] Mason, *Kingship and the commonweal*, 44.

[37] R. Mason, 'Covenant and commonweal: the language of politics in Reformation Scotland', in N. Macdougall (ed.), *Church, politics and society: Scotland 1408–1929* (Edinburgh, 1983), 97–126; 'Scotching the brut: politics, history and national myth in sixteenth-century Britain', in R. Mason (ed.), *Scotland and England 1286–1815* (Edinburgh, 1987), 6–84; 'Kingship, nobility and Anglo-Scottish union: John Mair's *History of Greater Britain* (1520)', *Innes Review*, 41 (1990), 182–222; 'Chivalry and citizenship'; 'Imagining Scotland: Scottish political thought and the problem of Britain 1560–1660', in R. Mason (ed.), *Scots and Britons: Scottish political thought and the Union of 1603* (Cambridge, 1994), 3–13; *Kingship and the commonweal*, 36–77.

[38] Q. Skinner, 'The origins of the Calvinist theory of revolution', in B. Malament (ed.), *After the Reformation: essays in honour of J. H. Hexter* (Philadelphia, 1980), 321; A. Broadie, *The circle of John Mair: logic and logicians in pre-Reformation Scotland* (Oxford, 1985).

Williamson has noted that Major wrote about improving agriculture 'and what might be remotely called social statistics: that is about the population of cities, wealth, and taxes'.[39] In terms which would have been familiar to Major's contemporaries, we should also see his *Greater Britain* as a treatise in theoretical political geography, both for the above reasons and because Major considered Scotland to be geographically divided between the Lowlands and the Gaelic-speaking Highlands. Unlike Boece, who praised the Highlanders for their qualities of endurance in their upland environment, Major saw in Highland Scotland only regional difference and cultural dissent. In those terms, his theoretical sensibilities, if such they may be called, had later practical consequences: 'his views lent intellectual responsibility to the ever-increasing hostility and contempt towards Gaeldom which would lead later in the century to James VI's attempts to reinforce Lowland hegemony by extinguishing Gaelic culture altogether'.[40]

For different reasons, there was much of the geographer in George Buchanan, or, perhaps more properly, much that is geographical in the opening sections to his last and greatest work, *Rerum Scoticarum Historia,* published in Edinburgh in 1582, the year he died. Born in 1506, Buchanan is now recognised as a great sixteenth-century humanist (see Figure 2.2). Educated at St Andrews (where he studied under John Major), Buchanan spent much time in France, it is possible he knew Boece in Paris and he certainly visited the Royal Cosmographer, André Thevet, before returning to Scotland in 1561. He was, variously, principal of St Leonard's College in St Andrews and, from 1570, tutor to the young James VI. In his *History,* Buchanan broadly followed the chronicle tradition of Fordun and of Boece. But on detailed linguistic grounds, he rejected both the Brutus and the Gathelus-Scota origin myths. In rejecting Scots' origin myths but accepting the historicity of Scotland's kings and in reasoning from the evidence of language, Buchanan was advancing a more critical historical scholarship than either Boece or Major.[41]

Buchanan begins his *History* with a detailed geographical description of Scotland. Among modern historians, only Ferguson has commented upon this, remarking how 'Buchanan's description of Scotland is, in fact, a remarkable *tour de force*, thorough, well organised and full of all kinds of acute observations', and terming it 'a brilliant comprehensive description of the geography of Scotland'.[42] Historians of geography have paid it no attention. In terms of contemporary thinking, however, and mindful of the political and ideological purposes to which chorographical descriptions were put, it is an

[39] A. Williamson, *Scottish national consciousness in the age of James VI* (Edinburgh, 1979), 97.

[40] Mason, *Kingship and the commonweal,* 55.

[41] Ferguson, *Identity of the Scottish nation,* 79–98. See also R. Bushnell, 'George Buchanan, James VI and neo-classicism', in Mason, *Scots and Britons,* 91–111; R. Mason, 'George Buchanan, James VI and the presbyterians', in Mason, *Scots and Britons,* 112–37; J. Burns, 'George Buchanan and the antimonarchomachs', in Mason, *Scots and Britons*, 138–58.

[42] Ferguson, *Identity of the Scottish nation,* 87.

2.2 The monument to George Buchanan in Killearn. Erected in 1788 from a design by James Craig, part of the inscription reads 'The monument was set up to the everlasting memory of George Buchanan, brave among the brave, learned among the learned and most wise amongst the wise'.

important contribution. Buchanan acknowledges a debt to Dean Monro's 'Description of the Western Isles', produced in 1549. He corrects some of Boece's geographical comments. He offers regional descriptions of parts of Scotland, commenting upon the relative state of natural resources and the agricultural economy of various counties: Lothian, for example, he considered 'far excels all the rest in the cultivation of the elegancies, and in the abundance of the necessaries of life'; the area between Badenoch and the North Sea is, he comments, 'a region – for a Scottish one – remarkably rich in the products of the water and of the land'.[43] Buchanan expresses an interest in Scotland's prehistoric antiquities both in speculating about their origins and, in recognising similar features in different parts of Scotland, as a matter of comparative regional geography. He takes issue with classical authorities including the Roman geographer, Mela Pomponius, on the etymology of Scottish island names. In his description of Highlanders, Buchanan offers both a geographical description of the region and a recognition of the regional emergence of the Highlands within the contemporary Scottish geographical and political consciousness.

All of these features mark Buchanan's geographical introduction to his *History* as a chorography. Quite what we should properly call it matters less, I suggest, than his view, and that of his contemporaries, as to its purpose. In one sense, this demands we recognise that chorography was an accepted part of historical enquiry. Geographical description was, in short, *necessary* for historical understanding. In another sense, Buchanan, like others, was valuing geographical enquiry for its own sake. Buchanan did have wider geographical interests. In his unfinished five-book cosmological epic, *Sphaera*, begun *c.*1555, Buchanan acknowledged the influence of Honter's *Rudimentorum cosmographicorum,* published in 1531; in Book III of the work (which is principally concerned with the geographical question of tropical habitability), Buchanan's account is, effectively, a debate with the authority of classical geographical accounts, notably of Aristotle and Posidonius.[44] Buchanan possibly used the works of ancient geographers in his teaching. He certainly influenced Andrew Melville, the Scottish churchman, to do so: Melville's attention to mathematics and geography as part of his teaching at Glasgow in 1574–7 – teaching which included use of Dionysius's *Geographia,* of Honter and the study of 'Historie, with the twa lights thereof, Chronologie and Chirographie' – followed Buchanan's guidance on the nature of a proper curriculum.[45] Geography was taught in Glasgow using these and other texts, appearing principally in the fourth year under the heading of cosmography, although it was

[43] George Buchanan, *The history of Scotland* (Glasgow, 1827), translated by James Aikman in four volumes, Vol. I, 21, 32.

[44] J. Naiden, *The* Sphera *of George Buchanan (1506–1582)* (Washington, 1969), 54–6, 113–14; J. Durkan, *Bibliography of George Buchanan* (Glasgow, 1994), 21–2.

[45] James Melville, *Mr James Melville's diary, 1556–1601* (Edinburgh, 1829), 38–9; T. McCrie, *Life of Andrew Melville* (Edinburgh, 1856), 7, 30–1.

taught in association with astronomy in the third year. Such instruction also obtained in Edinburgh where Robert Rollok, the first principal, was required to teach chorography upon his appointment in 1583.[46] Melville's own geographical work, the poem *Scotiae topographia,* later used as a part preface to Blaeu's maps of Scotland in the 1662 edition of his *Atlas novus,* was, for one eighteenth-century commentator, 'Buchanan's prose turn'd into elegant verse'.[47]

I am not claiming that Boece, Major and Buchanan should be seen as geographers, not least because they did not so term themselves. I do want to argue, however, that, in terms contemporaries would have understood, these men were in different ways thinking geographically and dealing with geographical issues: issues that had to do with the history of the idea of Scotland's national identity; with theoretical reasoning about crown control over what Scotland was then and had in the past been; and with the practical political management of the realm – with what Arthur Williamson terms the 'politics of civilization'.[48] Colin Kidd also notes that 'the chroniclers, humanists and antiquarians of late medieval and early modern Scotland had created a formidable ideology of Scottish nationhood'.[49] This was, I submit, not simply a historical enterprise, but the result of men whose writings helped constitute Scotland geographically. Neither Boece, Major nor Buchanan were alone in so doing. Buchanan's chorographical emphases are apparent, for example, in Bishop John Leslie's 1578 *History of Scotland,* with its attention to the 'Descriptione of the regiounes', the 'main tounes' and the 'fertilitie of the soile'.[50] From what we know of early Scottish libraries, many Scottish churchmen had works of geography in their collections: John Duncanson, Principal of St Leonard's College in St Andrews in *c.*1566, owned a copy of Sebastian Münster's *Sphaera mundi* (1546) as did James Stewart, a prior there; copies of Ptolemy's *Geographia* and of Mela Pomponius's *Cosmographia* (the Paris edition of 1513) were owned by the Bishop of Ross and by William Hay, successor to Hector Boece as Principal to King's College, Aberdeen.[51]

## *Geographical and chorographical descriptions, 1529–*c.*1680*

Hume Brown has documented twenty-two travellers to Scotland between about 1300 and 1680. Some commentators were more perceptive than others:

[46] J. Durkan and J. Kirk, *The University of Glasgow 1451–1577* (Glasgow, 1977), 315–17, Appendix K, 444; *Charters and statutes of the University of Edinburgh*, (1583).

[47] W. Nicholson, *Scottish historical library* (London, 1702), 18; McCrie, *Life of Andrew Melville,* 448, n.13.

[48] A. Williamson, 'Scots, indians and empire: the Scottish politics of civilization 1519–1609', *Past and Present,* 150 (1996), 46–83; 'George Buchanan, civic virtue and commerce: European imperialism and its sixteenth-century critics', *Scottish Historical Review,* 76 (1996), 20–37.

[49] Colin Kidd, *Subverting Scotland's past* (Cambridge, 1993), 29.

[50] E. Cody (ed.), *The history of Scotland, wrytten first in Latin by the most reverend and worthy Jhone Leslie* (Edinburgh, 1888).

[51] J. Durkan and A. Ross, *Early Scottish libraries* (Glasgow, 1961), 55, 56, 57, 93, 113, 149.

the remark of Aeneas Sylvius (the future Pope Pius II) during his tour of Scotland in about 1435 that 'Nothing pleases the Scots more than abuse of the English' recognised a vital element of contemporary Scottish identity.[52] Others, less discerning, are of no less interest in their own way. For example, James Howell (who also wrote descriptions of the Dutch under the name 'Owen Feltham') was the author of *A perfect description of the people and country of Scotland* (1648) in which, in addition to denigrating Scotland and the Scots as a whole – 'The Ayr might be wholesome, but for the stinking people that inhabit it. The ground might be fruitful, had they wit to manure it' – he reserved particular contempt for Scottish women: 'Pride is a thing bred in their bones, and their flesh naturally abhors cleanliness; their breath commonly stinks of Pottage, their linen of Piss, their hands of Pigs turds, their body of sweat, and their splay-feet never offend in Socks. To be chained in marriage with one of them, were to be tied to a dead carkass, and cast into a stinking ditch'.[53]

Of more significance, however, are the geographical descriptions of Scotland undertaken by Scots. A note of caution is necessary here. In documenting chronologically a variety of geographical descriptions, I am not supposing any straightforward intellectual genealogy. Similarly, it is not always possible to know much about their production, or about their audiences, or, for manuscript accounts, to know if they were intended for never-realised publications. Yet the fact that they exist at all is important: in that regard, these works of geographical description represent a richer tradition than has hitherto been supposed for Scotland, and, indeed, for Britain and early modern geography at this time. Such texts are historical *and* geographical: as Emerson put it in his 1652 poetical description of Orkney, 'Butt he that would a good historian bee/ Esteem'd and praise'd for full Geography/ Must show the length, the breadth, the situation/ The Lawes, Religion, Manners of a Nation'.[54]

'Jo. Ben's' 1529 'Description of the Orkney Islands' is probably the work of one John Benston (or Beinston), a clerk to the bishop of Orkney at that time. The work, in a Latin manuscript, was known to Sibbald. Translations have appeared from time to time, but little is known of the context to the work.[55] It contains, island-by-island, a chorographical description of the Orkneys with brief reference to customs, local etymology and agricultural produce. It shows stylistic similarities to the better-known 'Description of the Western Isles' by Dean Monro of the Isles in 1549, but there is no evidence that Monro drew upon the earlier work. Monro's 'Description' is chiefly a chorographical

[52] P. Hume Brown, *Early travellers in Scotland* (Edinburgh, 1978), 27.
[53] J. Howell, *A perfect description of the people and country of Scotland* (London, 1648), 3, 7–8.
[54] J. Emerson, *Poetical descriptions of Orkney* (Edinburgh, 1835), 7.
[55] A. Mitchell (ed.), *Geographical collections relating to Scotland made by Walter Macfarlane* (Edinburgh, 1908), Vol. III., viii–xiii, 302–24.

account of the Hebrides in which attention is paid to local antiquities and native belief. Mention is made of the clan chiefs and principal landowners. Monro's work was conducted by sea; in that respect, it shares something with Alexander Lindsay's *Rutter of the Scottish seas* (1540) and with the circum-navigation that year of Scotland's northern and western isles by James V. The account of this voyage, drawn up by the French cosmographer, Nicholas (sometimes Nicholay de Nicholay) d'Arfeville in 1583, has been seen 'as the record of a serious attempt to produce an accurate chart of the coasts of Scotland'.[56] This work may also have been a stimulus to the land-based mapping of Timothy Pont.

Several geographical descriptions, published and manuscript, appear in the 1630s. An anonymous 'Ane Descriptione of Certaine Pairts of the Highlands of Scotland' has been dated to 1630 but may be later.[57] This chorographical work offers a more detailed account than Monro, and focuses upon the rental value of parts of the Highlands as well as offering comments upon natural products and local beliefs. In style and organisation, it is similar to Sir Robert Gordon of Gordonstoun's 'Description of the Province of Southerland [Sutherland], with the Commodities thereof', which Gordon, a gentleman of the Privy Chamber of James VI and younger brother of the then Earl of Sutherland, included as part of his *Genealogy of the Earldom of Sutherland* (first published in 1813).[58] In its explicit attention to his family's lineage and, thus, with their connectedness to place, Gordon's work shares with those many contemporary English 'Gentlemen of the Shires' the concern to document their locality and themselves as part of that local geography.[59] There is no reason, however, to suppose Sir Robert was the author of 'Ane Descriptione'.

Of greatest interest in this period are the manuscripts entitled 'Topographical Descriptions relating to Scotland', brought together by Sir James Balfour of Denmilne, Lord Lyon King of Arms, between *c.*1632 and *c.*1654.[60] In places, it is difficult to distinguish on the manuscripts Balfour's hand from the later additions of Sibbald who used them, as we shall see he used so many other accounts, as part of his own geographical work. Sir James was, like his brother Sir Andrew – Sibbald's closest friend – a distant cousin to the Geographer Royal. There is no way of knowing if kinship rather than professional authority facilitated Sibbald's access. Balfour's 'Topographical Descriptions' emphasise the genealogies of the gentry and nobility within

[56] D. Moir (ed.), *The early maps of Scotland to 1850* (Edinburgh, 1973 edition), 19–23; I. Adams and G. Fortune (eds.), *Alexander Lindsay: a rutter of the Scottish seas* (Greenwich, 1980); J. Cameron, *James V the personal rule, 1528–1542* (East Linton, 1998), 239, 245, 251; A. Forte, 'Kenning be kenning and course be course. Alexander Lindsay's rutter and the problematics of navigation in fifteenth and sixteenth century Scotland', *Review of Scottish Culture,* 11 (1998–9), 32–45.

[57] Mitchell (ed.), *Geographical collections,* Vol. II., xxiii, 144–92.

[58] Sir R. Gordon of Gordonstoun, *A genealogical history of the earldom of Sutherland* (Edinburgh, 1813), Section I, 1–12; W. Fraser, *The Sutherland book* (Edinburgh, 1892), I., i–xiii.

[59] J. Gentry, 'English chorographers 1656–1695: artists of the Shire' (Ph.D, thesis University of Utah, 1985).

[60] National Library of Scotland [hereafter NLS], Adv. MS 33. 2. 27.

Scotland's counties as well as providing regional descriptions. It is possible, of course, that such geographical knowledge was compiled as part of Sir James' court position. If this is so, it was not the only such work. Together with Sir John Lawder, Lord Fountainhall, one of the senators of the College of Justice in Scotland, Balfour also produced, in manuscript, a 'Geographical Dictionary'.[61] Works of this type were to become increasingly common. In Balfour's work, the entries are mainly of Scottish place names and features, together with references to Camden's *Britannia.* Balfour was collecting Pont's map and manuscript materials at about the same time; he received support from the King to fund their publication, but the project was never realised.

In Balfour's collections, and, therefore, in his mind, geography was on the agenda as part of state knowledge. Unlike his cousin in the 1680s, we cannot unequivocally claim that Balfour was, as part of his professional position, using his geographical enquiries between about 1632 and 1654 as political instruments, although he did co-organise the 'symbolic geography' of Charles II's triumphal entry into Edinburgh in 1633. Yet it is probable both that he was and that he was not the only one, given that such knowledge would have been useful in understanding the disrupted nature of church and state politics in that period. In August 1641, Sir John Scot of Scotstarvet, director of the Chancery, had announced his intentions 'to have a description of our Shyredomes', by which, it is thought, he meant maps as well as written accounts.[62] In 1642, the General Assembly of the Church of Scotland instructed its constituent presbyteries 'to sett doun the descriptiouns of there severall paroches according to the alphabet [set of instructions] then given to the severall commissioners to deliver to there presbyteries and to report the same to the chancellorie'.[63] Scot was using the Church of Scotland to coordinate map and textual descriptions for use in Blaeu's 1654 *Atlas*. This the church did when it could: 'in its nine annual meetings in 1641–9 the general Assembly showed its interest in the *Atlas* project six times; and the three years in which it neglected the issue were those in which the political turmoil of the time reached peaks in open civil war (the Montrose campaigns in 1644 and 1645; the Engagement crisis in 1648)'.[64] Manuscript descriptions of several parts of Scotland from 1644, the essentially ecclesiastical survey of several south-east Scotland parishes in 1627, chorographical descriptions of Aberdeen by James Gordon in 1645, a description of Edinburgh in 1647, Scot's own travels in Fife in 1642–3 and several chorographical descriptions of sheriffdoms within south-west Scotland dating from 1649 must all be presumed the result of church and state involvement with geographical

[61] NLS, MS Acc. 10015.

[62] National Archives of Scotland [hereafter NAS], CH 2/154/2, ff. 39v-40v; D. Stevenson, 'Cartography and the kirk: aspects of the making of the first atlas of Scotland', *Scottish Studies*, 26 (1982), 1–12.

[63] SRO, CH 2/154/2, ff. 39v–40v, 44r–45r.

[64] Stevenson, 'Cartography and the kirk', 5.

enquiry.[65] It is to be regretted that the turmoil of the times prevented the wider uptake by presbyterial authorities of the 1642 Sir John Scot-General Assembly proposals for geographical descriptions.

We can state with some confidence, then, that in texts and manuscripts, Scotland was being 'given shape', in chorographical terms, from at least the second quarter of the sixteenth century. What Lynch has called 'the imagining of Scotland' from this period was, as he himself notes, strongly geographical.[66] The engagement with chronicle traditions and chorographical description marks, I suggest, something more than simply a *sensitivity* to geographical issues and to the regional differences making up the realm, something more than a self-awareness by certain Scots of their kingdom's bounds. They mark what I would want to see as a recognition of the *power* of geographical insight, of the importance of *thinking geographically* to an understanding of national identity. To be sure, this was not happening only in Scotland. There is, however, a particular expression of what I mean for Scotland in James VI's *Basilicon doron*. This work, first published in 1599, was intended as a manual of kingship for his son, Prince Henry, who had large numbers of geographers at his court.[67] Two statements within *Basilicon doron* lend weight to the more general claim I am making here: the king's suggestions for Scotland's emergent 'Highland problem'; and the importance for proper government of being geographically 'active' as a monarch.

For James, Highlanders were of two sorts: 'the one, that dwelleth in our maine land that are barbarous, and yet mixed with some shew of civilitie: the other, that dwelleth in the Iles and are alluterlie barbares, without any sorte or shew of civilitie'. For this latter group, James's instruction to his son (other than to think of Highlanders as 'Wolves and Wilde Boares') was to 'followe forth the course that I have begunne, in planting Colonies among them of answerable In-landes subjectes, that within shorte time maye roote them out and plant civilitie in their roomes'.[68] Controlling the realm was not just a Highland problem (the Borders were also singled out for attention), but the association between urbanisation, commercial development and colonisation – of geographical thinking as an aid to statecraft – is clear. James's advice to Henry noted that 'it will be a greate helpe unto you, to be wel acquent [acquainted] with the nature and humoures of all your subjects, and to know particularlie the estate of every part of your dominions'. 'I woulde therefore counsel you', continued James, 'once in the yeare to visit the principall parts

[65] Mitchell (ed.), *Geographical collections*, Vol. II, xxvi–li, 1–21, 192–200, 201–10, 224–305, 308–11, 469–508, 509–77, 578–613, 614–40; I. Whitaker, 'The reports on the parishes of Scotland, 1627', *Scottish Studies*, 3 (1959), 229–32.

[66] M. Lynch, 'A nation born again? Scottish identity in the sixteenth and seventeenth centuries', in Broun, Finlay and Lynch (eds.), *Image and Identity*, 82–104.

[67] Cormack, 'Twisting the lion's tail'; *Charting an empire*, 206–23.

[68] [King James VI], *Basilicon doron* (Edinburgh, 1599), 42–3.

of the cuntry ye were in; and (because I hope yee shal be King of moe cuntries then this) once in the three yeares to visite al your kindomes, not lippening [listening] to Viceroyes but hearing your self their complaints'.[69]

For kings no less than for geographers, this was a matter of trust and of reliance upon secure geographical information. Since, by 1603, the days of the itinerant monarch were over,[70] these circumstances only highlight further the importance of maps.

## 'Knowing one's bounds': Timothy Pont's chorographical survey of Scotland

It has been argued that in the period between 1546 and 1595, 'the map of Scotland developed from a crude outline . . . to a map full of detail'.[71] In these terms, there are several noteworthy maps in the half century from *c.* 1540. Alexander Lindsay's chart of 1540, known only from a version of 1583, which, like his *Rutter,* resulted from the navigation by James V of the northern and western isles as part of his attempted pacification of Highland chieftains. André Thevet published a transcription of Lindsay's work in 1587.[72] The *c.* 1566 map of 'Scotia', the first extant printed map of Scotland, derived from a 1546 map by George Lily of the British Isles, and was used as the basis to the map of Scotland included in Bishop John Leslie's 1578 *History.* The 1543 map by John Elder was borrowed by Gerard Mercator in 1564, and, in turn, by Abraham Ortelius in 1573. Skelton has argued that Boece 'exercised a considerable influence' upon the cartography of Scotland in this period in that George Lily drew upon Boece's 1527 work and later cartographers borrowed from Lily.[73] Elder is an interesting if shadowy figure. In about 1555, describing himself as 'John Elder, Clerke, a Reddschanke', that is, a Highlander (he was Caithness-born and educated at St Andrews, Aberdeen and Glasgow), Elder wrote to Henry VIII suggesting a union between the two kingdoms. His tract was, in effect, a preface, a plot in more than just the cartographic sense, to a geographical description of Scotland for the information of the English king.[74]

It is because mapping was understood to give shape to one's kingdom – to

[69] [King James VI], *Basilicon doron* (Edinburgh, 1599), 64–5; on the reception and translation of the work, see J. Wormald, 'James VI and I, *Basilicon doron* and *The trew law of free monarchies*: the Scottish context and the English translation', in L. Peck (ed.), *The mental world of the Jacobean court* (Cambridge, 1991), 36–54. On the political management of Scotland, see M. Lynch, 'James VI and the highland problem', in J. Goodare and M. Lynch (eds.), *The reign of James VI* (East Linton, 2000), 208–27; J. Goodare and M. Lynch, 'The Scottish state and its borderlands, 1567–1635', in Goodare and Lynch (eds.), *The reign of James VI,* 186–207.

[70] K. Brown, 'The vanishing emperor: British kingship and its decline 1603–1707', in Mason (ed.), *Scots and Britons,* 75.

[71] Moir (ed.), *Early maps of Scotland,* 9.

[72] M. Destombes, 'André Thevet (1504–1592) et sa contribution a la cartographie et a l'oceanographie', *Proceedings of the Royal Society of Edinburgh (B),* 72 (1972), 123–31. More generally, see B. Harley, 'Secrecy and silences: the hidden agenda of state cartography in early modern Europe', *Imago Mundi,* 40 (1988), 111–30.

[73] R. Skelton, 'Bishop Leslie's Map of Scotland, 1578', *Imago Mundi,* 7 (1950), 103–6.

[74] J. Elder, 'A proposal for uniting Scotland with England', in *Miscellany of the Bannatyne Club Vol. I* (Edinburgh, 1827), 1–18; P. Barber, 'A friend at a distance: Mercator and the mapping of Britain 1538–1595', in M. Watelet (ed.), *Gerardus Mercator Rupelmundanus* (Antwerp, 1994), 27–39.

know one's bounds – that Timothy Pont's manuscript maps and texts are so important. Pont's work has been the subject of considerable study.[75] He was the eldest son of Robert Pont, the Scottish churchman. He matriculated at St Andrews in 1580, graduated in 1583 and almost certainly was taught mathematical geography by Professor William Wellwood, Professor of Canon and Civil Law at St Andrews and a friend to John Napier. Robert Rollok was also taught by Wellwood.[76] Timothy Pont was appointed minister of Dunnet in Caithness in 1600 or 1601, and he is thought to have died between 1611 and 1614. At some unknown period and for reasons which are still unclear, but which have, I suggest, everything to do with that emergent geographical consciousness I am charting here, Timothy Pont undertook a mapping survey of Scotland. One manuscript only of Pont's work can be dated: to 'Sept et/ Octb:/ 1596'. From this, it has been argued that his survey took place between *c.* 1583 and 1596. In addition to his place in the wider context of geographical enquiry within contemporary Scotland, he has significance for several reasons: what he shows and how he shows it suggest that Pont was concerned to delineate, survey *and* to record the nature of the land; he was both topographer *and* chorographer, and not, in any strictly mathematical sense, a surveyor. Failed attempts to publish his work in the 1620s and 1630s suggest a contemporary recognition of its importance. Successful publication of his work, through the mediation of Robert Gordon of Straloch and his son, James Gordon of Rothiemay, as the source of the maps in volume five of Johannes Blaeu's 1654 *Atlas novus,* provides an enduring basis for the 'modern' outline of Scotland. His undertaking of the survey, and others' management and publication of his maps demonstrates for one context the social nature of, and the international networks sustaining, early modern geographical knowledge.

The majority of his manuscript maps are rough copies or working documents (see Figure 2.3). Pont does not use locational symbols consistently, place names are not spelled consistently, and there is no common standard employed to record buildings or settlements. Not all of Scotland is covered in the extant manuscripts (see Figure 2.4). Yet perhaps because of their unfinished nature, there is an immediacy to his maps. We can better sense something of the processes behind Pont's work: of the rigours of safe travel; of his needing to have a warrant for his own safe passage; of his having to have sufficient credibility as a transient stranger that others would trust him and his purpose. Such matters were vital: Richard Bartlett, a surveyor in Donegal in Ireland in 1602, was beheaded because the natives did not want their land 'to be written down'.[77] Pont was sketching the lie of the land in ways that sought

[75] This is summarised in J. Stone, *The Pont manuscript maps of Scotland: sixteenth century origins of a Blaeu atlas* (Tring, 1978), and in I. Cunningham (ed.), *The nation survey'd* (East Linton, 2001).

[76] NLS, Adv. MS 29.2.27.

[77] B. Friel, J. Andrew and K. Barry, 'Translations and a paper landscape: between fiction and history', *The Crane Bag,* 7 (1983), 121.

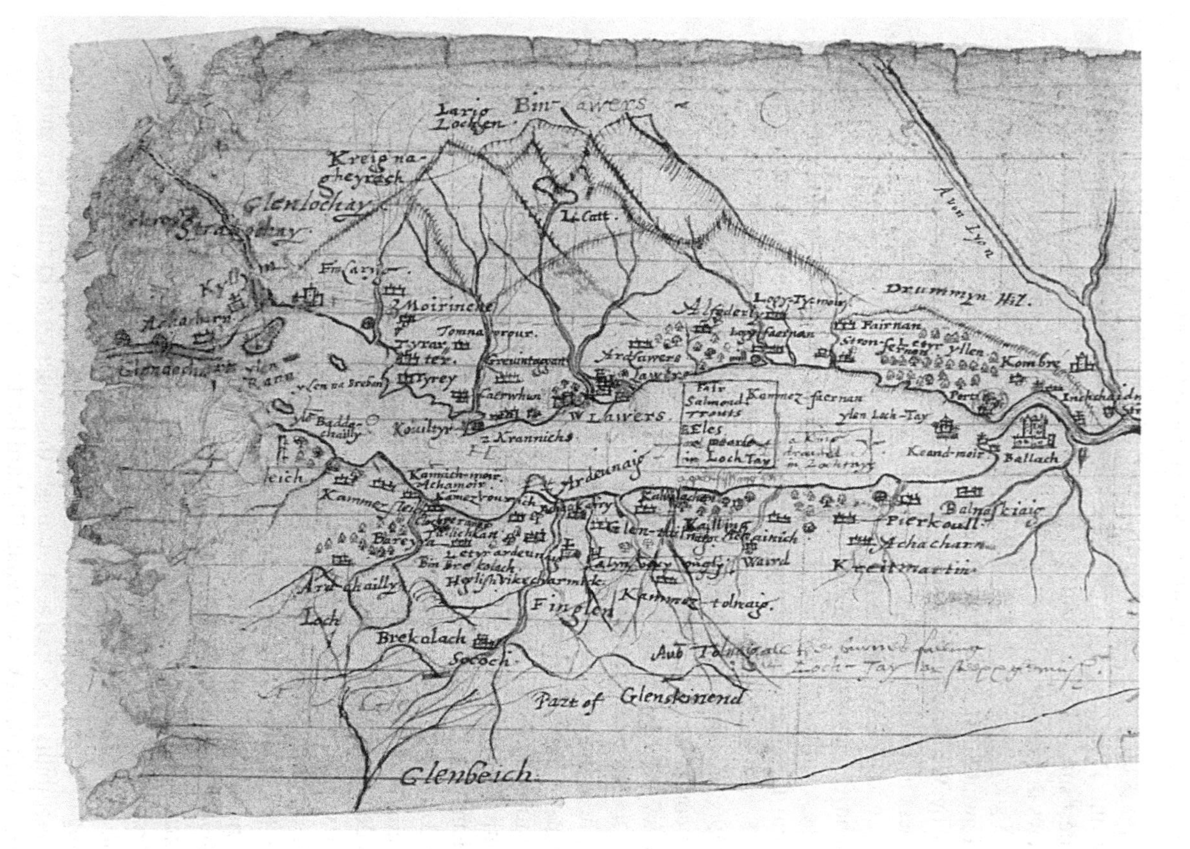


2.3 Timothy Pont's map of Loch Tay. The representation of the topography of Ben Lawers and the description of the 'Fair Salmond' as well as other products of the Loch are amongst the noteworthy features of Pont's chorographical concerns here.

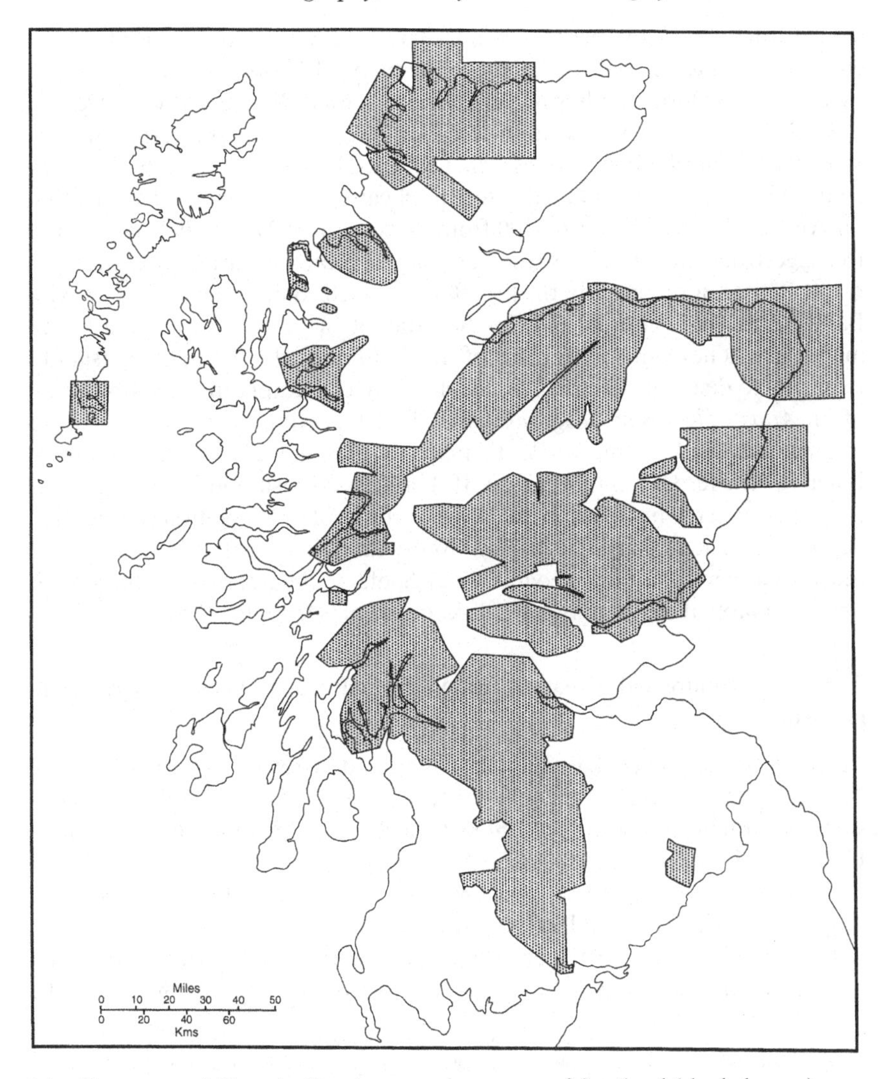


2.4 The extent of Timothy Pont's mapped coverage of Scotland (shaded areas).

its truthful delineation, perhaps in relation to the administrative needs of the church or as a terrestrial equivalent to the maps of Scotland's coast provided by Lindsay. As Goodare has it, Pont's concern was a 'landscape of power'.[78]

Sir Robert Gordon's dedication of 24 January 1648 in Blaeu's *Atlas* and the statement there that Pont's maps were neglected by his heirs have been used to confirm the view that Pont died before publication of his maps.[79] The fact that Sir Andrew Balfour received £100 from the king in 1629 is sufficient evidence to suggest that they were regarded as geographically accurate by contemporaries. This is supported by the fact that, by 1626, Blaeu was in contact with Balfour, through Sir John Scot of Scotstarvet, with a view to using Pont's maps: 'if you have any other maps of Scotland or of the surrounding islands, I beg you to deign to send them to me; for I am contemplating the publication of an *Atlas (Theatrum Geographicum)*'.[80] The Gordons were much involved in amending and adding to Pont's maps as they moved, via Scot, between Blaeu in Amsterdam and Scotland. If Timothy Pont should be confirmed as the map-maker who first provides the outline of a recognisably 'modern' Scotland, the fact that his work survived and, even, was begun at all, owed much to a contemporary recognition by Scots of the importance of geography and mapmaking to national self-knowledge (see Figure 2.5).

## Theories of empire and 'Greater Britain': the Union of 1603 as a geographical problem

The Union of Crowns of England and Scotland in 1603 and the accession of James VI to the throne of the 'United Kingdom' has been the subject of considerable attention, both by contemporaries and by modern historians.[81] Understood as historically significant, the Union of 1603 has been almost entirely ignored as a question of geography. Yet there are at least two compelling reasons why we should see it as such.

First, consideration of the contemporary reaction to the union shows that conflicting ideas as to its nature – the 'making of Britain' as, potentially, the

[78] J. Goodare, *State and society in early modern Scotland* (Oxford, 1999), 245.

[79] Stone, *Pont Manuscript Maps,* 7–9; 'Robert Gordon and the making of the first atlas of Scotland', *Northern Scotland* 18 (1998), 15–29.

[80] Moir (ed.), *Early maps of Scotland,* 37–53; Stone, 'Robert Gordon and the making of the first Atlas'; F. Emery,'The geography of Robert Gordon, 1590–1661, and Sir Robert Sibbald, 1641–1722', *Scottish Geographical Magazine,* 74 (1958), 2–12; J. Stone, 'Robert Gordon of Straloch: cartographer or chorographer?', *Northern Scotland,* 4 (1981), 7–22.

[81] Brown, 'The vanishing emperor'; B. Galloway and B. Levack, *The Jacobean Union: six tracts of 1604* (Edinburgh, 1985); M.Lee, *Great Britain's Solomon: James VI and I and his three kingdoms* (Urbana and Chicago, 1990); B. Levack, *The formation of the British state: England, Scotland and the Union 1603–1707* (Oxford, 1987); Mason, 'Scotching the Brut'; *Scots and Britons;* J. Robertson, 'Empire and union: two concepts of the early modern European political order', in J. Robertson (ed.), *A union for empire: political thought and the British Union of 1707* (Cambridge, 1995), 3–36; J. Wormald, 'James VI and I: two kings or one?',*History,* 68 (1983), 187–209; 'The creation of Britain: multiple kingdoms or core and colonies?', *Transactions of the Royal Historical Society,* 6 (1992), 175–94; D. Smith, *A history of the modern British isles 1603–1707: the divided crown* (Oxford, 1998).

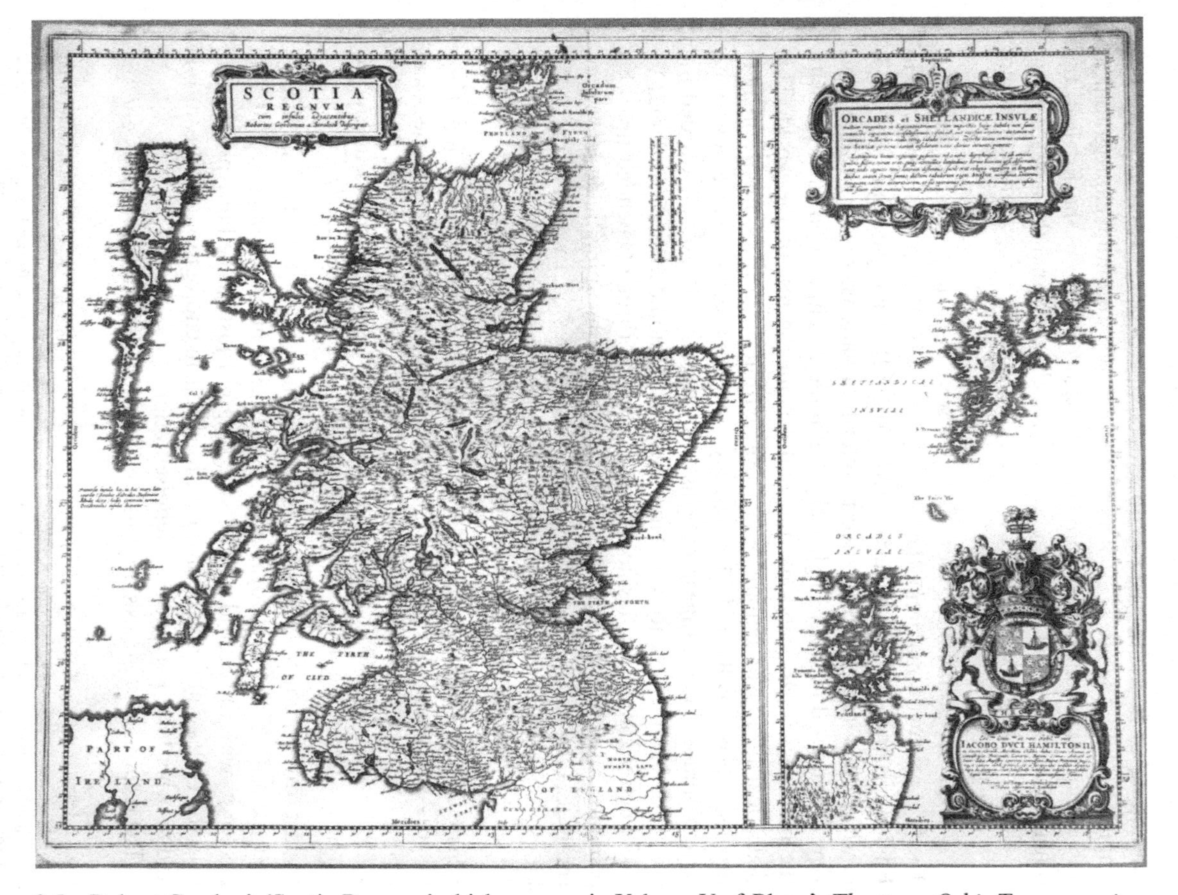


2.5 Robert Gordon's 'Scotia Regnum' which appears in Volume V of Blaeu's *Theatrum Orbis Terrarum, sive Atlas Novus,* published in Amsterdam in 1654.

'unmaking' of Scotland – had to do with competing claims about the idea of empire. Second, the presence of two ideologically sustaining chorographical and antiquarian chronicle traditions, one drawn upon by the English, the other by the Scots, had to be negotiated in such a way that the new kingdom would depend upon the creation of a new imperial genealogy for the 'Empire of Great Britain'.[82]

Two further political theories of empire were also understood at this time. The first was with imagining the kingdom as itself an empire, a view that, in England, is first apparent in Henry VIII's Act in Restraint of Appeals of 1533, in which the term 'empire' is used to refer to a national sovereign state. This view is sustained in many English chorographies.[83] The second has to do with the perceived separateness of that realm from the rest of the world. This dialectical notion of empire of a world within a world is important here because, theoretically and ideologically, contemporaries used 'empire' and 'union' almost interchangeably: 'This association of ideas . . . provided the conceptual framework for the Anglo-Scottish Union debate'.[84] As James VI knew better than most, his accession to a single throne depended upon his claiming, and others accepting, that an incorporating union made sense conceptually. One way in which this was represented by James was in terms of the king's own body: the two kingdoms should regard themselves as one wife to James the husband, one body to the king their head:

> What God hath conjoyned then, let no man separate. I am the husband, and all the whole Isle is my lawful Wife; I am the Head, and it is my Body; I am the Shepherd, and it is my flocke; I hope therefore no man will be so unreasonable as to thinke that I that am a Christian King under the Gospel, should be a Polygamist and husband to two wives; that I being the Head, should have a divided and monstrous Body.[85]

In part, this statement by James has to do with the place of a dual monarchy, then a question central to royal control in several European countries.[86] It also has to do with the central place of the king not just *as* the body-politic, but with the king's embodiment of the nation itself in chorographical illustration and in other forms of symbolic representation.[87]

The responses to James's proposal indicate a wide spectrum of conceptions

[82] Brown, 'The vanishing emperor'; W. Ferguson, 'Imperial crowns: a neglected fact of the background to the Treaty of Union', *Scottish Historical Review*, 53 (1974), 22–44; Helgerson, *Forms of nationhood*; Mason, 'Imagining Scotland'; S. Ellis, 'Tudor state formation and the shaping of the British Isles', in S. Ellis and S. Barber (eds.), *Conquest & union: fashioning a British state, 1485–1725* (London, 1995), 40–64; K. Robbins, *Great Britain: identities, institutions and the idea of Britishness* (London, 1998), 31–8; C. Withers, 'Geography, royalty and empire: Scotland and the making of Great Britain, 1603–1661', *Scottish Geographical Magazine*, 113 (1997), 22–32.

[83] Mendyk, 'Early British chorography', 435; R. Flower, 'Lawrence Nowell and the discovery of England in Tudor times', *Proceedings of the British Academy*, 21 (1935), 41–54.

[84] Robertson, 'Empire and union', 5. [85] Brown, 'The vanishing emperor', 61.

[86] P. Curtin, 'The environment beyond Europe and the European theory of empire', *Journal of World History*, 1 (1990), 131–50; Pagden, *Lords of all the world*; N. Phillipson and Q. Skinner (eds.), *Political discourse in early modern Britain* (Cambridge, 1993); Robertson, 'Empire and union'.

[87] This is to differ from Helgerson's view as outlined in his 'The land speaks', 56–65.

of union's relation with empire. Prominent amongst the enthusiasts was Francis Bacon. In his 'A Brief Discourse of the Happy Union of the Kingdoms of England and Scotland', Bacon drew upon natural and cosmological references in seeing kings and stable government as, like the sun and moon, celestial bodies always 'constant and regular, without Wavering or Confusion'. He also argued for the unique nature of the event:

> Your Majesty is the first King that had the Honour to be *Lapis Angularis,* to unite these two mighty and warlike Nations of *England and Scotland,* under one Soveraignty and Monarchy. It doth not appear by the Records and Memories of any true History, or scarcely by the Fiction and Pleasure of any fabulous Narration or Tradition, that ever of any Antiquity, this Island of *Great Britain* was united under one King before this day.[88]

Bacon's view of the unique position and authority of James is obvious. Yet appealing to antiquarian knowledges and historical traditions as means to verify kingly authority could not itself overcome the problem that James had no single lineage in which to place himself and a united Britain.

Bacon avoided privileging one national historical tradition over another by using analogies between natural processes and political circumstance: 'I set before your Majesty's Princely Consideration, the Grounds of Nature touching the Union with the Commixture of Bodies, and the Correspondence which they have with the Grounds of Policy, in the Conjunction of States and Kingdoms'.[89] The image of the sun and its central place in the firmament was cited. Reference was made to waters in nature: 'So we see waters and Liquors in small Quantity do easily putrify and corrupt: but in large Quantity, subsist long, by Reason of the Strength they receive by Union'.[90] The point of importance is not the accuracy of these and other claims but their ideological appeal to the *naturalness* of the Union, to the naturally-ordered place of the king in overseeing his new political *imperium.* Bacon's emphasis on 'natural Unions' and upon the 'Conditions of Perfect Mixture' allowed him to promote the idea of a political continuum in which James' new empire, Britain, could be situated in relation to the old one, of Rome. What Bacon thus termed 'the naturalisation of Scotland' was historically rooted *and* natural. It was also politically expedient. Scotland's past alliance with France and Ireland's connections with Spain had exposed these countries as 'a Temptation to the Ambition of Forreigners: (now) their Approaches and Avenues are taken away'. A new Greater Britain was almost a heaven on earth:

> For Greatness, Mr Speaker, I think a Man may speak it soberly, and without Bravery, That this Kingdom of *England,* having *Scotland* united, *Ireland* reduc'd, the Sea Provinces of the Low- Countries contracted, and Shipping maintain'd, is one of the greatest Monarchies, in Forces truly esteem'd, that hath been in the World. For certainly the Kingdoms here on Earth, have a resemblance with the Kingdom of Heaven.[91]

[88] F. Bacon, *Reasons for an union between the kingdoms of England and Scotland* (London, 1705), 3.
[89] Bacon, *Reasons for an union,* 3–4. [90] Bacon, *Reasons for an union,* 4–5.
[91] Bacon, *Reasons for an union,* 27.

Others, like Sir Thomas Craig in his 1605 *De unione regnorum Britanniae tractatus*, saw a new united Britain as a means to rival and oppose the Spanish Empire. Hispanic imperial ideology at that time adopted a version of Roman imperial theory, stressing military glory and the conversion, rather than the naturalisation, of 'the Other'.[92] Craig considered the *Hispanicum imperium* the most powerful in existence, and a model to be followed by Britain. Others were like the churchman Robert Pont, Timothy's father, numerologist and friend to John Napier upon whose cosmological thinking Pont drew, who argued in his 1604 *De unione Britanniae* that union of kingdoms demanded full union.[93] Yet others saw the union as a matter of alchemy, just as earlier Dee had sought the aid of natural magic as a political instrument in promoting the Elizabethan vision of a 'Brytish Empire'. In separate elements (England and Scotland) the union was, for John Thornborough, 'of repugnant qualities', but united kingdoms would 'agree in one form, as in a *Medium* uniting and mixing them together'.[94]

In such ways, the geographical making of the 'Empire of Great Britain' was justified and realised through theories of union and of empire in which national identity was made from within and defended from without. It depended, too, upon reconciliation of those competing chronicle traditions discussed earlier. It does not matter that such origin myths could not be substantiated. What matters is that they were believed as part of contemporaries' geographical imagination. What also mattered was that neither *mythomoteur* alone could be mobilised as historical justification for the new geographical entity that was Greater Britain. In the event, greater reliance was placed upon the more strictly Elizabethan theories of *imperium* and upon English mythic traditions, both because of threats to English Common Law and from an English fear of too many Scots in positions of authority. If, in political terms, 'the attempt to establish a 'perfect' union was dead in the water' by 1607–8,[95] and the longer term consequence for Scottish political and economic interests was what has been considered 'provincial relegation',[96] royal patronage of imperial and British themes was nonetheless essential. William Cornwallis' *The miraculous and happie union of England and Scotland* (1604), Anthony Munday's pageant *The triumphes of re-united Britania* (1605), George Marcelline's 1610 *The triumphs of King James the First,* and, notably, John Speed's *The theatre of the empire of*

[92] Pagden, *Lords of all the world,*

[93] Robert Pont, *De unione Britanniae* (Edinburgh, 1604); A. Williamson, 'Number and national consciousness: the Edinburgh mathematicians and Scottish political culture at the union of the crowns', in Mason (eds.), *Scots and Britons,* 187–212; *Scottish national consciousness,* 30, 103, 181.

[94] John Thornborough, *The joyful and blessed reuniting of the two mighty and famous kingdomes, England and Scotland into their ancient name of Great Brittaine* (Oxford, 1604); Cormack, *Charting an empire,* 3–4, 229; R. Deacon, *John Dee: scientist, geographer, astrologer and secret agent to Elizabeth I* (London, 1968).

[95] J. Young, 'The Scottish parliament and national identity from the Union of the Crowns to the Union of the Parliaments, 1603–1707', in Broun, Finlay and Lynch (eds.), *Image and identity,* 105–43.

[96] The term is from A. MacInnes, *Charles I and the making of the Covenanting movement 1625–1641* (Edinburgh, 1991), 22.

*Great Britaine* (1611) all reflect the making of 'Greater Britain' as, variously, the result of monarchical authority and empire, a geographical union rooted in a common past and displayed in triumph.

## Symbolising identity: Scotland and Britain display'd, 1604–1661

For matters of national and imperial identity to mean anything to others demanded that the geographical materials on which they were dependent were constituted not just in map and text but also in other more symbolic ways. In the first part of the seventeenth century, and spectacularly in London in 1661, geographical meaning was made in the court and on the street.

### *Geography's theatre: court masques and the representation of Britain, 1605–1634*

Between 1604 and 1640, over fifty court masques were presented to the Jacobean and Caroline courts. Such masques, vehicles for the veneration of the King and his courtiers, were also means to the representation in staged form of issues of identity and empire.[97]

The idea of Britain as itself an empire, and one set apart from the world for a special destiny, is the central theme of Ben Jonson's 1605 *Masque of blacknesse*. *Blacknesse* is about royal and courtly magnificence and the place of a united Britain as an imperial nation. The play takes its title from the central conceit of the queen and her ladies disguised as 'blackamoors', then a popular form of courtly disguise. Jonson casts the ladies as 'Aethiopians', dwelling by the River Niger, which flows, in Jonson's erroneous geography, into the Indian Ocean. The masque proceeds by the 'dark ladies' experiencing a vision in which they are instructed to seek a land whose name ended in —— *tania*, where, they are told, their complexion would be refined and their beauty made ideal. Ocean, a character, explains that the vision referred to 'Britannia'. The moon goddess appears and affirms themes of geography, national identity, and kingship:

> BRITANNIA, which the triple world admires,
> This isle hath now recovered for her name;
> . . . this blest isle
> Hath wonne her ancient dignitie, and stile,
> A world, divided from the world.

[97] D. Howarth, *Art and patronage in the Caroline courts* (Cambridge, 1993); D. Lindley (ed.), *The court masque* (Manchester, 1984); *Court masques* (Oxford, 1995); S. Orgel, *The illusion of power* (Berkeley and London, 1975); G. Parry, *The golden age restor'd: the culture of the Stuart court 1603–1642* (Manchester, 1981); *The trophies of time* (Oxford, 1995); R. Strong, *Art and power: Renaissance festivals 1450–1650* (Woodbridge, 1984); T. Marshall, *Theatre and empire: Great Britain on the London stages under James VI and I* (London, 2000); J. Gillies and V. Vaughan (eds.), *Playing the globe: genre and geography in English renaissance drama* (London, 1998).

Here is one expression of that dominant theory of *imperium* embracing internal union and separateness from the rest of the world. Further, the king was seen – was *embodied* – in cosmological terms. His single, unmonstrous body was vested with an authority simultaneously monarchical, scientific and celestial in order to 'make' the new Britain, itself described as being:

> Rul'd by a Sunne, that to this height doth grace it:
> Whose beames shine day, and night, and are of force
> to blanche an Ethiop, and revive a Cor'se.
> His light sciential is, and (past mere nature)
> Can salve the rude defects of every creature.[98]

Several themes are at work simultaneously. Britain is recovering her name through political union. The New World, albeit not very accurately portrayed in Jonson's imaginative geography, is being made sense of in a racialised discourse itself the consequence of geographical encounter. Britain's 'ancient dignities' are being restored and used as the basis to a new empire, while contemporary differences between peoples are slighted. Britain is ruled by a king whose 'sciential light' of knowledge may be considered analogous to divine wisdom; a light, moreover, that is powerful enough to change skin colour in order that all subject peoples appear the same.

Thomas Carew's *Coelum Britannicum* of 1634 was likewise designed to symbolise Stuart rule over the united kingdom. At the start of the masque, 'there began to arise out of the earth the top of a hill, which . . . grew to be a huge mountain: . . . about the middle part of this mountain were seated the three kingdoms of England, Scotland and Ireland, all richly attired in regal habits appropriated to the several nations, with crowns on their heads, and each of them bearing the ancient arms of the kingdoms they represented'.[99] On the mountain sat 'the Genius of Great Britain', like the other symbols a female suitably dressed, bearing fruits and corn as emblems of the peace and richness of Charles' reign. Stars and the figure of Atlas were used to affirm these neo-Platonic ideals. In such productions, the British court affirmed its importance in ways others have shown of European courts – as a centre for patronage of natural knowledge and artistic endeavour.[100] And it used geographical ideas, in symbolic and emblematic form, to do so. This was not the only site so used.

## *Geography in the street: triumphal processions in 1605, 1633 and 1661*

Throughout early modern Europe, the royal progress or 'magnificence' upon entry to a city affirmed through spectacle the political authority vested in the

[98] Lindley, *Court masques*, 6. [99] Strong, *Art and power*, 167.

[100] R. Westfall, 'Scientific patronage: Galileo and the telescope', *Isis*, 76 (1985), 11–30; M. Biagioli, *Galileo, courtier: the practices of science in the culture of absolutism* (Chicago, 1993); P. Findlen, 'Controlling the experiment: rhetoric, court patronage and the experimental method of Francesco Redi', *History of Science* 31 (1993), 35–64.

monarch. The triumphal entry of James VI and I into London in 1604 sought to endow him with a British imperial heritage in order to confirm the union and support moves to establish colonies overseas: London, indeed, was the new Rome, the Thames as the Tiber receiving her Augustus.[101]

James' entry into the geographical centre of the new British empire was through seven triumphal arches, especially designed and constructed as theatrical 'symbols' to be 'read' and interpreted. On the first arch, the Londinium arch, the British monarchy was symbolised, seated beneath the two crowns of England and Scotland, holding a globe inscribed '*Orbis Brittanicus, divisos ab orbe*' (Britain as a world within a world; see Figure 2.6). This phrase, originally from Virgil, had been used by Giordano Bruno to celebrate Elizabeth I as Astraea, the goddess of justice, presiding over sacred waters. It was employed in 1604 by Ben Jonson, who made additional reference to Virgil, to show how James was usher to a new Golden Age. The third arch was entitled 'Nova Felix Arabia', a reference to the newly-created geographical entity of Britain as the new Arabia, traditionally in ancient geographical and chorographical accounts the fertile land of peace, and home of the Phoenix, symbol of virtue and of immortality. The fifth arch was the New World arch, topped by a large rotating globe. As James approached the arch, an interpreter explained that the globe of Britain had been moving awry since the death of Elizabeth, but, with the arrival of James as the new Brutus, England and Scotland, and, thus, the empire of Britain, would move in a right and orderly fashion again.

The 1604 triumphal procession should be seen as spectacular emblematic geography. James was being integrated with imperial *mythomoteurs* in order to make a new Britishness. London was both the new Rome and the new Jerusalem. Britain was the new-born centre of virtue and courtly knowledge about the world. Contemporaries considered the event cosmologically: the king's progress through the streets was seen as the sun's movement in the heavens and other natural emblems, understood as royal signifiers, were used as political instruments.[102]

Scotland had had triumphal processions since 1503, but the above imperial messages were most clearly evident in the 1633 state entry into Edinburgh of Charles I.[103] William Drummond of Hawthornden, the poet who coordinated the event in association with Sir James Balfour, constructed four arches throughout the capital: the Arch of Welcome; of Caledonia; of Genealogy; and of Destiny. The first arch represented Edinburgh. At this arch, local citizenry

[101] T. Harrison, *Arches of triumph* (London, 1604); V. Hart, *Art and magic in the court of the Stuarts* (London, 1994); J. Nichols, *The progresses, processions and magnificent festivities of King James the First* (London, 1828); Parry, *Golden age restor'd*; Strong, *Art and power*.

[102] Hart, *Art and magic*; J. Williams, *Great Britain's Solomon: a sermon preached at the magnificent funerall, of the most high and mighty King James, the late King of Great Britaine* (London, 1625).

[103] I. Campbell, 'James IV and Edinburgh's first triumphal arches', in D. Mays (ed.), *The architecture of Scottish cities: essays in honour of David Walker* (East Linton, 1997), 26–33.

2.6 The New World Arch, from Thomas Harrison's *Arches of Triumph* (1604).

presented the keys of the city to the king, himself seen as a national safeguard: 'Presenting you *Sir,* (who art the strong key of this little world of Great Brittaine)'.[104] The Caledonia arch had a painted scene showing: 'in land-skip [landscape], a country wild, full of Trees, Bushes, Bores, white kine'. As the king stopped at this arch, he was confronted by two women representing the Genius of Caledonia and the Genius of Nova Scotia. The latter symbolised Scotland's nascent colonialism. In 1621, James VI and I had granted a royal warrant to the Scot Sir William Alexander to colonise land between New England and Newfoundland, to be called Nova Scotia.[105] Together, the two arches affirmed the incorporation of Scotland within Britain and of overseas possessions within the emergent British empire.

In being thus painted and symbolised, Scotland was being figuratively represented in the 1633 triumphal entry in ways which demanded that its wild landscapes and uncultivated peoples be brought to order in much the same way as the Americas. Scotland was not, admittedly, a rich new colony, however much she was rhetorically and visually so imagined. She had 'no Serean fleeces, Peru gold, Auroras gemmes, nor wares by Tyrians sold'. As a land of ancient historical traditions, however, she had perhaps greater riches: 'Yet in this corner of the World doth dwell/With her Pure Sisters, *Truth, simplicitie*: Heere banish'd *Honour* beares them company'. At the Arch of Genealogy, actors playing Scottish kings put Charles' dominion in historical context, in order to demonstrate the 'haill number of the Kings that ever rang [reigned] in Scotland from fergus the first to his awin Majestie'.[106] Others played a number of 'ancient Worthies of Scotland', including Boece, Major and Buchanan. The Arch of Destiny employed cosmological devices to position the king as a celestial body around which all other planets orbited.[107]

The coronation pageant for Charles II, at the restoration of the monarchy in 1661, likewise emphasised monarchy, nationhood, union and empire. The event was organised as a street spectacle by John Ogilby, the Scottish-born chorographer and translator. He is best known for his 1675 *Britannia*, a chorography and book of road maps, based on circulated questionnaires and undertaken with the assistance of John Aubrey, Gregory King and Robert Hooke, which was of considerable influence upon later mapping projects. Ogilby had trained as a dancer, had danced in court masques (to his own permanent injury), was a one-time Master of the King's Revels and founder of the theatre in Ireland, and, from 1649, achieved fame and royal patronage with

[104] W. Drummond, *The entertainment of the high and mighty monarch Charles King of Great Britaine, France and Ireland, into his ancient and royall city of Edinburgh, the fifteenth of June, 1633* (Edinburgh, 1633), 4.

[105] J. Martin, 'A study of the entertainments for the state entry of Charles I, Edinburgh, 15 June 1633' (MA thesis, University of Edinburgh, 1990); on topographical poetry and plays being used to 'locate' the king, see also A. Mill, *Mediaeval plays in Scotland* (Edinburgh, 1927), 8–87; H. Shire, *Song, dance and poetry of the court of Scotland under James VI* (Edinburgh, 1969).

[106] Drummond, *Entertainment of the high and mighty monarch,* 8.

[107] NLS, Adv. MS 15. 2. 17, 'Sir James Balfour's Order of King Charles Entering Edinburgh, 1633'.

a number of beautifully-produced translations of Virgil and Homer and of great atlases, chiefly drawn from the work of others.[108] Trained as a choreographer, active as a chorographer: Ogilby was just the man to know the right steps to take in staging symbolic geographies.

> On the South Pedestal [on the first of four triumphal arches], was a Representation of *Britains* Monarchy, supported by *Loyalty,* both Women. *Monarchy* in a large purple Robe, adorned with Diadems and Scepters, over which she had a loose mantle edged with blue and Silver Fringe, resembling Water, on her Mantle the map of Great Britain, on her head *London,* in her right hand *Edinburgh,* in her left *Dublin.*[109]

At this arch, the character 'Rebellion' spoke, and was driven off by 'Monarchy' and 'Loyalty'. The second, naval, arch was used to laud Charles as ruler of the seas, an important claim given debates then about jurisdiction over the oceans. The living tableaux was also about empire in other ways. Pedestals in the upper storey of the arch:

> were adorn'd with living Figures, representing *Europe, Asia, Africa,* and *America,* with Escutcheons and Pendents, bearing the Arms of the Companies Trading Into those parts. *Europe* a Woman arm'd antique, on her Shield a Woman riding on a Bull, and at her foot a Cony. *Asia,* on her head a Glory, her Stole of Silk, with several forms of Wild Beasts wrought on it. *Africa,* a woman, in her Hand a Pomegranate, on her head a Crown of Ivory and Ears of Wheat, at her Feet two Ships laden with Corn. *America,* Crown'd with Feathers of divers Colours; on her Stole a Golden River, and in her hand a Silver Mountain.[110]

On the upper part of the arch, entablatures represented Charles I and his Prince, Charles II: 'In his hand viewing, the Soveraign of the Sea, the Prince leaning on a Cannon, over above the Cornish [corniche] between the two Celestial Hemispheres an *Atlas* bearing a Terrestial Glob, and on it a Ship under Sail.' In the four niches within the arch were four women representing 'Arithmetick, Geometry, Astronomy, and Navigation'.

Such pomp and ceremony was about power manifest in geographical knowledge and in its figurative display. National identity was being made and displayed emblematically as, in turn, geographical knowledge was part of that array of rationalising discourses bringing the world to order. As Knowles notes of the 1661 event, the depiction of arithmetic, geometry, astronomy and navigation marked a 'notable shift in emphasis from the *studia humanitatis* to techniques of calculation well-adapted to the new golden age, in which the Royal Society flourished'.[111] As we shall see, such practices were to be widely

[108] K. van Eerde, *John Ogilby and the taste of his times* (London, 1978).

[109] John Ogilby, *The king's coronation* (London, 1685), 2; see also, for a fuller account of the theatrical management of this event, Ogilby, *The entertainment of his most excellent Majestie Charles II in his passage through the city of London to his coronation* (London, 1662).

[110] Ogilby, *The king's coronation,* 4.

[111] R. Knowles (ed.), *The entertainment of his most excellent Majestie Charles II in his passage through the city of London to his coronation* (Introduction to the facsimile text of Ogilby (1662)),33.

used by geographers in Scotland, as both practices and practitioners reflected and directed the growth of the new sciences from the later seventeenth century.

**Conclusion**

In discussing the nature of identity in sixteenth- and seventeenth-century Scotland, Lynch has argued that 'it seems clear – although relatively little has been made of the phenomenon – that a widespread, multi-layered exploration of Scotland's identity had emerged by the last quarter of the seventeenth century.'[112]

This chapter has shown geographical knowledge to be a part of that exploration. Geographical knowledge as undertaken by Scots in this period partly drew upon classical accounts and chronicle traditions in order, often, to reject them; but, from the 1520s, it principally took the form of 'descriptions' and chorographical accounts designed to 'situate' Scotland historically. In combination with such textual survey, mapmaking was also important in giving 'shape to the nation', notably in Pont's survey and in its incorporation in Blaeu's *Atlas Novus.* What should be termed 'symbolic' or 'emblematic' geographies were also used to present Scotland's place in a greater Britain, both at the Union of 1603 and in the years following. The making of geographical knowledge in these terms took place amongst a handful of men, located either in the universities or in royal courts. Such making was not alone an academic affair: Scotland's representation on stage and street involved artists, playwrights and poets.

In these terms and recognising that patronage was as important for scholarly disquisition as it was for geography's public theatre, it is appropriate to think of the production and consumption of geographical knowledge within Scotland before about 1680 as the work of a very few small groups, perhaps even of single individuals, many with European connections, rather than by any consciously constituted network of scholars. Even so, we may with reasonable confidence state that, by about 1540 if not before, geographical knowledge was a *necessary* part of the processes by which kingship was understood – part, then, of that set of practices by which Scotland as a nation and a kingdom were 'self-fashioned'. Geography was taught in Scotland's universities, certainly by 1570, chiefly as mathematical geography and as chorography. Although it is likely that Pont's work in the 1580s was prompted by the needs of church and state to know Scotland for reasons of ecclesiastical and political administration, this cannot be confirmed. The first certain signs of geography being employed in relation to what contemporaries would have seen as political self-knowledge date from the 1630s and 1640s.

Two general claims follow from this evidence. The first has to do with the

[112] Lynch, 'A nation born again?', in Broun, Finlay and Lynch (eds.), *Image and identity*, 89.

significance for different disciplines of our taking seriously the complex nature and purpose of geographical enquiry in the early modern period. In drawing upon a range of work by historians, I hope also to have contributed a geographical perspective to what we know of the making of national identity in Scotland at this time. For historians of early modern geography, this evidence provides one reason why Britain should not be seen as an undifferentiated whole and, importantly, why the terms 'England' and 'Britain' should not be conflated. Further, recent work on performance and embodiment within geography, and related calls for attention to be paid to music and 'soundscapes' to counterbalance what some consider geography's predominant focus on visual representation, would do well to think historically and consider context more carefully.[113] If the evidence from triumphal processions is to be believed, we should not see too firm a distinction between sight and sound in presenting images of power, national identity and geographical meaning. To the intellectual historian, geographical knowledge was part of Renaissance learning of the humanist curriculum in Scotland's universities and of the slow move to 'modernity', facts which are considered in more detail in later chapters.

My second claim has to do with what I have here called the emergence of 'geographical thinking' in early modern Scotland. The emergence of a strong and self-aware geographical consciousness – a sense of national identity *as geographically defined and constituted* – is, of course, not the same thing as saying that there was a clear sense of geography, however formally understood, as an agreed *means* to that national self-awareness. I am arguing, however, that certain people in Scotland were thinking geographically in several ways well before the appointment of an official Geographer: in terms of the Highlands as a regional political problem; in regard to the 'external shape' of the nation from the 1540s and its 'internal shape' from the 1580s; and, from the 1630s, in regard to geographical variation in the state of the nation.

[113] S. Smith, 'Soundscape', *Area,* 26 (1994), 232–40; 'Beyond geography's visible worlds: a cultural politics of music', *Progress in Human Geography,* 21 (1997), 502–29; my argument is accepted, tacitly anyway, in F. Driver, 'Visualising geography: a journey to the heart of the discipline', *Progress in Human Geography,* 19 (1995), 19–34.

# 3

# Geography, credibility and national knowledge, 1682–1707

This chapter examines the connections betweeen geography and national knowledge in the quarter century between Sir Robert Sibbald's appointment as Geographer Royal and the union of Parliaments. In relation to the changing nature of science in this period, three main themes are explored. The first considers Sibbald's vision for geography, the methods he used and the social networks through which he operated. The second discusses the idea of 'competing geographies' by examining the different sorts of and means to geographical knowledge undertaken by several of Sibbald's contemporaries. The third discusses geography teaching in Scottish universities in the late seventeenth century and the production of geographical knowledge in other ways and sites, including Scotland's failed attempts at colonialism and the impact of that failure as part of the move to union in 1707. In short, this chapter attempts to show how and why the last twenty years or so of the seventeenth century was a crucial period in the making of geographical and scientific knowledge in Scotland and to show, in particular, how geographical knowledge, variously arrived at, was central to contemporaries' understanding of the nature of their nation.

The three central issues are not discrete. Amongst several themes, one issue in particular – credibility – connects them. Credibility and the problem of ensuring reliable natural knowledge have been key questions in several studies of science in this period.[1] In the context of its time, Sibbald's appointment as Geographer Royal conferred upon geography *intellectual credibility,* as it also conferred social authority upon Sibbald. However differently contemporaries

[1] M. Hunter, *Science and society in Restoration England* (Cambridge, 1981); *Science and the shape of orthodoxy: intellectual change in late seventeenth-century Britain* (Woodbridge, 1995), 101–19, 151–66, 169–80. S. Shapin, *A social history of truth: civility and science in seventeenth-century England* (Chicago, 1994); '"A scholar and a gentleman": the problematic identity of the scientific practitioner in early modern England', *History of Science* 29 (1991), 279–327; S. Shapin and S. Schaffer (eds.), *Leviathan and the air-pump: Hobbes, Boyle and the experimental life* (Princeton, 1985); P. Dear, '*Totius in Verba*: rhetoric and authority in the early Royal Society', *Isis,* 76 (1985), 145–61; F. Fernandez-Armesto, *Truth: a history and a guide for the perplexed* (London, 1997).

practised the term, the fact that geography was so recognised is a matter of moment: historiographically for geography in general and particularly for matters of national knowledge in Scotland. Sibbald's geography should not simply be seen as some sort of 'official' account. What was important about his appointment was the social *and* intellectual value others placed upon the subject, not directly the manner he brought to its doing. Yet credibility is also a matter of conduct: of *how*, not alone *who*. The fact that Sibbald received such a warrant meant that he had to undertake geographical understanding in ways which would not betray others' faith, either in the subject or in himself. Sibbald had to secure and to maintain *personal credibility* and he had to establish *methodological*, or, perhaps more properly, *epistemic credibility*. For Sibbald, that demanded he consider earlier chorographical traditions, that he demonstrate the utility of geographical enquiry to an understanding of the Scotland of his day, and that he consider the nation's future potential.

Such matters were embodied in other ways and in other people. What was at issue in the work of John Adair, Martin Martin and others in their methods – mapping, speaking to people and transmitting that knowledge to others – was also credibility of a personal *and* an epistemic nature. For such men to be taken seriously, their chosen methods for undertaking geographical knowledge had to be capable of adding to that knowledge and capable, too, of being transmitted as useful information to often distant audiences.[2] Such questions about the reliability of different methods and of practitioners in competition with one another, are, of course, applicable to other times and places. They are particularly apparent, however, in the relationships between these men, just as debates about empire and union prompted by the failure of Scottish imperial and commercial ambitions in Darien in 1699 highlight concerns about what was thought of as Scotland's *national credibility*, using that term loosely to denote the sense that contemporaries held of Scotland at the moment of union with England.

## Sibbald's geographies: the production of credible geographical knowledge

Sibbald was born in 1641 in Edinburgh and died there in 1722.[3] His lifetime covered a difficult and contentious period in Scottish history, one character-

[2] Shapin, *A social history of truth*, 243–309; D. Lux and H. Cook, 'Closed circles or open networks?: communicating at a distance during the scientific revolution', *History of Science* 36 (1998), 179–211.

[3] The principal source for Sibbald's life is his autobiography which is both incomplete – he wrote it in middle age, stopping in 1692, and is written with a degree of selective omission – and extant in a number of forms. The best is Francis Paget Hett (ed.), *The memoirs of Sir Robert Sibbald (1641–1722)* (London, 1932), and I have used Sibbald's 'Memoirs of my Life', EUL, Laing MS III.535 (unpaginated) and the copy in the National Library of Scotland: NLS, Adv. MS 33.5.1. These served as the basis for James Maidment's *Remains of Sir Robert Sibbald, KNT, MD, containing his autobiography, memoirs of the Royal College of Physicians, portions of his literary correspondence, and an account of his MSS* (Edinburgh, 1837) and for Hett's *Memoirs*. Both MSS are transcripts of Sibbald's lost original: a conjectural, but probably accurate history of the Sibbald biography is provided by Hett (1933), 1–7.

ised by friction between crown, parliament and church, of parliamentary union, of religious persecution, and, in the 'Seven Ill Years' of the late 1690s, by widespread economic hardship and local famine.[4] It was, paradoxically, also a time of relative expansion in the universities, of a rise in patronage for the arts and sciences, including medicine, and a period marked, for some, by the adoption of Newtonianism and the expression of claims to natural knowledge as useful, to the promotion of natural philosophy and to Scotland.[5]

Sibbald was both Scot and European. He was educated at Edinburgh University, and at the universities of Leiden, Paris and Angers. At Leiden, he studied under Christian Marcgraf, brother to the author of *The natural history of Brazil* (1648), a fact that, for Mendyk, prompted Sibbald's interests in natural phenomena.[6] Sibbald studied widely: anatomy, surgery, chemistry, botany, medicine and the earth sciences. Sibbald's intentions on return to Edinburgh in 1662 were modest: 'I had from my settlement here, a designe to informe myself of the subjects of the naturall history this country could affoord, . . . so I resolved to make it part of my studies to know what animalls, vegetables, minerals, metalls, and substances cast up by the sea, were found in this country, yt might be of use in medicine, or other artes useful to human lyfe, and I began to be curious in searching after ym and collecting ym'.[7] He was influenced in this respect by Sir Andrew Balfour, with whom, in 1667, he established what became the Edinburgh Botanical Garden. By 1679–80, Sibbald had founded a medical-*virtuoso* club, which became, from December 1681, the Royal College of Physicians of Edinburgh (RCPE). Like the botanic garden, the foundation and success of the RCPE owed much to the Duke of York's patronage. Sibbald was also supported by the Earl of Perth from 1678, it was through his offices that Sibbald was knighted in 1682, and, on 30 September 1682, appointed Physician to the King and Geographer Royal.

Sibbald's commission was to 'publish the naturall history of ye Country, and the geographical description of the kingdome'. Knowledge of natural productions, their 'virtues and uses', was important to scientific advance, of potential benefit to the nation and would 'honour and glory' Scotland.[8]

[4] A. Simpson, 'Sir Robert Sibbald – founder of the College', in R. Passmore (ed.), *Proceedings of the Royal College of Physicians of Edinburgh: ter-centenary congress* (Edinburgh, 1982), 59–91; I. Whyte, *Agriculture and society in seventeenth-century Scotland* (Edinburgh, 1979), 179, 233–4, 246–50, 255–8.

[5] H. Ouston, 'York in Edinburgh: James VII and the patronage of learning in Scotland, 1679–1688', in J. Dwyer, R. Mason and A. Murdoch (eds.), *New perspectives on the politics and culture of early modern Scotland* (Edinburgh, 1980), 133–55; C. Shepherd, 'Newtonianism in Scottish universities in the seventeenth century', in R. Campbell and A. Skinner (eds.), *The origins and nature of the Scottish Enlightenment* (Edinburgh, 1982), 65–85; R. Emerson, 'Sir Robert Sibbald, Kt, the Royal Society of Scotland and the origins of the Scottish enlightenment', *Annals of Science*, 45 (1988), 41–72.

[6] S. Mendyk, *Speculum Britanniae: regional study, antiquarianism, and science in Britain to 1700* (Toronto, 1989), 215; 'Scottish regional historians and the *Britannia* project', *Scottish Geographical Magazine*, 101 (1985), 165–73.

[7] Hett, *Memoirs*, 65.

[8] EUL, MS Laing III.535. Sibbald's 1682 warrant of appointment appears as Appendix I of C. Withers, 'Geography, science and national identity in early modern Britain: the case of Scotland and the work of Sir Robert Sibbald (1641–1722)', *Annals of Science*, 53 (1996), 29–73.

Sibbald's warrant of appointment was a royal proclamation of the benefits to Scotland of an utilitarian and empiricist geography. Understanding further Sibbald's purpose and how he undertook his geography demands attention to his plans for a Scottish atlas, to his use of circulated queries and to the social network of geographical practitioners and contributors upon which he relied.

## *Sibbald's Scottish Atlas: geography as national knowledge*

Sibbald's 1683 proposals for the 'Scotish [sic] Atlas, or the description of Scotland Ancient and Modern',[9] make clear that he intended a two-volume work. Volume one, *Scotia antiqua* (to be published in Latin), was to embrace the historical development of Scotland, the customs of the people and their antiquities. The second, English-language volume, *Scotia moderna,* would describe the country's resources as a work of contemporary chorography, on a county-by-county basis. Sibbald's *An account of the Scotish atlas* (1683) (Figure 3.1) is worth quoting from at length as it outlines his vision for geography. It is one of the clearest expressions in the late seventeenth century – possibly for any period – of the contemporary purpose and value of geographical knowledge.

Among the many sciences, which persite [sic] and adorn the mind of man, *Geography* is worthy of all praise; because it both affordeth matter of much delight, and likewise is of much use for the life of man. This World being a Theater, whereupon each act their Part, and by the several Personages, which they represent, very much demonstrate the marvellous Wisdom, Power and Goodness of GOD who hath contrived all for the perfection of the Universe his great Work, and by his never erring Providence bringeth all about to the accommplishment of his good will and pleasure: Man, who is the lesser World and Abridgement of the Greater, cannot but find the advantages of *Geography*, by which we see all the parts of this great Machine, even which are most remote from us, and look upon these who are absent, as if they were present with us. What delight must it give us, when we learn of the Situation of Countreys, Towns and Rivers, as well as if we had viewed them with our own eyes! It is certain that most of men have a great desire to travell, and it is certainly an inexcusable fault to be ignorant of what concerneth our own Countrey: Yet many, because of their poverty or want of health, are deprived of these Advantages. And therefore we are much beholden to *Geography*, by which all the face of the World is exposed to us, and we can even sitting at home view the whole Earth and Seas, and much sooner pass through them in our studies, than Travellers can do by their Voyages; and so may, without the hazard of being infected by the vices of Forreiners, improve our minds and reap all other Advantages from them. Not to mention here, how the Policy of each place, the studies of *Theology,* of *Natural Philosophy, History* and several Arts may be improved by this knowledge. I shall only instance in *Merchandising, Navigation* and the Practice of *Medicine.*

And for *Merchandising* (which with *Navigation* is the great support of corporations)

[9] NLS, Adv. MSS. 15.1.1, 15.1.2 (MS 15.1.1a is the maps, collected by Sibbald as part of the never published 'Atlas').

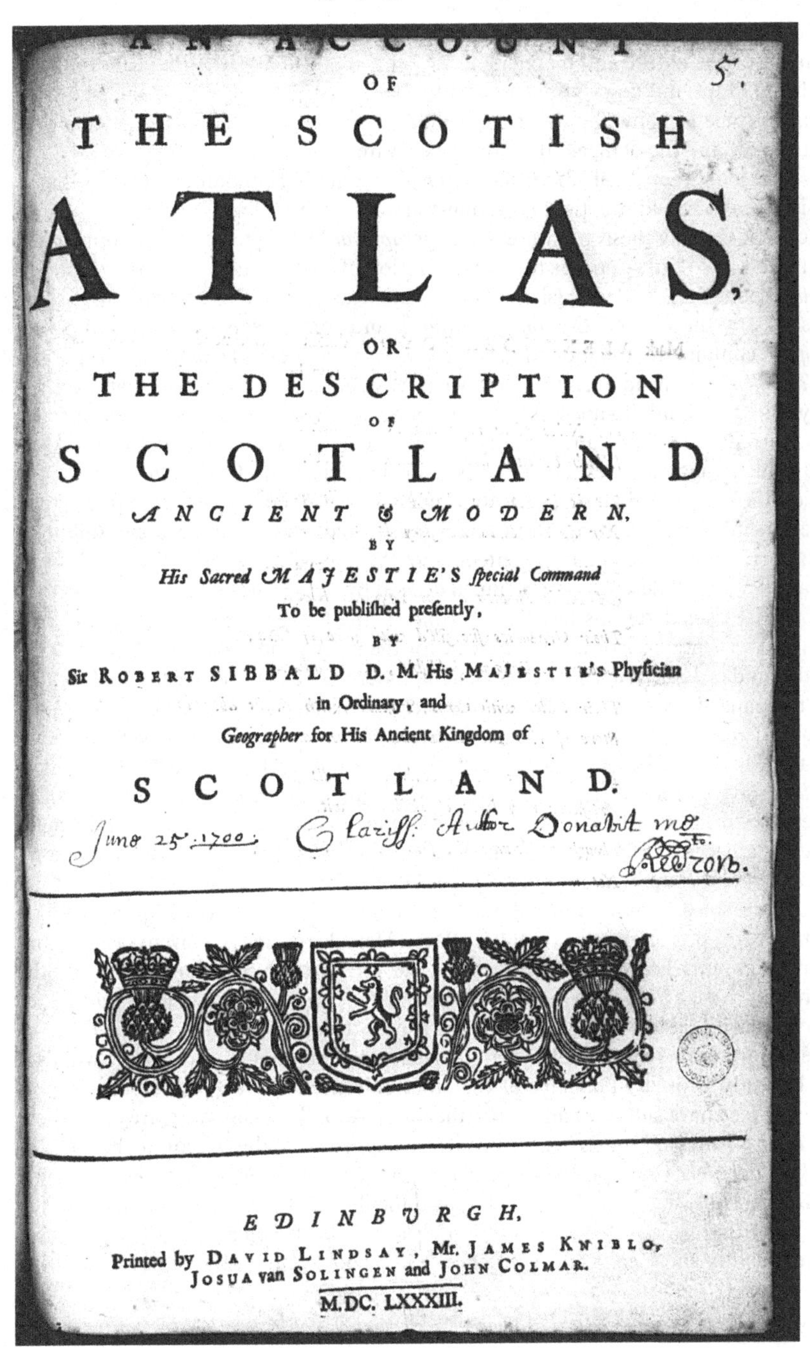

AN ACCOUNT
OF
THE SCOTISH
ATLAS,
OR
THE DESCRIPTION
OF
SCOTLAND
*ANCIENT & MODERN,*
BY
*His Sacred* MAJESTIE'S *special Command*
To be publiſhed preſently,
BY
Sir ROBERT SIBBALD D. M. His MAJESTIE's Phyſician
in Ordinary, and
*Geographer* for His Ancient Kingdom of
SCOTLAND.

*EDINBURGH,*
Printed by DAVID LINDSAY, Mr. JAMES KNIBLO,
JOSUA van SOLINGEN and JOHN COLMAR.
M.DC. LXXXIII.

3.1 The title page to Sir Robert Sibbald's *An Account of the Scotish Atlas* (1683).

it can hardly be entertained without the knowledge of the Countreys and their Products, of the nature and manners of the Inhabitants and Rodes that lead to them, which the Maps and descriptions joyned to them inform us of; and by the Maps the Mariners come to know the situation of the Countreys they intend for, the Rout they are to follow and the Dangers they are to be aware of.

As for the Practice of *Medicine, Hippocrates* hath abundantly proven, that a Physician, who would practise aright, must first know the place.

Now considering these great uses of *Geographical* Descriptions, many complained that there was so little done, as to the Description of our Countrey: For the *theater of Scotland* published by Bleau, [sic] for all its Bulk (except it be the Description of some few shires by the learned *Gordon of Straloch*, and some sheets of his of the *Scotia Antiqua*), containeth little more than what *Buchanan* wrote, and some few Scrapes out of *Cambden*, who is no friend to us in what he writeth. And in respect that there are many Islands around this ancient Kingdom of *Scotland,* and many impetuous and contrary Currents and Tides, and in several places the Coast is full of Rocks, or Banks of Sand, which ought to be exactly described for the security of Trade. And because the face of the Country, by the peace and quiet we enjoy under his *Majestie's* happy Government, is quite changed from what is was of old; and now for Stately Buildings, Parks, fertile Fields, we begin to contend with the happiest of our Neighbours: therefore a new and full description was much desired by all ingenious persons.

For these considerations His *Sacred Majesty*, the Father of his Country, and gracious and wise provisor of all that may be for the profile and honour of it, hath commanded and ordained, by His Letters patent under the Seal, Sir Robert *Sibbald* one of his Phisicians in ordinary and His *Geographer* for His Ancient Kingdom of *Scotland,* to publish an exact description of *Scotland Ancient and Modern,* with the Maps. And, that the Maps may be exact and just, the Lords of His *Majesties* Privy Council in *Scotland* gave Commission to *John Adair,* Mathematician and skilful Mechanick, to survey the shires. And the said *John Adair* by taking the distances of the several Angles from the adjacent Hills, hath designed most exact Maps, and hath lately made an *Hydrographical* Map of the River of *Forth* Geometrically surveyed. Wherein, after a new and exact way, are set down all the Isles, blind Rocks, Shelves and Sands, with an exact Draught of the Coasts, with all its Bayes, Head lands, Ports, Havens, Towns, and other things remarkeable; the Depths of the Watters through the whole *Firth,* with the Courses from each Point, the prospect and view of the remarkable Islands, Head-lands, and other considerable landmarks. And he is next to survey the shire of *Perth*, and to make two Maps thereof, one of the south-side, and another of the North. He will likewise be ready to design the maps of the other Shires, that were not done before, providing he may have sufficient allowance therefore. And, that these, who are concerned, may be the better persuaded thereto, there is joyned with this Account the Map of *Clackmanan Shire* taken off the Copper-Plate done for it, where may be seen not only the Towns, Hills, Rivers and Lochs; but also the different face of the grounds, which are Arable, and which Moorish and by convenient Marks you may know the Houses of the Nobility and Gentry, the Churches, Mills, woods and Parks &c.[10]

[10] Sir R. Sibbald, *An account of the Scotish Atlas, or the description of Scotland ancient and modern* (Edinburgh, 1683), 1–2.

Sibbald further outlined the geographical nature of his projected 'Atlas' in 1683 in the opening sections of his *Nuncius Scoto-Britannus,* itself prefaced to his 1684 *Scotia illustrata, sive prodromus historie naturalis* [*Scotland illustrated, or an introduction to the natural history*]. For a variety of reasons – too little cash, too much material and too little focus, personal disputes, Sibbald's loss of personal credibility in adopting Catholicism – Sibbald's intended 'Atlas' was never published. Yet it is clear from manuscript material that much was undertaken by him and on his behalf. What we have in the 1683 *Nuncius* and the 1684 *Scotia illustrata* are the only published expressions of Sibbald's geographical thinking on Scotland as a whole, allowing that several regional volumes were later printed.[11] In one sense, Sibbald was concerned to provide what we might now see as a medical chorography, with information about different diseases, their approximate geographical distribution, and the local cures sought in medicinal simples and other plants. Perhaps reflecting the emphasis of his 1682 warrant, with its focus on enriching 'the said kingdom, by keeping in the money which is sent abroad',[12] Sibbald was at pains to stress the virtues of domestic over foreign medicines.

In another sense, Sibbald was both a 'geographical modern' *and* someone who continued earlier traditions. Like many natural philosophers, he thought about the relationships between ancient claims and the observations made by contemporaries about their world. He wrote and thought in terms of that established distinction between 'special geography', with its roots in chorography and its emphasis upon given regions, and 'general geography', understood as world survey, in which regional descriptions were placed in comparative context. Sibbald knew and drew from the works of Ptolemy and other classical geographers, as well as upon ancient authorities like Pliny and Seneca. His library (in so far as we know it following the sale in 1708[13]), shows him to have been familiar with Bochart's *Geographia sacra* (1674), with the works of Varenius, Cluverius, and Keckermann, with Münster's *Cosmographiae universalis* (1544) and his 1558 *Rudimenta mathematica,* and with the works of Hakluyt, Kircher and Newton. Sibbald noted of the last two in particular: 'And since Kircher's *Subterranean World* was in great repute, as was Varenius' *General Geography* edited by Isaac Newton, wherein the general characteristics of the earth are set forth, I have inserted many

[11] Sir R. Sibbald, *The history, ancient and modern of the sheriffdoms of Fife and Kinross: with descriptions . . . of the firths of Forth and Tay . . . with an account of the natural products of the land and waters* (Edinburgh, 1710); *The history, ancient and modern, of the sheriffdoms of Linlithgow and Stirling . . . with an account of the natural products of the land and water* (Edinburgh, 1710); *The description of the isles of Orkney and Zetland . . . published by S. R. S. M. D.* [from the MS of Robert Monteith, 1633] (Edinburgh, 1711). For a list of Sibbald's publications, see C. Withers, 'Sir Robert Sibbald (1641–1722)', *Geographers' Biobibliographical Studies,* 19 (1997), 12–21. [12] EUL, MS Laing III.535.

[13] F. Emery, 'The geography of Robert Gordon, 1580–1661, and Sir Robert Sibbald, 1641–1722', *Scottish Geographical Magazine,* 74 (1958), 3–12; Simpson, 'Sir Robert Sibbald', 89–91. See also the anonymous *Bibliotheca Sibbaldiana* (Edinburgh, 1722) – a catalogue of Sibbald's texts prepared at his death.

things from them'.[14] Sibbald was, then, well aware of wider geographical thinking and used it, where necessary, to help make sense of geographical enquiry in and of Scotland.

In a further sense, Sibbald used his geographical enquiries as a 'historical modern'. I mean by this that he drew upon geographical work and past claims about Scotland's geography as part of contemporary enquiries into Scotland's historical identity. Ferguson has documented what he calls the move 'towards Enlightenment and a new view of history' in the later seventeenth century, a move distinguished by an insistence amongst historical scholars of verification from the sources.[15] For Scotland, the emergence of this new and critical sense of scholarship – itself bound up with advances in empirical methodology in science – was apparent in challenges to Boece and Buchanan. These were evident, for example, in William Lloyd's *An historical account of church government* (1684), in which Lloyd, the Bishop of St Asaph, dismissed the claims of the Scottish chroniclers over the lineage of Scotland's kings.

The details of this historiographical engagement have been documented elsewhere.[16] If, in general terms, we should accept that 'the old version of Scottish history derived from Fordun, as developed by Boece and Buchanan, was still accepted by leading Scottish intellectuals in the 1680s',[17] it is clear that Sibbald used geographical work to consider these questions. He drew together 'Manuscripts Containing the Generall and Speciall Geographie the Natural Histories and State Government of the Kingdome and Church of Scotland'.[18] The material relating to general geography – Scotland in context – he used in answer to the Bishop of St Andrews, who had asserted that Scotland was without its own historical traditions in comparison to England and other European nations. In doing so Sibbald makes reference to the work of Buchanan, Maule of Panmure's unpublished 'De Antiquitate Gentis Scotorum' of 1609; his argument is most apparent in his tract 'A defence or Vindication of the Scottish Historians, wherein the ancient Race of the Scottish Kings, ther ancient Possessions in this Island of Great Britain, and the Antiquity and Dignity of the Scottish Church are asserted: and the Objections of the Bishop of St Asaph are answered By Sir Robert Sibbald, his Majesties Physitian and geographer, and President of the Royal College of Physitiones of Edinburgh', which was widely referred to by contemporaries. Sibbald was situating the historical identity of the nation as part of contemporary geographical concerns. In order to do so, he had to rely on others.

[14] Sir Robert Sibbald, *Vindiciae Scotiae illustratae, sive prodromi naturalis historiae Scotiae, contra prodromomastiges* (Edinburgh, 1710), 5.

[15] W. Ferguson, *The identity of the Scottish nation: an historic quest* (Edinburgh, 1998), 145.

[16] Ferguson, *Identity of the Scottish nation,* 140–72.

[17] Ferguson, *Identity of the Scottish nation,* 154. Sir Robert Gordon of Straloch confessed his shame at the 'manifest fictions' of Boece, however, as early as 1649: see Bishop William Nicholson, *Scottish historical library* (London, 1702), 75, 107.

[18] NLS, Adv. MS 33.1.16, ff.1–2, 9, 9v *et seq.;* Adv. MS 33.5.15, 34.2.8.

*Sibbald's 'Queries': establishing credibility through the word of others*

The *Scotia Illustrata* was in part based on returns to queries circulated by Sibbald. Twelve main questions were asked, with additional queries for five specified 'target groups' (see pp. 78–9). The questions posed were typical of the time, and share much, for example, with those of Petty, Plot, Ogilby and Boyle. What is strikingly apparent is the emphasis on the social order: the nobility and the gentry being quizzed on *their* memorable exploits, the extent of their economic wealth, and so on. Here, I suggest, is geographical understanding – national self-knowledge – emerging as a form of political enquiry through survey, through the qualitative measurement of a nation's social order, and through chorographical description embracing natural history and antiquarian knowledge.

In a follow-up to his *Advertisement,* he had printed that same year (1682) a broadside entitled *In order to an exact Description,*[19] which was directed towards respondents of the 'General Queries'. Here Sibbald was providing an instruction guide for completion of his queries: 'In Order to an exact Description', he wrote, 'remember to confine your self within one Parish at a Time.' 'Omit nothing', he cautioned, 'that can any Way contribute to the Knowledge of the present Geographical face of our Country'.[20] He even provided a standard entry as guidance.[21]

What is clear is that geography's credibility as a means to national knowing was being established by use of a particular method at a time when just such questions about the rigour of empirical procedure were taxing many natural philosophers.[22] The epistemic credibility of circulated queries depended upon reliable informants. The reliability of such people did not have to be established or verified. It was, simply, understood: 'Participants "just knew" who a credible person was. They belonged to a culture that pointed to gentlemen as among their society's most reliable truth-tellers, a culture that associated gentility, integrity, and credibility'.[23]

Sibbald's sense of who such people were is clear from the categories he used in his 1682 *Advertisement*. This evidence allows us to recognise the detail asked for by Sibbald in his use of circulated queries as a means to geographical knowledge. It also shows the social order upon which such geographical knowledge rested and through which it was channelled. Sibbald paid attention to the leading groups within Scottish society because he considered such people, by virtue of their social position, to be reliable sources

[19] NLS, Crawford Deposit, Crawford MS, MB 277.
[20] NLS, Crawford Deposit, Crawford MS, MB 277, 2.
[21] The full version of this broadside appears as Appendix II of Withers, 'Geography, science and national identity', 68–9.
[22] Shapin, *Social history of truth,* 243–309; *The scientific revolution* (Chicago, 1996), 65–118.
[23] Shapin, *Social history of truth,* 241.

for its geographical description. Epistemic credibility was a matter of social worth: of Sibbald himself and of those upon whom he presumed.

*Sir Robert Sibbald's Advertisement of 1682 and his General Queries with a view to a Geographical Description of Scotland*[24]

WHEREAS His Sacred MAJESTY, by His Patent, hath constituted *Sir Robert Sibbald,* one of his *Physicians* in Ordinary, His *Geographer* for His Kingdom of SCOTLAND, and Commandeth and Ordaineth him to publish, the Description of the *Scotia Antiqua, & Scotia Moderna,* and the natural History of the Products of His *Ancient Kingdom* of SCOTLAND. These are earnestly to entreat all persons, that they would be pleased freely to communicate their *Answers* to these following *Queries*, or any of them, Directing them to the said Sir *Robert Sibbald* at his Lodging at *Edinburgh,* or to Master *James Brown* at his house in *Harts-closs*, who is Deputed by the said Sir *Robert Sibbald*, to receive and Registrate them; or to *Robert Mean, Post-master* at *Edinburgh*, to be sent to any of them: Withall specifying in their Letters, the place of their Habitation, that they may be again Written to, if Occasion should require; And an Honourable Mention shall be made of them in the Work, according to the importance of the Information.

*General Queries, to which Answers are desired.*

I What Nature of the County or place? And what are the chief products thereof?

II What Plants, Animals, Mettals, Substances cast up by the Sea, are peculiar to the place, and how Ordered?

III What Forrests, Woods, Parks? What Springs, Rivers, Loughs? With their various properties, whether Medicinal? With what Fish replenished, whether rapid or slow? &c. The rise of the Rivers, and their Emboucheurs?

IV What Roads, Bayes, Ports for shipping, and their Description? And what *Moon* causeth High-Water? What Rocks, and sholes on their Coast?

V What Ancient Monuments, Inscriptions, graved and figured Stones; Forts and ancient Camps? And what Curiosities of Art are, or have been found there?

VI What great Battels have been there fought, Or any other Memorable Action or Accident?

VII What peculiar Customs, Manners or Dispositions the Inhabitants of each Country or Town have among them?

VIII What Monasteries, Cathedrals, or other Churches have been there, and how named?

IX What places give, or formerly have given the Title to any Noble-man? As also, what ancient Seats of Noble-Families, are to be met with?

X What the Government of the County is? whether Sheriffdom, Stewartry, or Baillery?

XI What towns of Note in the County, especially Towns Corporate? The names of the Towns both Ancient and Modern? Whether they be Burrows Royal, of Regality or Barony? The Magistracy of Towns Corporated, when Incorporated? And by whom built? With the return of Parliament -Men? The Trade of the Town? How inhabited, and their manner of Buildings? What publick or Ancient Buildings? Their Jurisdiction? *&c.*

[24] NLS, MS s.302 b. 1 (21) [1682]; see also *Bannatyne Club Miscellany,* 3 (1855), 371–80.

XII In what Bishoprick each County or any part thereof is? Who is Sheriff, Stewart or Baily? And who commands the *Militia*? What Castles, Forts, Forrests, Parks, Woods, His MAJESTY hath there? *&c.*

*To the* NOBILITY
What Sheriffdomes, Bailliries, Stewartries, Regalities, Baronies and Burrows thay have under them? What Command of the Militia? What special Priviledge, Dignity and Heritable Command they have? The Rise of their Family, Continuance, and their Branches? What Forrests, Woods, Parks, Loughs, Rivers, Mines, and Quarries they have? What Fishing? *&c.* What Harbours they have? What their Titles are? What Memorable Actions raised or Aggrandized their Family? *&c.*

*To the* CLERGY
What their Priviledges and Dignities are? Their Erection? The Bounds of their Diocese? Their Chapter? The number of their Parishes in their Diocese? Their Jurisdiction, their Foundations for publick and pious Uses, their Revenues? What Lands hold of them? Their Houses? *&c.*

*To the* GENTRY
What the rise of their Family, their Priviledge and Dignity? What Baronies and Burrows under them? What Harbours, what Forrests, Woods, Parks? Their Houses, the Description and Names of them? The Chief of the Name and the Branches? The memorable Expolits done by them, and the Eminent Men of the Name? Their Heritable Command and Jurisdiction? *&c.*

*To the Royal* BURROWS
Of what Standing? The Constitution of their Government? their Priviledges, jurisdiction and its Extent? their publick Houses, Churches, Forts, Monuments, Universities, Colledges, Schools, Hospitals, Manufactures, Harbours? What their Latitude and Longitude is? *&c.*

*To the* UNIVERSITIES *and* COLLEDGES
What Standing they are of? their priviledges, Jurisdiction and its Extent, their Constitution? The Number of the Professors, their Names, what they teach? Their Salaries, Foundations, and their Founders? Their revenue and Dependencie? Their Houses, Churches and Chappels, edifices and Monuments? Their Libraries, Curious Instruments? The account of the famous Men bred there or Masters there? What are the Observations of the Masters or Students, that may be for the Embellishment of this Work?

It is clear from Sibbald's *Memoirs* that the project was personally costly and, to some degree, that shortages of funding underlay Sibbald's failure to publish his two-volume work. Writing in 1692 of the King's command a decade earlier to produce 'the geographicall description of the kingdom', he reflected how the doing of it 'was the cause of great paines and very much expense to me and how he only ever got 'one yeer's payment' for it.[25] In a letter in 1707 to

[25] Hett, *Memoirs,* 74–5.

Robert Wodrow, Sibbald wrote 'I have been more as these threttie [thirty] years past preparing the Geographicall description' and that 'if the Government give due incouradgement, it may be putt to the presse soone'. He had, he continued: 'all the originall mapps and surveys and descriptions of Mr. Pont, the Gordons and others, . . . and several mapps never printed. I give account of the natural products, especially the mynes from the MSS of those employed in working them, and there is ane account of all the ancient monuments and of the considerable actions join'd, and of the chief seats of the nobility and gentry. I not only viewed many, but I also have from the intelligent men resident in the places, the account of the countries'.[26]

The picture we should hold of Sibbald is not of the field-based geographer. He did not actively engage in direct observation, even allowing for his claim to have 'viewed many' of the houses of the nobility and others. Rather, Sibbald was the compiler of other's accounts, a personal 'centre of calculation' situated at the hub of a network of correspondents upon whose word he was reliant. By a close reading of Sibbald's manuscripts, notably his 'Repertory of Manuscripts', it is possible to piece together his network of contributors. There is, however, no complete list of respondents. William Nicholson, Bishop of Carlisle and a friend of Sibbald, noted only that 'a great many more short Descriptions, Observations, Traditional reports, &c. were sent in to the said Sir Robert, from most of the Counties in Scotland, in Answer to such Queries as he had transmitted to the Learned Men of those Parts; and he [Nicholson] listed twenty – some of the most considerable'.[27] In his 'Repertory' Sibbald noted, 'I applyde myself to procure descriptions from those who resided in the Several Countreys and were most capable to doo them and then I wrott myself'. He also noted: 'I shall first give ane account of thes I pleased to be done for me and after that, of what I wrott myself relating to the Speciall Geographie of this Country'.[28] There then follows a list of sixty-five names detailing, by name of respondent, by place and by length of manuscript, their contribution to the geographical survey of Scotland. This list, which can be supplemented by references elsewhere in Sibbald's papers and cross-checked against Nicolson's 1702 enumeration, points both to the audience for his work and, importantly, to the sources for geographical knowledge of Scotland in the late seventeenth century. It is here presented in the Appendix (pp. 256–62).

I am not suggesting that the identified respondents should all be thought of as geographers or, even, that their responses together amount to what we might understand as a consistent body of geographical knowledge, despite Sibbald's strictures to that effect. I am claiming, however, that this material illustrates what I will call the *constitution* of geographical knowledge. I mean

[26] Maidment, *Remains of Sir Robert Sibbald,* 36 (a letter of 11 November 1707 to Wodrow).
[27] William Nicholson, *Scottish historical library* (London, 1702), 19–20. [28] NLS, Adv. MS 33.3.16, f.14v.

by this that it is possible to see here the social and epistemic bases to geographical enquiry and authorship, to understand the presumed credibility upon which such knowledge was based and, at the same moment, to 'witness', as it were, the making of geographical knowledge about one's nation as a particular form of understanding.

Some of Sibbald's respondents drew upon earlier geographical and chorographical work, dating mainly from the 1630s and 1640s (Appendix, nos. 9, 11, 21). Most replies were written as a direct response to Sibbald's circulated 'Queries'. Dr George Archibald's 'Account of the Curiousities at Dumfries' as part of a larger 'General Description of the Stewartry of Kirkcudbright' (Appendix, no. 13) was entitled 'An answer to the Doctor his Advertisement'.[29] Others simply replied 'Answers to the General Queries', as did William Dundas in writing of Caithness, or Mr John Ouchterlony of Guynd in Forfarshire in his 'Information for Sir Robert Sibbald' (Appendix, nos. 5, 23).[30] Closest in form to the style requested by Sibbald is Bishop Fraser's 'Ane Answer to Sir Robert Sibbald's queries for the Isles of Tirrey [Tiree], Canna, Colle and Icolumbkill [Iona] lying within the Sherrifdom of Argyll and the Bishoprick of the Isles'.[31] Sibbald's cousin, Sir Andrew Balfour, younger brother to Sir James upon whose own earlier collection of topographical descriptions Sibbald also drew, was used as an intermediary by some respondents. 'I beseech you let Sir Robert Sibbald know that I have received his Queries, & wish that I could send him such information as he desires', wrote one Dr Miln of Inverurie in March 1683. 'But this place & ye country hereabouts affords but little matter for such a design: yet I shall not be wanting in [making] what comes to my knowledge worthy [of] his notice'.[32] To judge from the returns, no material was ever forthcoming from Dr Miln.

Several of Sibbald's correspondents were themselves reliant upon local knowledge and networks of 'intelligent persons' in their local area. This is evident for Sir George Mackenzie, Earl of Cromartie, reporting on the western isles (Appendix, no. 3), and is most clear in the constitution of geographical knowledge from and about the northern isles (Appendix, nos. 26, 27, 45–56). In this sense, for some parts of Scotland, Sibbald's position as a 'centre of calculation' was replicated at a local scale.

Certain of Sibbald's respondents were actively engaged in geographical work: William Geddes in Caithness; Alexander Pennecuik; Robert Edward; the Wallaces in the Orkneys (Appendix, nos. 5, 8, 24, 42, 43). James Wallace senior was a Puritan natural historian, a former physician in Sumatra and considered by his contemporaries 'well vers'd in Antiquities'.[33] The younger Wallace published *A description of the isles of Orkney* in 1693, based upon his

[29] NLS, Adv. MS 33.5.15, ff. 224–5.
[30] NLS, Adv. MS 33.3.16, f.15.
[31] NLS, Adv. MS 34.2.8, ff.197–197v.
[32] NLS, Adv. MS 33.5.15, f.14.
[33] E. Gibson (ed.), *Camden's Britannia* (London, 1695), col.1073; Mendyk, *Speculum Britanniae*, 221.

father's papers, to which Sibbald attached an essay on the Thule of the Ancients. Wallace's *Description* emphasised the natural history of the coastlines: not just in a focus on types of species, but from a concern to *measure* what was there in terms of economic capacity as, for example, inlets likely to make good harbours and safe fishing grounds. The chorographical accounts of Robert Gordon, who, with his son, James, was influential in ensuring Pont's maps were published (see pp. 53–6), produced several regional descriptions which Mendyk considers 'more realistic and geographical than most landscape interpretation' in mid-seventeenth-century England.[34] The involvement of John Adair and Martin Martin (Appendix, nos. 31, 39 and 58), is considered further below.

In these ways, then, Scotland was being geographically constituted through local geographical enquiries, written (in intention anyway) to a common plan, with the purpose of providing a description of the nation. In those terms, Sibbald was a political and natural arithmetician like John Graunt and Gregory King, and a chorographer like Robert Plot. It is anachronistic to concur with Kidd's view that Sibbald was easing 'the birth of Scottish sociology', yet it is clear that through Sibbald's geographical work the nation was coming to know itself historically.[35] Sibbald's geography was the result of networks of informants, some of whom were geographically active. It was a practice that incorporated historical material as part of contemporary survey. In several instances, the reports of his informants stress the utility of their knowledge not just as *local* knowledge, but as part of a bigger inductive project to benefit the nation.

### *'Useful to human lyfe': Sibbald's geography, institutionalised natural knowledge and the state of the nation*

In 1684, Sibbald was elected president of the RCPE and first professor of medicine at Edinburgh University. Influenced by his patron, Sibbald became a Roman Catholic in 1686. The conversion shocked his contemporaries, led a mob to storm his house in Edinburgh and Sibbald to resign his presidency of the RCPE and flee to London. There he met Robert Boyle and Hans Sloane, amongst other members of the Royal Society, and was elected a fellow of the College of Physicians. He also repented his conversion to Catholicism, and, on return to Edinburgh in 1686, rejoined the Protestant church. He was, however, slighted in verse for his conversion by his fellow physician, the Newtonian Archibald Pitcairne, most notably in a satirical poem 'A Catholick

[34] Mendyk, *Speculum Britanniae*, 214.

[35] C. Kidd, *Subverting Scotland's past: Scottish Whig historians and the creation of an anglo-British identity, 1689–c.1830* (Cambridge, 1993), 112; D. Dickson, 'Science and political hegemony in the seventeenth century', *Radical Science Journal*, 8 (1979), 7–38. David Allan, *Virtue, learning and the Scottish enlightenment: ideas of scholarship in early modern history* (Edinburgh, 1993), 12.

Pile to purge out Christianity, Prescribed by Doctor Sibbald'.[36] Sibbald, in short, lost personal credibility and was out of the country at an important time in relation to his plans for geography in and of Scotland.

It is difficult to know if this fact affected contemporaries' faith either in Sibbald or in his work but, together with the variability of returns to his queries, it is unlikely to have directly helped. Yet Sibbald brought his geographical activities to bear upon Scotland's economic condition. His manuscript 'Description of Scotland' pays attention to soil types and to different agricultural geographies.[37] Similar interests inform his 1699 pamphlet on 'Provision for the poor in time of dearth and scarcity', written in response to widespread harvest failure in Scotland at that time, and even his 1701 'natural history of fishes', in addition to descriptions of 'aquatick animals' found in Scotland's rivers and firths, comments on potential improvements in navigation and the connections between profitable fisheries and national well-being.[38]

This attention to the present state of the nation and to its future condition is most evident in Sibbald's political agricultural tract entitled 'A Discourse Anent Improvements may be made in Scotland for Advancing the Wealth of the Kingdom', written in 1698.[39] This work is in three parts: the present state of the nation; 'the means by which the wants of the countrey may be made up wherein it is discussed of Acts, trades, navigations, & improvement of land, and colonies'; and on improving Scotland's fisheries 'and the advantages the Nation may have yr by'. The focus on the state of the nation was to look to utility and to the future; as Sibbald put it, 'What is wanting to make the people in all those places happy'.[40] He also noted that 'Colonies may improve this Countrey',[41] but was otherwise silent on Scotland's colonial ventures at this time.

Sibbald's concerns with natural knowledge as useful knowledge are also evident in his plans for an intended Royal Society of Scotland. Between 1698 and 1701, Sibbald twice proposed the establishment of a Royal Society of Scotland, a body to mirror the Royal Society of London and the Dublin Philosophical Society, both of which concerned themselves with geographical knowledge. In his 'Overture presented to the Commissionary and Severall of the Nobility & Gentrie for erecting the Royal Society of Scotland for improving of Usefull Arts', Sibbald proposes a 'Society for the improvement of Arts useful for Humane Lyfe, especcially Husbandry, Gardenry, Medicine, and the knowlege of Natural things that are the products of this Country, and may be useful for Mechanick Arts, and Matters of Trade yr in'.[42] Emerson's analysis

[36] NLS, MS 2257, f. 3. [37] NLS, Adv. MS 15.1.1, ff. 16v–17. [38] NLS, Adv. MS 33.5.16, ff. 1, 23–72.

[39] NLS, Adv. MS 33.5.16, ff. 76–93v; Whyte, *Agriculture and society in seventeenth-century Scotland,* 206, 208, 209, 211. [40] NLS, Adv. MS 33.5.16, ff. 77, 80v. [41] NLS, Adv. MS 33.5.16, f. 87.

[42] NLS, Adv. MS 33.5.19, f. 490; see Appendix I of Emerson, 'Sir Robert Sibbald, Kt, the Royal Society of Scotland', 67–9.

of this intended body has shown how Sibbald and his fellow proponents should be seen as Baconians, concerned with empirical information about the nation and with its improvement: 'They had rejected traditional wisdom and were prepared to base their claims on carefully scrutinized empirical findings'.[43] The fact, however, that Newtonian beliefs were being institutionalised by university teachers of mathematics and natural philosophy, by leading doctors in Scotland's universities and to some extent by teachers of geography in Scotland's universities, means that we should not see Sibbald's Baconian notion of natural knowledge as the only one at work in Scotland in the 1690s.

Neither should we lose sight of the wider networks in which he worked. Sibbald saw himself, from 1685–6 at least, as a source of information to London's Royal Society. In 1693, he wrote to Hans Sloane to say that if the fellows would welcome a copy of Wallace's work on the Orkneys, he would provide it.[44] In 1707, a quarter of a century after his geographical enquiries were formally begun, he solicited that Society's views on his geography, asking whether they might assist in its publication (partly because Scotland's printing presses were then taken up with printing literature on the Union). The fact that they did not does not diminish the point about the wider context in which Sibbald operated and through which so much of his geographical work was carried out. By that time, his own capacities were weakened: 'My age now makes me unfitt for the search of yes [these] products, yett my curiosity continues and I ame mightily satisfied to find yt you [Hans Sloane and the Royal Society] advance the natural history so much'.[45] What is also the case is that others, differently pursuing geographical knowledge, were also involved with the Royal Society.

## Competing geographies, testing credibility

One recent discussion of the 'Scientific Revolution' has noted that 'in the seventeenth century natural philosophers were confronted with differing repertoires of practical and conceptual skills for achieving various philosophical goals and with choices about which ends they might work to achieve. The goal was always some conception of proper philosophical knowledge about the natural world, though descriptions of what that knowledge looked like and how it was to be secured varied greatly'.[46]

Several other conceptions of natural and geographical knowledge were apparent in Sibbald's time. James Paterson's *A geographical description of Scotland,* for example, published in 1681 and in expanded editions in 1685 and 1687, was essentially an almanac with tide tables, dates of fairs and so on, all,

[43] Emerson, 'Sir Robert Sibbald, Kt, the Royal Society of Scotland', 64.

[44] Royal Society, EL.S2.6, f. 1 (letter of 7 February 1693); EL.S2.3, f. 1 (letter of 7 February 1693), and EL.S2.3, f. 1 (letter of 11 July 1693).

[45] EUL, MS Dc.8.35, f. 33.

[46] Shapin, *The scientific revolution,* 117.

as the preface states, 'Exactly Calculate and formed, for the use of all Travellers, Mariners, and others, who have any Affairs, or Merchandizing in this Kingdom of Scotland'. Paterson was an Edinburgh mathematician: his work advertises him 'at the Sign of the Sea Cross-staff and Quadrant'. He also wrote *The Scots arithmetician, or arithmetic in all its parts* (1685) and, in the same year, *Edinburgh's true almanack, or a new prognostication for the years 1685–1692*. Similar work was being done in Glasgow by James Corss, author of a Glasgow almanac in 1662 and, in 1666, of *Practical geometry*; Corss had earlier taught geography in Edinburgh.[47]

In considering the 'differing repertoires' of geographical knowledge, my concern here is with three other men. The three are Robert Wodrow (1679–1734), ecclesiastical historian, Glasgow University Librarian between 1697 and 1701, and, from 1703 until his death, parish minister; Martin Martin (*c.* 1660–1718), a native Highlander educated at Edinburgh and Leiden universities who was involved in reporting upon the Hebrides; and John Adair (1660–1718), 'mathematician and mapmaker' as Sibbald termed him, who, from 1681, was funded by the Privy Council to map the nation. Consideration of their work highlights both the contested nature of geographical knowledge and suggests how such understanding was part of wider contemporary concerns to do with the observation of nature and with the credibility of geographical knowledge.

Wodrow maintained a voluminous correspondence with persons including Sibbald, James Sutherland, Keeper of the Botanic Garden in Edinburgh and others in the Scottish Highlands, Ireland, London, the Netherlands, Darien in Panama, Virginia and Guinea.[48] Wodrow was no field scientist. He acted, like Sibbald, at the centre of a network of people who, in being judged reliable, were called upon for their knowledge of nature. Wodrow's interest as a *virtuoso* was, to borrow Eamon's phrase, with the 'rehabilitation of curiousity' about natural products.[49]

Partly on behalf of the antiquarian and natural historian Edward Lhuyd, Wodrow sent 'Queries and things to be done in the western Highlands' to the ministers there. His focus upon 'the curious', upon rare and unusual features and events, stemmed both from the need to know about such things (rather than with their presumed utility) and from a concern that such local natural knowledge might, upon closer enquiry, be scientifically well founded. His intention may also be interpreted as an attempted accommodation of such local knowledge not only within the classifications of authoritative writers such as John Ray, but also within a view of nature he saw as divinely ordained.[50] He asked in his 'Queries' how Highland names of plants accorded

[47] M. Wood (ed.), *Extracts from the records of the burgh of Edinburgh* (London, 1940), 93.
[48] L. Sharp (ed.), *Early letters of Robert Wodrow 1698–1709* (Edinburgh, 1937).
[49] W. Eamon, *Science and the secrets of nature: books of secrets in medieval and early modern culture* (Princeton, 1994), 314–15. [50] Sharp (ed.), *Early letters of Robert Wodrow*, 188.

with 'Mr Ray's Dictionarium trilingue', about local charms, and about 'the peculiar games and customes observed on set dayes throughout the year, or any other fashions peculiar to the Highlanders', noting to one correspondent how 'Mr Edw. Lhuyd has 3 sheets of the customes and rites of the Highlands quhich he procured from some correspondent in Scotland'.[51]

Wodrow's Highland correspondents were nearly all Church of Scotland ministers. His reliance upon such men is understandable; there was no one else in the region who might be judged reliable or capable of finding things out and reporting back. Such men, Gaelic-speakers themselves, drew upon native vernacular knowledge as legitimate knowledge. That was quite acceptable to Wodrow. His request, for example, to have 'ane accompt of the opinions of the vulgar Highlanders touching [concerning] the adderstones, toad-stones, cockneastones, mole-stones &c',[52] was in keeping with a more widely-held wish to know about the natural products of a region unknown to most natural philosophers and is illustrative of *virtuosi* interest in natural knowledge and the popular understanding of it.[53] As a Baconian, Wodrow encouraged his correspondents to provide examples from which he could construct a general picture of Scotland's geography and natural history and the workings of the world in general. As he wrote in October 1700 to James Paterson, author of the *Geographical description,* 'I think we have not as yet a sufficient stock of experiments & tryalls to build our hypotheses on'. *All* information was useful: he commented to another respondent, 'Accompts from hills and mountains, moss or more, bank and syke, sea or shore, books, stones, coins, charters – in short anything rather than nothing will be acceptable'.[54]

Like Sibbald, Wodrow was prepared not only to accept such knowledge but also to trust his informants. Unlike Sibbald, Wodrow did not place a significance upon the gentlemanly or elite status of his respondents. Wodrow's emphasis was with local 'folk' knowledge. In those terms, he was according a place for what some historians of science would consider as 'the humble' playing a part through native knowledge in the promotion of natural learning.[55] Like Sibbald, Wodrow's network connected others interested in Scotland's geography. In a letter of 9 November 1702 to Mr Lachlan Campbell in Kintyre, Wodrow noted 'I send you some queries of Mr Adaires, quhich I hope you will communicate with the curious and observing gentleman, ministers and others, and let me have your oun and their answers to them nou and then as you have occasion'.[56]

[51] Sharp (ed.), *Early letters of Robert Wodrow,* 237. On the relationships between Lhuyd and Wodrow and Lhuyd's travels in the Highlands, see J. Campbell and D. Thomson, *Edward Lhuyd in the Scottish highlands, 1699–1700* (Oxford, 1963), 16, 53, 56–8, 64–5, 95.

[52] Sharp (ed.), *Early letters of Robert Wodrow,* 160.

[53] Eamon, *Science and the secrets of nature,* 301–18, 319–50; Hunter, *Science and the shape of orthodoxy,* 101–19.

[54] Sharp (ed.), *Early letters of Robert Wodrow,* xxiv.

[55] Eamon, *Science and the secrets of nature*; Hunter, *Science and the shape of orthodoxy.*

[56] Sharp (ed.), *Early letters of Robert Wodrow,* 237.

In 1694, Adair printed and circulated 'QUERIES in Order to a True Description; And an Account of the Natural Curiousitys, and Antiquities'.[57] Although most of Adair's papers are now lost – no replies to his queries have been located – his intentions in this respect mirror Wodrow's. (It is uncertain whether Wodrow is referring to Adair's 1694 'Queries' in his letter of 1702.) Some questions were antiquarian in nature. Most were directed at local knowledge and at the curious, the aberrant: 'Information is desired about any strange Appearances, that have happended in the Air . . .'; whether 'any uncommon substances have been digged from under, or found above ground'; or 'If Men or Women have been attended, by any thing not common'.[58] Such chorographical information, like that solicited by Wodrow, was certainly local knowledge, but it was not knowledge which easily lent itself to inclusion in wider explanatory reasoning as did Wodrow's, or had an economic utility as did Sibbald's. As a form of geographical enquiry, Adair's 'Queries' were in contrast to his work as a mapmaker and surveyor, although they were quite consistent with contemporaries' interests in natural wonders. What united Sibbald, Wodrow and Adair was their reliance upon the word of people 'out there' in providing what were judged credible facts; the acceptability and credibility of such knowledge depended upon the acceptability and credibility of one's informants.

### *Martin Martin and the geographical credibility of the local*

For Mendyk, Martin Martin was 'perhaps the most famous and the most important of all of Sibbald's collaborators'.[59] This is arguable. Martin did provide Sibbald with accounts of Skye, St Kilda and other Hebridean islands (see Appendix, nos. 39, 58). But he was not just an agent for the Geographer Royal. Martin was involved with Sibbald and others in ways which reinforce the points I have been making here about credibility and the social networks generating late seventeenth-century geographical knowledge. Martin was a source of information about the western Highlands and islands for Edward Lhuyd. Martin was, as Sibbald recognised in a letter to Sloane, a 'local', able, by virtue of language and if given the right support, to gather trustworthy facts about the Hebrides: 'He was borne in the Isle of Sky, was Gobernour to ye Chieffs of ye Clans in ys isles and heth yt interest and favour with them, they will doe for him what they will doe for no other. yr. Language is his Mother Language, and he is well acquainted with yr Maners and Customes and is the person here most capable to Serve the Royall Societty in the accounts of what relateth to ye description of ys Isles'.[60]

Martin had been reporting to the Royal Society for at least three years by

[57] J. Adair, *QUERIES in order to a true description, and an account of the natural curiousitys, and antiquities* (n.p., 1694). [58] Adair, *QUERIES in order to a true description*, 2.
[59] Mendyk, *Speculum Britanniae*, 218. [60] EUL, MS Dc.8.35, ff.9–10.

the time of Sibbald's letter. In December 1695, Martin had written to a friend in Edinburgh noting that the President of the Royal Society had 'oblidged me to send all my observations' to that body, and that Viscount Tarbat, Sir George Mackenzie – himself a source of geographical information for Sibbald (Appendix, no. 3) – was to write to them in support of Martin.[61] In a letter of 1 August 1696 to David Gregory, Professor of Astronomy at Oxford, Martin discussed enquiries he had made about the geography of Hebridean Scotland under the title 'Observations in the North of Scotland':

> Before I came from London Octr. last a ffriend [sic] of yours advised me to wait upon Mr Charleton, who promised to consider me above the Common rate, if I should bring him any Curiosities from our Western Islands, which I did not omit in the course of my short Travels in *Skie Harries* South and North *Uist* and some Isles. Dr Gordon thereafter made my acquaintance with Dr: Sloane, who expects great things from our Western Isles. I promised to inform him of any natural observations I could learn. . . . If he with yourself will influence the Royal Society to satisfy what you determine for my future encouragement, to prosecute a further Enquiry into the Natural History of all the Western Isles of Scotland, If this Essay be Satisfying, it will engage them to give me the encouragement that is necessary for such an undertaking.[62]

In October 1697, Martin published his 'Several Observations in the North Islands of Scotland' in the Society's *Philosophical Transactions.* Twelve items were briefly discussed, ranging from the seasonality of egg-laying amongst sea birds, Highland medicinal practices, the coincidence of respiratory disorders on St Kilda with the arrival of the estate chamberlain, and the transient blindness of a Harris man who lost and regained his sight with each new moon and was known locally as the '*Infallible Almanack*'.[63] In a letter to Sloane, Martin wrote that he had about 100 'Curiosities' from Skye for the Royal Society 'and near the Same number of Natural Observations'.[64] In a letter of 16 November 1697 to Hans Sloane from Charles Preston (later to be Professor of Botany at Edinburgh), Preston noted how 'our friend Mr Martin' was on his way to London with collections for the attention of the Royal Society: 'he spares no pains in collecting things'.[65] Martin was, then, a geographical field agent for the Royal Society as well as for Sibbald, directly observing and reporting upon things and transmitting specimens and facts about unknown parts of Britain.

His travels resulted in two main works: *A late voyage to St Kilda, the remotest of all the hebrides, or western isles of Scotland,* published in 1698 and

[61] NLS, MS 1389, f. 84, Martin to Alexander Macleod, 14 December 1695.

[62] Royal Society, LBC 11 (2), 160, Martin to David Gregory, 1 August 1696. The 'Charleton' here referred to is not the physician Walter Charleton but the collector William Courten (1642–1702) who also went by the name 'Charleton': see C. Gibson-Wood, 'Classification and value in a seventeenth-century museum: William Courten's collection', *Journal of the history of collections,* 9 (1997), 61–77.

[63] *Philosophical Transactions of the Royal Society,* October 1697, Number 233, 727–29.

[64] Royal Society, LBC 12.334, Martin to Hans Sloane, 4 September 1697.

[65] Royal Society, LBC 12.338, Charles Preston to Hans Sloane, 16 November 1697.

dedicated to Charles Montague, then President of the Royal Society; and, in 1703, *A description of the western isles of Scotland.* At one level, these are works of local description. They illustrate Martin's interest in 'remarkable curiousities' and include attention to local knowledge: plant names, the healing use of certain stones or waters, Hebridean customs, the portentous interpretation of migratory bird behaviour. In the *Preface* to his 1698 work and again in 1703, Martin stressed the value of local knowledge: 'The modern itch after the knowledge of foreign places is so prevalent that the generality of mankind bestow little thought or time upon the place of their nativity'.[66] To most people, even to most Scots, the Hebrides and St Kilda were then as unknown as the Americas or other parts of the New World. As Martin stated, 'Foreigners, sailing thro the Western Isles, have been tempted, from the sight of so many wild Hills . . . and fac'd with high Rocks, to imagin the Inhabitants, as well as the Places of their Residence, are barbarous; . . . the like is suppos'd by many that live in the South of Scotland, who know no more of the Western Isles than the Natives of Italy'.[67] Martin's comments, however, were not just about the value of knowing one's locality.

At another level, Martin is expressing a belief that Hebrideans' knowledge merited the same philosophical attention as knowledge gained by travellers abroad. Martin is stressing the credibility of his informants. 'The inhabitants of these islands', he wrote, 'do for the most part labour under the want of knowledge of letters and other useful arts and sciences; notwithstanding which defect, they seem to be better versed in the book of nature than many that have greater opportunities for improvement'.[68]

Martin's words and actions in directly observing nature highlight the importance of language to geographical enquiry. Martin is unusual, as chorographer, geographical agent and native Highlander, in being part of the culture he wrote about. Unlike geographers overseas, he did not, in his practical enquiries and communication with the natives, have to overcome the hermeneutic problem of ascribing meaning through translation to the subjects under review. In order to convince distant others, notably within the Royal Society, of the credibility of his narratives and, thus, of Hebrideans' worth as imparters of natural knowledge, he had to represent Gaelic as a legitimate language through which to assess the nature of Nature. His remarks amount to a moral and philosophical judgement about the credibility and worth of such people, of whom he was one. They are a claim about the need to treat such knowledge as epistemically credible, as Martin himself sought personal and intellectual credibility.

Martin's narratives do not demonstrate that 'language of mathematical plainnesse' which Thomas Sprat and the Royal Society sought in descriptions

[66] M. Martin, *A late voyage to St Kilda* (London, 1698), Preface; *A description of the western isles of Scotland* (London, 1703), 62. [67] M. Martin, *A description of the western isles* (London, 1716 edition), ii.

[68] Martin, *A description of the western isles* (1703 edition), 63–4.

of natural and social phenomena.[69] Martin admitted to 'Defects . . . in my Stile and Way of Writing' and confessed that 'he might have put these papers into the hands of some capable of giving them, what they really want, a politer turn of phrase'.[70] In terms of his wider credibility, Martin's geographical encounters demanded – indeed, *depended upon* – a linguistic shift from the spoken Gaelic of his informants to the written English of his texts. Yet knowing and using Gaelic, a language not then associated with civility and certainly not with the conduct of geographical enquiries,[71] was essential to his credibility *and* to the epistemological claims that underlay his work. Martin's credibility rested, then, upon proof of his own encounter with nature and the trustworthiness of his local informants. He did this by use of their language, by reference to their moral virtue and by reference to their intrinsic inability to deceive.

There is nothing related to the following Account, but what he vouches to be true, either from his own particular Observation, or else from the Constant and harmonious Testimony that was given him by the inhabitants; and they are a sort of People so Plain, and so little inclined to Impose upon Mankind, that perhaps no place in the World at this day, knows such Instances of true primitive Honor and Simplicity; a People who abhor lying tricks and Artifices, as they do the most poisonous Plants, or devouring Animals.[72]

In a letter of 27 October 1697 from Charles Preston to Hans Sloane, Preston wrote:

I have met frequently with Dr Sibbald & Mr Martin who informs me he has Lately travelled through Some of the western Isles . . . and has made Severall Curios observations of the natural products of those Isles and more particularly of St Kilda which Island it seems is undescribed by Buchanan or any other author it is in the Latitude of 58° & Some minutes. They have no trade no communication with any other people save those that come with the Steward once a year to exact the masters dues. There are about 200 soules upon it: they live upon the product of the Isle Sea Fowl is a great part of their dyet and thier manner of Catching them is pretty odd.[73]

Preston's letter highlights further the place of language in considering credibility and trust in others. In his use of the rhetoric of latitudinal measurement, Preston is attempting to give a location for St Kilda in a way that reflects the place of mathematics in geographical and natural knowledge. He is, as it were, attempting to site intellectually the geographically unsighted and is doing so for someone sharing neither the immediacy of the meeting with Martin nor Martin's own vision and description of the island. If his rhetoric

[69] Shapin, *Social history of truth*, 122–3, 199, 211–14; A. Goldgar, *Impolite learning: conduct and community in the republic of letters, 1680–1750* (New Haven, 1996).
[70] Martin Martin, *A voyage to St Kilda* (Glasgow, 1818 edition), iv.
[71] C. Withers, *Gaelic in Scotland 1698–1981: the geographical history of a language* (Edinburgh, 1984).
[72] Martin, *A late voyage to St Kilda* (1698 edition), 2.
[73] RS, EL.P1: 101, Charles Preston to Hans Sloane, 27 October 1697.

is, in one sense, mathematical, it is in another, and simultaneously, one of credulousness about what he hears from Martin about this unknown world. How could he express himself so as to be believed about what – upon first encounter in print or conversation or direct observation – was unfamiliar and strange, even inexpressable? Much of what Martin would have told Preston and Sibbald and reported to the Royal Society may well have struck them as 'pretty odd'.

This trope of astonishment and a sense of the marvellous in considering the nature of human society and natural productions was common in contemporary accounts of New World geographies: the whole of nature was a matter of wonder.[74] The languages of wonder and of incredulity stem, of course, from the difficulties of ensuring commensurability between what one knew and had seen and what one had heard reported or had read of the unknown. There is every reason to expect it of narratives detailing the wondrous unknown 'Old World'. And being believed was important if others had competing claims to public support and credible knowledge.

### *Martin Martin, John Adair and competing geographies*

Martin undertook at least two Hebridean voyages in the company of John Adair. This is clear in a letter from Preston to Sloane of 13 October 1698: 'Mr Adair . . . is not as yet returned from his voyage to the western Isles, . . . Mr Martin went in company so that I doubt not they will return freighted with a Large Cargoe of natural curiosities'.[75] In a later letter that year to Sloane, Preston noted how Adair 'seems resolved to give a short history of his voyage apairt from his other work & to publish it very suddainly, Mr Martin who was with Mr Adair some part of the voyage is also arrived but I have had no time to discourse [with] him'.[76] Relationships between Martin and Adair were not good: Sibbald noted in a letter to Sloane that Adair treated Martin 'scurvily'.[77] The cause of the tension between the two was, I suggest, rivalry over different sorts of geographical knowledge, getting published and the fact that both men, differently at work in the same area, sought the support of the Royal Society.

Adair had received 'incouragement and assistance' from the Privy Council since 4 May 1681 when he was commissioned to 'take a Survey of the whole Shires in the Kingdom, and to make up Mapps thereof'.[78] These maps were wanted both by Moses Pitt for his multi-volume *Great atlas*, and by Sibbald as part of the Scottish *Atlas.* Initial funds for Adair's work proved inadequate.

[74] S. Greenblatt, *Marvellous possessions: the wonder of the new world* (Oxford, 1991); Goldgar, *Impolite learning*, 117–42; L. Daston and K. Park, *Wonders and the order of nature, 1150–1750* (New York, 1999).
[75] RS, EL.P1: 102, Charles Preston to Hans Sloane, 13 October 1698.
[76] British Library, Sloane MS 4060, ff. 125–6 [Undated but the BL catalogue dates it at 22 November 1698].
[77] EUL, MS Dc.8.35, ff. 9–10. [78] SRO, RH 14/203 (15 June 1686).

Adair solicited further support, and on 15 June 1686, an Act was passed 'In Favours of John Adair, Geographer, for Surveying the Kingdom of Scotland, and Navigating the Coasts and Isles thereof'. Funding was to be forthcoming from a tonnage levy applied differentially upon Scottish and foreign ships: it was never sufficient.[79]

The Act is important, however, for the values it articulates both about mathematical and survey work and for the capacity of Adair to undertake such work:

> [E]xact *Geographical* Descriptions of the several Shires within this Kingdom, will be both Honourable and Useful to the Inhabitants; and the *Hydrographical* Description of the Sea-Coast, Isles, Crieks, Firths, and Lochs, about the Kingdom, are not only Honourable and Useful, but most necessary for Navigation . . . The want of such exact Maps, having occasioned great losses in time past: and likewise, thereby *Forraigners* may be invited to Trade with more security on our Coasts.[80]

The Act re-affirmed Adair's credibility in the eyes of contemporaries as a geographical practitioner of a certain sort. We should not doubt his public esteem: Adair bore, at one time or another, the titles 'Surveyor of the Shires', 'Hydrographer Royal', 'Geographer to the King', 'Geographer for Scotland', and 'The Queen's Geographer in these Parts' [Scotland], and was made a burgess of three towns in Scotland.[81] In these terms and given this Act's warrant of mapmaking as a geographical practice and instrumental craft, Adair mirrors for Scotland that utilitarian impulse apparent for mapmaking in Europe and in England in the work of men such as the Cassinis, Ogilby, Moxon and Hooke[82] (Figure 3.2).

The Act also contains a clue to understanding the relationship between Adair and Martin. The Act made provision that the Sheriff of each county would 'appoint one or two knowing men, in each Paroch, to go alongst with the said *John Adair,* when he is actually Surveying the same, to design unto him the particular Places of each Paroch, for the more exact performance of the said Work'.[83] No sources exist by which we can confirm Martin as such a local man. But we may infer as much from the fact that Martin began his

[79] C. Withers, 'John Adair (1660–1718)', *Geographers' Biobibliographical Studies,* 20 (2000), 1–8.

[80] SRO, RH 14/203 (15 June 1686).

[81] *Calendar of treasury books* (London, 1969), Vol. XXVIII (1717), part II, 170 (4 March 1714): (a list of proposed custom's establishments in Scotland 'examined and is signed by John Adair, the Queen's Geographer in Scotland'). In Rotterdam on 1 September 1687, Adair signed a commission to his wife using the designation 'Geographer to his Sacred Majesty', SRO, Register of Deeds, DAL, LXVII, 1279. See also Withers, 'John Adair', and J. Moore, 'Scottish cartography in the later Stuart era, 1660–1714', *Scottish Tradition,* 14 (1986–87), 28–44.

[82] Moore, 'Scottish cartography'; D. Buisseret (ed.), *Monarchs, ministers and maps: the emergence of cartography as a tool of government in early modern Europe* (Chicago, 1992); J. Konvitz, *Cartography in France 1660–1848: science, engineering and statecraft* (Chicago, 1987); E. Taylor, 'Robert Hooke and the cartographical projects of the late seventeenth century (1666–1696)', *Geographical Journal,* 90 (1937), 529–40; H. Wallis, 'Geographie is better than divinitie', in N. Thrower (ed.), *The compleat platt-maker* (Chicago, 1986), 1–43; P. Barber, 'Necessary and ornamental: map use in England under the later Stuarts, 1660–1714', *Eighteenth-Century Life,* 14 (1990), 1–28.

[83] SRO, RH 14/203 (15 June 1686).

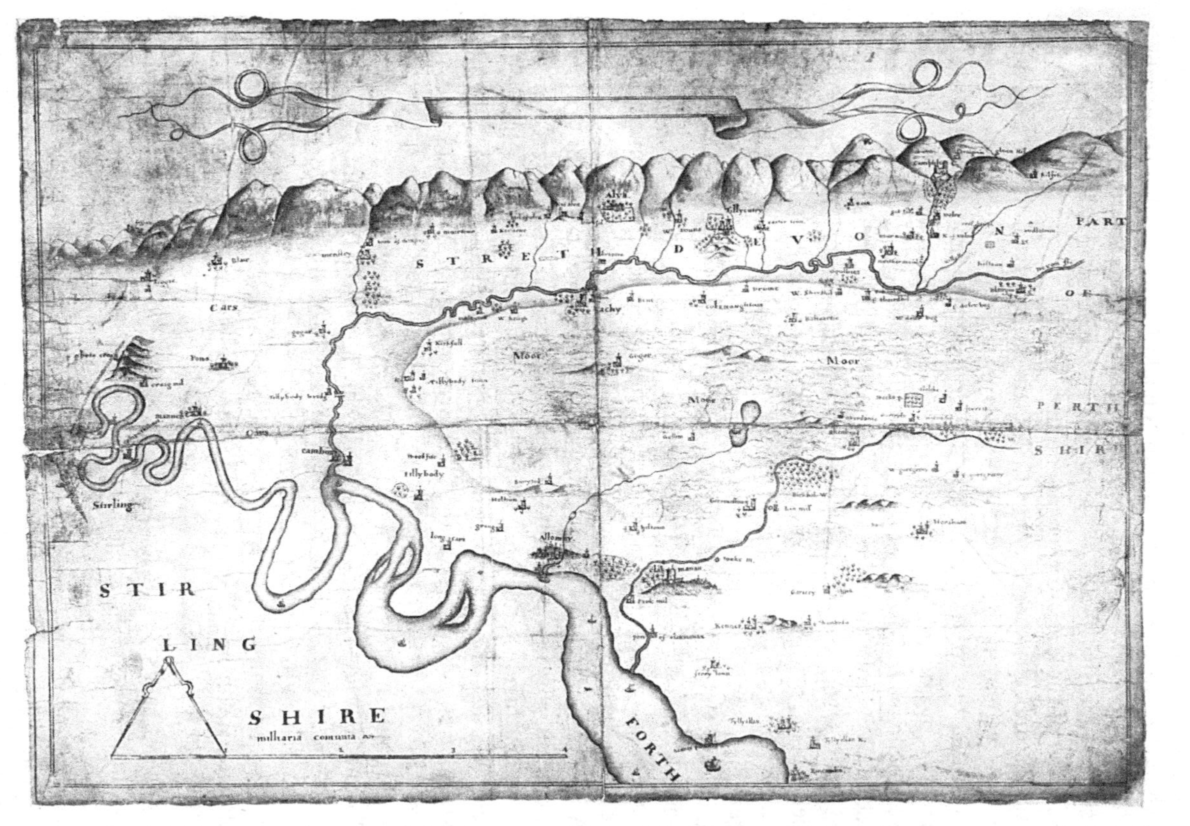


3.2 John Adair's manuscript map of Clackmannanshire, 1681. Adair's qualities and style were regarded highly by contemporaries in England as well as by Sibbald and others in Scotland.

travels in the area about 1695, from Sibbald's letter to Sloane describing Adair's attitude to Martin, and by the support given to Martin by Sir George Mackenzie, the Sheriff over those islands and west mainland areas that were the subject of these competing geographies.

Adair's difficulties were not with Martin alone. His work was seriously restricted by funding shortages, despite the tonnage levies passed in his favour. He resented the contractual restrictions imposed by Sibbald and spoke of Sibbald's 'envy, malice and oppression' towards him. In 1693, Sibbald confessed to Sloane that Adair was slow: 'I cannot find yt any progress heth been made in Mr Adairs his topography of Scotland thes eight years elapsed'.[84] And Adair's work was further hindered by the claims of John Slezer. Slezer, Chief Engineer with the army in Scotland since 1671, had published his *Theatrum Scotiae* in 1693: the work, of topographic views of the nation (see Figure 3.3), was also to have involved Sibbald, but they fell out over contractual matters.[85] Slezer intended a further work: 'Of Ancient Scotland, and its Ancient People'. The Scots Parliament, in approving Slezer's work, added Slezer's name to the beneficiaries of the tonnage levies: Privy Council records note on-going disputes between Adair and Slezer.[86]

Adair may have been slow in his work simply because of the nature of instrumental survey at that time as well as because of his insistence on purchasing instruments and contracting engravers from Holland.[87] But he, like others, was also soliciting the support of the Royal Society. In November 1687, Adair had shown 'several of his curious maps of Scotland made from his late survey' to the Royal Society, and, at their meeting of 2 November 1687, had informed its members 'that to the westward of the isles of Scotland called the Hebrides there was an island, which had been frequently seen from the land of Argyle's jurisdiction: but it was not known, that any person had been upon it'.[88] In a letter of 20 December 1698, by which time Martin and he had sailed to this unknown island – St Kilda – John Adair made clear his prolonged involvement in mapping the Hebrides and other parts of the Scottish coast:

These 3 Years by Gone I have been taken up in making Maps and Descriptions of the Sea-Coast, and Isles of this Kingdom, for the use of Seamen, so that what Observation I made in Natural Learning by the way, could not well be Soon published. . . . I am resolved to be there early next Spring, and use all means to get the Survey of the West Isles finished, and so soon as possibly published, for I could not have believed that any

[84] Royal Society, EL.S2.6, f.1, Sibbald to Hans Sloane, 7 February 1693.

[85] K. Cavers, *A vision of Scotland: the nation observed by John Slezer, 1671–1717* (Edinburgh, 1993).

[86] Withers, 'John Adair', 2–3, 5–6.

[87] He contracted James Moxon as his engraver, brother to the mapmaker and geographical author Joseph Moxon. Adair's wife sold most of Adair's instruments at his death to help pay debts incurred by his mapping: Withers, 'John Adair', 3–4.

[88] T. Birch, *History of the Royal Society* (4 volumes, London, 1756–57), Vol. IV, 550, 552 (2 November and 9 November 1687).

3.3 Glasgow from the south, from John Slezer's *Theatrum Scotiae* (1693). This work – whose costs of production effectively bankrupted Slezer – is an important early example of the topographic visualisation of Scotland so common from the later eighteenth century.

Land especially at the Doors, and where there are so many Excellent Roads and harbours, had been so falsely laid down, as this is in all Maps I have seen.[89]

Here is first-hand testimony of the conduct of geographical knowledge, of its time-consuming nature, and of the pressures upon practitioners – effectively, Adair was a government employee – to make their efforts public. It is evidence, too, of the inaccuracy of earlier efforts, including Blaeu; the inadequacies of his 1654 *Atlas* and later editions had been one stimulus to Sibbald in 1682.

In summary, two different means to geographical knowledge were being practised in the same region at the same time by two different men. Both Adair and Martin relied upon personal observation. Unlike Sibbald and Wodrow, both undertook their enquiries whilst enduring the rigours of travel. Both were connected with the Royal Society, and both worked for Sibbald. For Martin, geographical knowledge involved the collection of natural productions and the documentation of local social customs in ways which lent moral value to his informants yet prompted a sense of incredulity from some of his audience. Adair sought in his maps to reduce local variety to a replicable geometry and to correct the errors of previous delineations so as to provide a 'true Shape' which would honour both the nation and the mapmaker (Figure 3.4 a and b). Yet, despite official support, he could not publish his material: he lacked funds, there was so much to do. On the other hand, Martin continued to be supported through the patronage of Sir George Mackenzie and others.

Adair's 1698 manuscript 'A Journal of the Voyage made to the North and West Islands of Scotland' was never published and is now lost. His principal work, *Description of the sea coast,* which mapped the eastern seaboard of Scotland (with which parts he had begun in 1681), was published in 1703, the same year as Martin's account of the western isles. Unpublished maps by Adair of parts of the western seaboard survive, 'Survey'd Navigated and Designed by John Adair', but they stand as the legacy of a particular, underfunded and largely unfulfilled geographical practice.

## Sites of knowledge and the public engagement with geography

The promotion of geographical knowledge by the state in Scotland did not have institutional expression. Sibbald's house in Edinburgh's Canongate may be seen as a collecting centre and site of scrutiny for the returns to his queries, but geography did not have a 'house of experiment' in which its practitioners could meet and debate as did those involved in the promotion of experimental natural philosophy in late seventeenth-century England.[90] Scotland's natural knowledge was the subject of some attention in the Edinburgh Botanical Garden, which, from its foundation in 1667, became a centre for

[89] RS, LBC. 12.69–72, John Adair to Hans Sloane, 20 December 1698.
[90] S. Shapin, 'The house of experiment in seventeenth-century England', *Isis,* 79 (1988), 373–404.

research into and the display of nature's products. It was first run by James Sutherland, who became Professor of Botany at Edinburgh University in 1695 and, in 1699, the King's Botanist and Keeper of the Royal Garden at Holyrood, and who was author, with Sibbald's assistance, of *Hortus medicus Edinburgensis.*[91] A further site was the cabinet of natural history, also established by Sibbald with Sir Andrew Balfour. This was to become, upon Balfour's death in 1699, the basis to Edinburgh University's Natural History Museum: Sibbald's speech at its foundation emphasised the utility in displaying materials which 'contribute to the use & ornament or pleasing of Human Lyfe'.[92] This 'Compleat Cabinet', as Sibbald termed it, was largely of Balfour's own making and consisted, like those of many *virtuosi,* of fossils, plant and animal specimens and *materia medica.* Edward Lhuyd, who visited it in the company of Sibbald and James Sutherland, was dismissive of the Scottish content in a letter of 15 December 1699 to Robert Lister: 'The Museum Balfourianum wch ye College of Edenborough lately purchased, consists of such exotics as are to be seen in most other collections; and contains very little of ye product of their own countrey excepting some fossils and a few marine animals.'[93] The clearest evidence for the situated production of geography comes from the universities.

*Teaching sites: geography in the universities, c.1680–1711*

In Edinburgh on 16 November 1670, Mr George Sinclair was given licence by the town's council 'to profess several usefull sciences', which, in addition to exploring the 'strange and wonderfull effects and causes . . . of The Pneumaticall or air pump' and the study of 'Hydrostaticks', included 'Mathematicks, Geometrie, Astronomy, Special and Theoricall Geography'.[94] He was one of a number of learned men in the universities then incorporating geography in their teaching. Geography lectures were given in Edinburgh University as part of philosophy courses in 1679 and 1680 by John Wishart, variously Professor of Humanity and of Philosophy between 1653–54 and 1667 and, again, from 1672 to July 1680. Surviving 'Dictates on Geography' taken by John Cranstoun, one of Wishart's students, show his lectures to have been concerned with cosmography, with regional descriptions of Great Britain with 'Scotia' figuring centrally, and with debates on Cartesian natural philosophy.[95] In King's College Aberdeen, the regent and, later, sub-principal,

[91] A. Cunningham, 'The culture of gardens', in N. Jardine, J. Secord and E. Spary (eds.), *Cultures of natural history* (Cambridge, 1996), 38–56; H. Fletcher and W. Brown, *The Royal Botanic Garden in Edinburgh 1670–1970* (Edinburgh, 1970), v, xiii–19; A. Grant, *The story of the University of Edinburgh,* (London, 1884), Vol. I, 218–20.

[92] Sir Robert Sibbald, 'A speech dedicating Balfour's museum', EUL, MS Laing III, 535, f.1.

[93] R. Gunther, *Early science in Oxford: volume XIV life and letter of Edward Lhuyd* (Oxford, 1945), 418–19.

[94] Wood, *Extracts from the records of the burgh of Edinburgh, 1665–1680,* 92–3.

[95] EUL, MS DK.5.27, ff. 57–70.

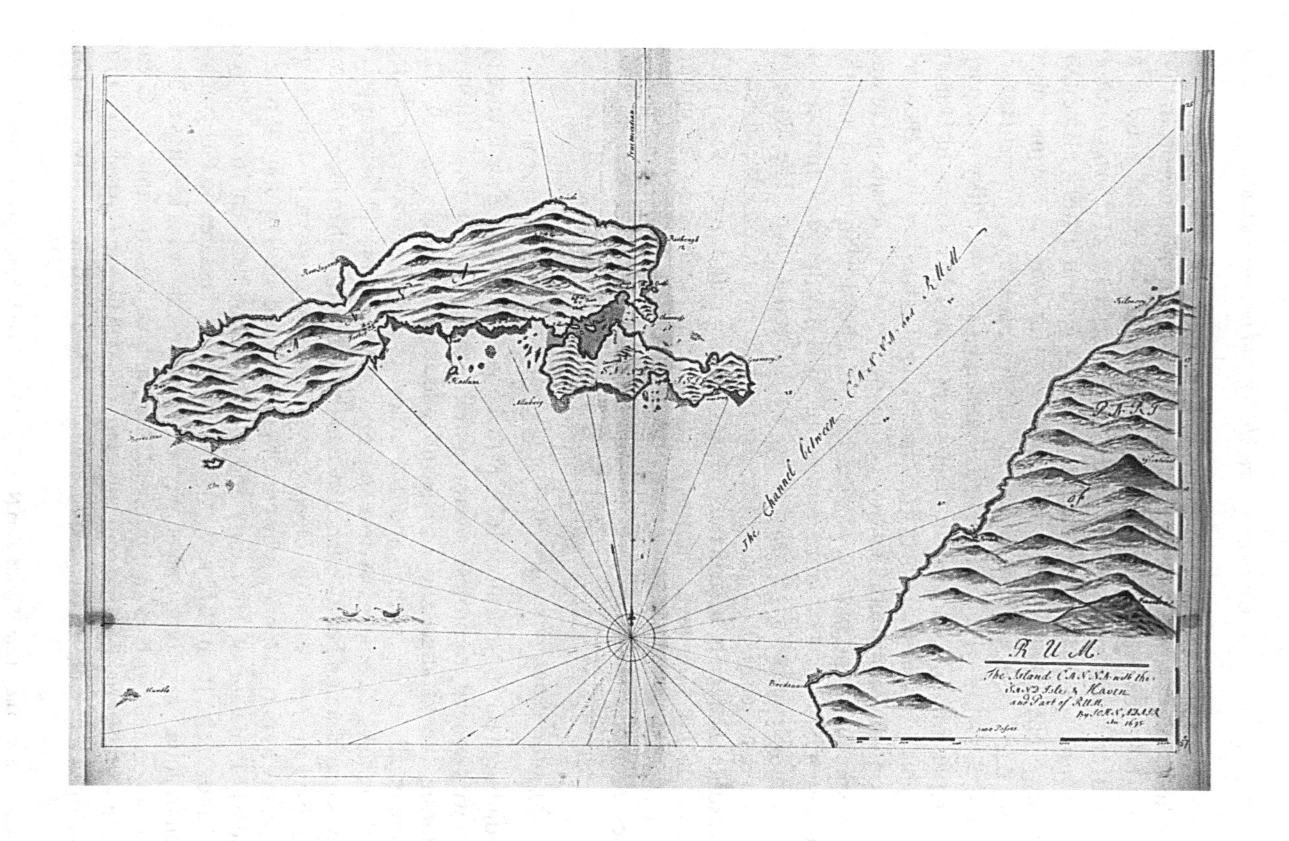
R U M.
The Island CANNA with the
SAND Isle & Haven
and Part of RUM.
The Channel between

When Mr Adair graduate the meridian of this Chart he had certain
by design'd it to begin at the parallell passing through the South of Rum
and to end Northward at the Parallell of the Maiden-Rock on the West
Coast of Skye: By the Meridian graduation of this Map the Island
Canna measures 20 Miles from West to East & 5 to 3 Miles from North
to South, also the Difference of Latitude of that part of Rum in this Chart
is 20 Miles of the Meridian: But by the annex'd Scale Canna measures 4½ Miles of a degree allowing 1222 passes to a Mile; That
part of Rum in the Chart measures 3¼ —

Martin a Native of the Western Isles in his descripton of these
Isles says Cannay is 2 Miles from South to North & 1 from East to West
Rum 5 Miles from South to North & 1 from East to West —

Note, the Graduation of the above Map would answer to
that of Rum Island. & part of Skye mls. —

3.4 The island of Canna from John Adair's unpublished 'Atlas'. The note to the right hand corner of the map (3.4A) dates this map to 1695. The text (3.4B) shows Adair discussing the different means to accurate measurement of the island and, by implication, to Scotland's west coast as a whole. Note the reference to Martin Martin.

William Black, taught geography and cosmography in 1692–3, and structured his geographical material as a series of 'propositions' to be debated and solved.[96] Surviving notes from Alexander Maclennan's classes with the regent George Skene show that Skene's geography lectures, delivered between 1701–4, were also delivered as practical propositions.[97] In St Andrews James Gregory, the Professor of Mathematics in 1684–5, gave 'some succinct institutions in Astronomy and Geography' in 1685, but nothing is known of their content.[98]

Together with what we have seen of geography teaching in Scotland's universities in the late sixteenth century, several points arise from this evidence. The first is historiographical: the current history of academic geography in Scotland understood as an institutionalised history with origins in formal departments in the later nineteenth or early twentieth century[99] must be replaced by recognition of geography's longer-running place in the academy, just as Cormack has extended our understanding of England, and others have shown of Russia.[100] Over two hundred years before being formally sited in departments, geographical knowledge as a taught subject was a compound enterprise with links to astronomy and cosmology, geometry, philosophy, the humanities and mathematics.

The second point is contextual and more problematic. It is not easy to know the extent to which Wishart's focus on Cartesianism in his geography classes in Edinburgh in the 1680s, or Black's practical geography, either ran counter to or supported what we know of the adoption of Newtonianism in Scotland. Texts such as Honter's Aristotelian *Cosmographicorum* were used in teaching 'geographie' to final-year students in Edinburgh in the 1620s.[101] Surviving lists of graduates' theses in cosmology, even allowing for their variability over time and for different universities, provide important clues. Apart from Edinburgh, which, between 1616 and 1626, witnessed a short-lived move to embrace Copernican theory, cosmological teaching in the universities largely followed traditional Aristotelian-Ptolemaic lines until the 1640s. From then until about 1670, it is possible to discern an engagement with the ideas of Copernicus and Tycho Brahe, for example, but, whilst discussed, they were always rejected. For Russell, 'The real watershed separating old and new cosmology occurred

[96] Aberdeen University Library [hereafter AUL], MS K 153, ff. 164r–184r–v and ff. 188r–206v; R. Emerson, *Professors, patronage and politics: the Aberdeen universities in the eighteenth century* (Aberdeen, 1992), 24–6, 136, 143.

[97] AUL, MS 2092 'Tractatus Geographicus', ff. 4r–9v; Emerson, *Professors, patronage and politics,* 72, 73, 90, 93, 104–5, 126, 128–9, 143, 146.

[98] Bodleian Library, Ashmolean MSS (1813), f. 243, 3 February 1685.

[99] E. Lochhead, 'Scotland as the cradle of modern academic geography in Britain', *Scottish Geographical Magazine,* 97 (1981), 98–109.

[100] L. Cormack, *Charting an empire: geography at the English universities, 1580–1620* (Chicago, 1997); D. Shaw, 'Geographical practice and its significance in Peter the Great's Russia', *Journal of Historical Geography* 22 (1996), 160–76.

[101] A. Morgan (ed.), *University of Edinburgh: charters, statutes, and acts of the town council and the senatus 1583–1858* (Edinburgh, 1937), 62.

about 1670 with the sudden appearance of mechanistic scientific philosophy, based on Descartes and Boyle, at St Andrews (1668), Marischal College [Aberdeen] (1669) and Edinburgh (1670). By 1682 Cartesian cosmology was completely dominant at Edinburgh and almost equally so at the other universities.' Further, 'Newton's cosmology was taken up with alacrity at both Edinburgh and St Andrews very soon after the publication of *Principia mathematica* [1687], and was evidently being taught at all universities by about 1710'.[102]

It is possible to suggest, then, that the Baconian geographical endeavours of Wodrow, Martin, Adair and Sibbald were paralleled in the universities, in the teaching of geography, by a theoretical engagement with current cosmological thinking. Shepherd's work on the Scottish universities' curricula and the Edinburgh-led adoption of Newtonianism would support this contention.[103] Geography's teaching, whilst never a major part of the curriculum, was in keeping with that advocacy of mechanistic natural philosophy that distinguished the 'Scientific Revolution' at this time. But this general claim about what we might think of as 'the map of geographical knowledge' in late seventeenth-century Scotland must recognise local difference and individual circumstance. In Edinburgh, for example, John Wishart at one time proclaimed the followers of Copernicus 'mad or weak in the head'.[104] His earliest lectures were solidly Aristotelian, but by the later 1660s his lectures refer favourably to Descartes and Boyle amongst others, and by 1679 his final lecture notes record disapproval of Aristotle and make reference to Newton's theory of light published in the *Philosophical Transactions* of 1672, 1675 and 1676. In Aberdeen, William Black only slowly embraced the heliocentric system between 1690 – when he agreed it could be accepted as a hypothesis 'unless it should be contrary to Scripture' – and his ready adoption of it by 1705.[105] In Glasgow, it is likely that Robert Simson, Professor of Mathematics from 1711–61, at once embraced Newtonianism: he had many geographical texts in his library and was a correspondent with the Cambridge mathematician James Jurin, who, with Newton, had produced an edition of Varenius' *Geographia generalis* with additional mathematical and geographical appendices.[106]

A point may also be made about audiences. The work of Sibbald and Adair could be considered geographical knowledge for an elite few, albeit that their interest in Scotland's geography was shared and facilitated by so many. The same group provided the intake to university classes and, as clerics or medical

[102] J. Russell, 'Cosmological teaching in the seventeenth-century Scottish universities, part 1', *Journal of the history of astronomy,* 5 (1974), 122–32, and 'Part 2', 145–54: quotation from page 152.

[103] Shepherd, 'Newtonianism in Scottish universities', 65–75, 'Philosophy and science in the arts curriculum of the Scottish universities in the seventeenth century' (Ph.D. thesis, University of Edinburgh, 1975).

[104] Russell, 'Cosmology in Scottish universities', 128.

[105] Russell, 'Cosmology in Scottish universities', 150.

[106] W. Warntz, 'Newton, the Newtonians and the *Geographia Generalis Varenii*', *Annals of the Association of American Geographers,* 79 (1989), 165–91.

men, made up Scotland's emergent 'scientific community' in this period.[107] That community was also a social one. Although there is not the detailed evidence for Scotland to parallel Stewart's work on Restoration science in London, in which he sketches for people like James Brydges, later First Duke of Chandos, what might be seen as his local geography of public knowledge as he moved between Garraway's coffee house in Exchange Alley, Gresham College and the meetings of the Royal Society,[108] it is likely that such local itineraries between sites of information exchange characterised, say, Sibbald's life in Edinburgh.

### *Geography as public knowledge*

There is limited evidence of public classes in geography before 1707. James Corss in Edinburgh in April 1658 advertised the teaching of 'Arithmetique Geometrie Astronomie and all uther airts and Sciences belonging thereto as horometrie Planimetrie Geographie Trigionometrie'.[109] Two other matters also merit comment: the gentlemanly Scot abroad, and the publication of works of geography designed both to describe the world in whole or in part and, thus, to instruct people in geography.

Although travellers to Scotland increased during the later seventeenth century, most did not travel to the north and west Highlands and islands; a fact, of course, that made Martin's work, Lhuyd's travels and the queries of Wodrow and Sibbald so important to the Royal Society and others. For Scots abroad, it is possible to claim that European travel and education, to Leiden especially for medical men, was a geographical experience. Sir Andrew Balfour, for example, spent fifteen years in France and Italy. His posthumous pamphlet *Letters writen to a friend, containing excellent directions and advices for travelling thro' France and Italy* (1700) mirrors Robert Boyle and others' emphases upon the benefits of travel and assurances about credibility:[110] 'What Sir *Andrew* delivereth here, is what he observed, experienced, and handled himself, and none need question the relations in any Circumstance; He was a most candid and Ingenious Gentleman'.[111]

For those who could not travel, geography books helped bring the world to them. I would not want to take geography texts out of context here, nor to over-play public interest in them. Despite the numbers of such works,[112] it is almost impossible to know who bought them and how they were read and used. Mayhew has discussed something of these issues for English geography from 1660. His concern is with works of *geography* – in distinction to my

[107] J. Christie, 'The origins and development of the Scottish scientific community, 1680–1760', *History of Science*, 12 (1974), 122–41.
[108] L. Stewart, 'Public lectures and private patronage in Newtonian England', *Isis*, 75 (1986), 47–58.
[109] Wood (ed.), *Extracts from the records of the burgh of Edinburgh*, 93.
[110] Hunter, *Science and society in restoration England*, 53.
[111] *Letters writen to a friend by the learned and judicious Sir Andrew Balfour* (Edinburgh, 1700), viii.
[112] O. F. G. Sitwell, *Four centuries of special geography* (Vancouver, 1993).

broader interest in the texts and practices of *geographical knowledge* – with geography understood as 'a coherent body of knowledge about a clearly-defined object, namely the situation of places on the earth and the content of those places in natural and human terms'.[113] In those strictly definitional terms, late seventeenth- and early eighteenth- century Scotland did not have an indigenous geographical book publishing industry, nor, with the partial exception of men like James Paterson and perhaps some of Sibbald's correspondents, did Scotland have many authors of geography books. Such activities were centred in London.

Some exceptions are, however, worth noting. Patrick Gordon, son of James Gordon of Rothiemay, published his *Geography anatomiz'd* in London in 1693. Gordon's work was intended for the children of the gentry. His prefatory claim 'that among all other parts of ingenuous Learning, Geography hath this peculiar property and excellency, that it is easily attainable by persons of any Age or Sex, and no less useful to persons in all stations, and of all professions; insomuch that no person whatsoever can be said to be compleatly accomplisht without some competent knowledge and understanding of it' repeats a rhetoric shared by many authors, before and since, who, after all, had to make money by their writing. But the fact that Gordon wrote it at all and that, variously corrected and enlarged, *Geography anatomiz'd* ran to twenty editions by 1760 is indicative of a certain public interest in geography.[114] So, too, is Matthias Symson's *Geography compendiz'd* (1702), written for the Marquis of Duglass, which focused upon the utility of geography as earth description. It was the first such book published in Scotland. Gawin Drummond's *A short treatise of geography* was published in 1708 'for the use of schools'.[115]

If there was little formal geography publishing before 1707 in Scotland, the evidence of other texts – in combination with those practices and works discussed above – suggests an awareness by the literate public at least of the geographical condition of Scotland. This is particularly true of works devoted to Scotland's agriculture which, like Sibbald's unpublished 'Description of Scotland', were perhaps prompted by the economic circumstances of the 1690s. John Reid's *Scots gard'ner* (1693), written 'for the climate of Scotland', James Donaldson's *Husbandry anatomised, or an enquiry into the present manner of teiling and manuring the ground in Scotland* (1697), and the anonymous *The countryman's rudiments, or an advice to the farmers of East Lothian how to labour and manure their grounds* (1698) may be cited here.[116] In those

[113] R. Mayhew, 'The character of English geography *c.*1660–1800: a textual approach', *Journal of Historical Geography*, 24 (1998), 385–412.

[114] Sitwell, *Four centuries of special geography*, 261–66; B. McCorkle, 'The maps of Patrick Gordon's *Geography Anatomiz'd*: an eighteenth-century success story', *The Map Collector*, 66 (1994), 11–15.

[115] Sitwell, *Four centuries of special geography*, 547–8. Drummond's book was advertised as such in *The Edinburgh Courant*, 16–18 June, 1708.

[116] J. Watson and G. Amery, 'Early Scottish agricultural writers (1697–1790)', *Transactions of the Highland and Agricultural Society of Scotland*, 43 (1931), 60–85.

terms, and including Symson (1702) as part of my argument, I would argue that, in addition to official support for different practices of geographical knowledge and the teaching both of geography in the universities, there was by 1707 a developing, if limited, public awareness both of geography within Scotland and of the state of the nation as a geographical matter. Such awareness was heightened by those ideas of empire that underlay the Darien scheme and parliamentary union with England.

**Scotland's imperial geographical imagination: empire, union and the Darien scheme**

The Union of 1707 was the subject of attention by contemporaries and has been so for modern historians. For Scotland, the union is explained by economic poverty, political expediency and the effects of long-term religious and political factionalism. For England, dynastic crises over the Hanoverian succession, the increasingly independent foreign policies of the Scots Parliament, the threat of a possible Jacobite restoration and the need to secure the nation's northern borders at a time of war with Spain all demanded that a resolution to the instability of Scottish affairs be found.[117]

The union was also the consequence of Scotland's failure to establish an empire. The collapse of the Darien colony in Panama in 1699 and 1700 is seen by most historians as one contributory element amongst many in the move to union. The Darien scheme may also be considered a geographical matter for two reasons. Darien was rooted, as Armitage shows, in theories of empire and in recognition of the site's geographical importance in terms of world trade: 'The argument that to straddle the junction of two oceans was the key to the empire of the seas was, of course, the heart of the strategic justification for Scotland's expedition to the isthmus of Panama: to plant a colony in Darien would be to hold the world's *entrepot* between east and west.'[118] In relation to other contemporary geographical concerns, the Darien colony represents, as an indigenous colonial enterprise, a further expression of geographical awareness by the state and a matter of popular concern at Scotland's geographical identity.

Plans for a trading colony at Darien in the Panama isthmus in central America have their origins in debates within the Privy Council in February 1681 to establish a Scottish trading colony 'in some place of America'.[119] By

[117] J. Robertson (ed.), *A union for empire: political thought and the British Union of 1707* (Cambridge, 1995); 'An elusive sovereignty: the course of the union debate in Scotland, 1698–1707', in Robertson (ed.), *A union for empire,* 198–227; Sir J. Clerk of Penicuik, *History of the union of Scotland and England*, D. Duncan (ed.), Scottish History Society, Fifth Series, Vol. 6 (Edinburgh, 1993); P. Hume Brown, *The Union of 1707* (Glasgow, 1907); J. Young, 'The Scottish parliament and national identity from the union of crowns to the union of the parliaments, 1603–1707', in D. Broun, R. Finlay and M. Lynch (eds.), *Image and identity: the making and re-making of Scotland through the ages* (Edinburgh, 1998), 105–42.

[118] D. Armitage, 'The Scottish vision of empire: intellectual origins of the Darien venture', in Robertson, *A union for empire,* 108.

[119] SRO, PC 12/7/13A, Memorial in favour of a Scottish colony, 28 February 1681.

1699, the Scots Company Trading to Africa and the Indies – later simply known as either the Company of Scotland or Darien Company – had been established. There is no doubt that this was *Scottish* colonialism. As one contemporary noted, 'Scotland is amongst the last of the Nations in Europe in setling forrain plantations, tho few can propose more advantage that way Because the Nation affords many subjects of manufactor and abounds in men which is the greatest riches as well as strength of a Nation'.[120] Contemporary reaction was initially mixed. A leading proponent was the London-based Scottish merchant, William Paterson, who had been involved in 1694 in founding the Bank of England. Paterson, who had not visited Darien, nevertheless recognised its strategic value: 'Thus this door of the Seas and Key of the Universe with anything of a reasonable management will of course enable its proprietors to give Laws to both oceans and to become arbitrator of the commercial World'.[121] Others, London Scottish merchants and traders in Scotland, were cautious both of Paterson's advocacy and of the scheme itself, particularly given that the proposed site would conflict with Spanish interests in the region. James Gregory, the mathematician, sought on the other hand to promote a 'Navigation and Writing School' by which young men might be trained in mathematics and other practical skills given that Scotland had 'now a fair propsect of a considerable foraigne trade to the Indies, Africa, etc'.[122]

King William's initial support turned, however, to outright opposition once it became apparent that England's trading monopolies would be harmed: further restrictions were placed by the King on Scots seeking funding abroad. All the capital was raised in Scotland. Subscription lists show financial support was forthcoming from across Scotland with about half coming from the nobility and gentry and one third from the merchants.[123] In that regard, as well as because of contemporary distrust of England, Darien was a matter of national importance. The eponymous '*Philo-Caledon*' [Archibald Foyer] wrote in his *A defence of the Scots settlement at Darien* (1699) of 'the whole Kingdom of *Scotland* being more zealous for it, and unanimous in it than they have been in any other thing for forty or fifty years past'.[124] Others thought likewise. The anonymous author of a short pamphlet on Darien, printed and sold in Edinburgh's coffee houses in 1699 noted: 'The Project of our Scots African Company, being a Design of so great Import, That upon the Success thereof, the Honour and Happiness of the Nation, does in a great Measure depend'.[125]

120 SRO, GD 45/1/161, Memorial on behalf of the Scots Company Trading to Africa and the Indies, 1699.

121 NLS, Adv. MS 83.7.3, f. 44v.

122 J. Gregory, 'A Prospect for a Navigation and Writing School by the E. India Company of Scotland', EUL, MS. Dc. 1.60, ff. 751–4.

123 NLS, Adv. MS 83.1.8v (Subscription list for the Company of Scotland, 1696).

124 Anon, ['Philo-Caledon' = Archibald Foyer], *A defence of the Scots settlement at Darien* (Edinburgh, 1699), 24.

125 *A short account from, and description of the isthmus of Darien, where the Scots collony are settled with a particular maps of the isthmus and enterance to the river of Darien. According to our late news, and Mr Dampier, and Mr Wafer* (Edinburgh, 1699), 1.

The first ships – the largest of which were the symbolically-named *Saint Andrew*, *Caledonia* and *Unicorn* – sailed in July 1698, and reached Darien in October. The new colony was established as Caledonia, with its main settlement New Edinburgh protected by Fort St Andrew (Figure 3.5). Initial prospects were good. In December 1698, the colonists' council wrote to the company in Scotland to report on the bounteousness of the site – 'In fruitfulness this Country seems not to give place to any in the Commercial world' – and to stress the patriotism of the colonial venture:

[We] shall always account it our greatest honour to expose our persons . . . in promoting this hopeful design, as not only promising profitt and glory to the Company . . . but as being the likeliest means that ever yet presented towards enabling our Countreymen to revive, recover and transmitt to posterity the virtue, lustre and wonted glory of their renown'd ancestors. And to lay a foundation of wealth and Security and greatness to our mother Kingdom for the present and succeeding ages.[126]

The younger James Wallace, a correspondent of Sibbald, was amongst the Darien colonists and kept a journal in which, amongst other things, he sought to understand the strangeness of this new world by thinking of it in terms of Scotland's geography: Caribbean islands were likened to the Bass Rock, for example, yet 'this place [Caledonian Bay in Darien] affords legion of monstrous Plants, enough to confound all the Methods of Botany ever hitherto thought upon'.[127]

The colony soon failed. The king instructed English colonies in the New World not to trade with the Scots; the Spanish objected to their presence; disease killed many; no supplies or further support was forthcoming from Scotland. By July 1699, Darien was deserted: surviving colonists made for Jamaica and New England and suffered further high mortality on their voyage.

If more extremely than Martin Martin's contemporary travels to St Kilda, the Darien affair also illustrates how the facts of geographical distance influenced secure information about unknown places. News of the failure reached Scotland only after a second expedition had set out (on 24 September 1699), although rumours – which were discounted – had been received by company directors on 19 September 1699. We know the fate of the first colonists and the reactions of the second expedition from Borland's *History of Darien*, published in 1779, but written in 1700 and largely based on his own experiences of the second expedition: 'The Author of these following Memoirs having been an Eye-Witness of many of the tragical passages of Providence, and exercised with a share of many of the calamities that befel his country-men abroad, in the wilderness and on the sea, was the more capable to give an

126 SRO, GD 406/1/6489, Letter from the Council in Caledonia to the Company in Scotland, New Edinburgh, 28 December 1698.

127 G. Insch (ed.), *Papers relating to the ships and voyages of the Company of Scotland trading to Africa and the Indies 1696–1707* (Edinburgh, 1924), 69–77: quote on p. 75.

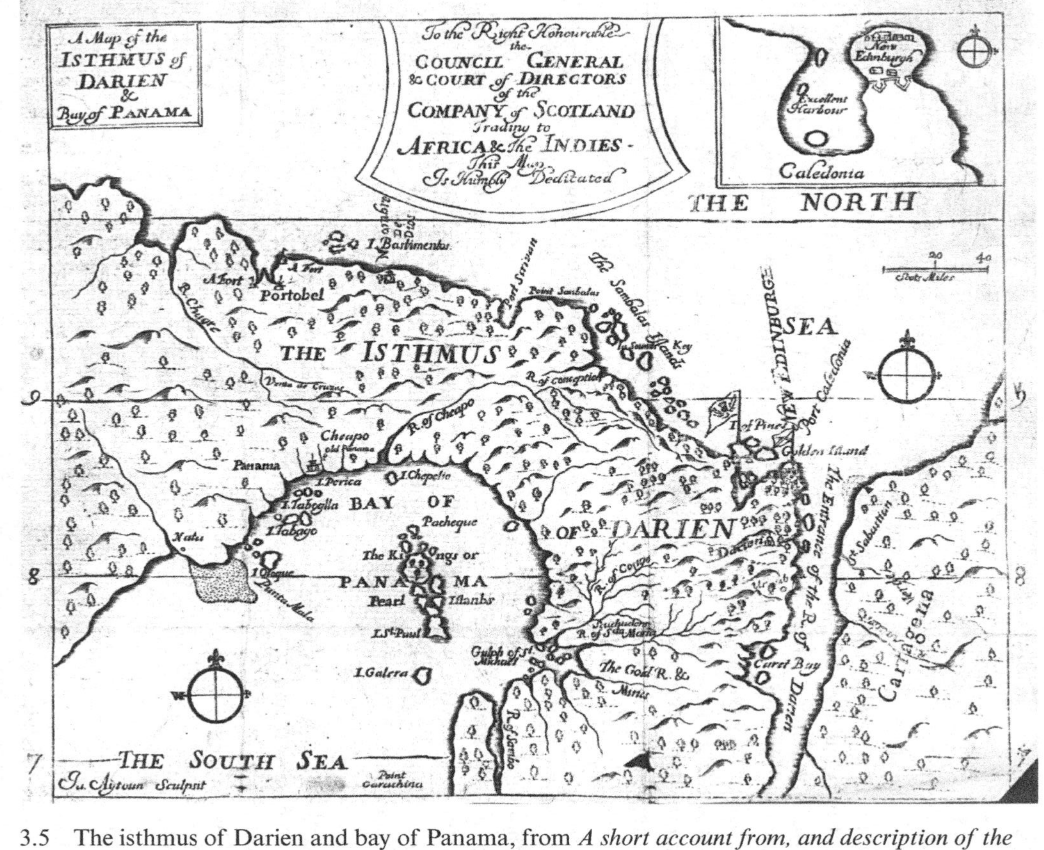


3.5 The isthmus of Darien and bay of Panama, from *A short account from, and description of the isthmus of Darien* (1699). Note the inset showing the settlement described as 'New Edinburgh' and, on the main map, the flags with the iconic thistle and saltire.

account of particulars relating to the design of Caledonia. And what passages he did not see himself, these he relates as he had them delivered by credible persons, who were Eye-witnesses of them when they occurred'.[128] Borland, who sailed on the inaptly-named *Company's Hope*, noted simply: 'Upon our arrival in this new world, we met with a sorrowful and crushing-like dispensation, for expecting here to meet with our friends and countrymen, we found nothing but a waste, howling wilderness: the colony deserted and gone, their Huts all burned, their Fort most part ruined, the ground which they had cleared adjoining to the Fort all overgrown with shrubs and weeds'.[129] The second party was finally driven off by Spanish troops in May 1700.

Contemporary reaction in Scotland stressed the affront done to Scotland's national honour and credibility more than the financial hardship caused. Pamphlets such as *The people of Scotland's groans and lamentable complaints* treated the failure of Scotland's imperial ambitions at Darien as the latest in a series of English insults.[130] The Parliament of Scotland condemned English interference and the General Assembly of the Church of Scotland called for a national day of fasting and remembrance to record 'the loss of many of our Countrey men; and even some of our dear and worthie Brethren in the Ministery, and of a great part of the nation's treasure and substance'.[131] Yet Darien was a failure of native administration with important consequences. As Armitage notes:

> The failure of the attempt did not lead them [Parliament and the Company in Scotland] to demand access to the English commercial empire, but rather to bemoan the abuse of their *imperium* at the expense of English trading interests. It also narrowed the field of possible solutions to the unequal distribution of power enshrined in the Union of the Crowns. . . . It is in this sense that . . . the Darien venture is historically important. Only by recovering the dilemmas it was intended to resolve, the avenues its failures closed off, and the new possibilities it created, can we understand how the Scots in 1707 came to calculate that their vision of empire could only be realised by Union, and within a British state.[132]

Darien was also geographically important: if not a failure of the contemporary Scottish geographical imagination, it certainly shared with other work at that time a failure to be realised in full.

## Conclusion

It has been claimed that 'by the end of the seventeenth century the language of one science, geography, routinely pervaded the language of all the

[128] F. Borland, *The history of Darien* (Glasgow, 1779), 5.
[129] Borland, *History of Darien*, 29.
[130] 9 SRO, PA 7/17/1/21A, *The people of Scotland's groans and lamentable complaints* (Edinburgh, 1700).
[131] SRO, PC 13/3, Act of the General Assembly for a Fast Day, 4 March 1701.
[132] Armitage, 'The Scottish vision of empire', 117–18.

sciences'.[133] This is arguable. This chapter has suggested that a distinction should be made between several things. It is possible, firstly, to consider the use of geographical rhetoric – the 'map of geographical knowledge', the 'location' of science, the 'situated nature' of knowledge making in the public sphere and so on. Secondly, we should recognise the values placed by contemporaries upon what was understood by them as geography. Thirdly, we should note that wider range of practices which, at any given moment, constituted geographical knowledge.

What is not arguable is that geography was recognised by the state and by the crown as a key means to national knowledge in and of Scotland from the early 1680s and that different means to that end were undertaken: mapping and accurate survey in the work of John Adair from 1681; chorographical enquiries based on circulated queries and utilitarian natural history in Sibbald's work from 1682. Both men, in different ways, were concerned, literally and figuratively, to map the state of the nation, to measure its present capacity and, for Sibbald particularly, to consider the future geographies of Scotland. Sibbald and Adair were geographers *and* they were civil servants in Scotland's move to 'modernity', providing, by intention anyway, geographical knowledge of the nation supported both by official warrant of their credibility and by government monies. In those terms, Sibbald's appointment as Geographer Royal represents a new departure for the status of geography as a credible intellectual pursuit for Scotland and for Britain; albeit we should, as Emerson has noted, see the work of Sibbald and his friends 'as a continuation of the efforts of men such as Timothy Pont, Robert Gordon of Straloch, James Gordon of Rothiemay, and others who had sought to map and describe Scotland'.[134] And as Emerson also notes, the fact that Sibbald was concerned to use geographical information for Scotland's improvement marks him as an early figure of the Enlightenment.[135] For these reasons, the significance of Sibbald and of Adair and of their purposive ends is not diminished by the fact that much of their work was unfinished: after all, the *Atlas* was never realised; returns to their queries were not forthcoming from everyone, funds for both were insufficient from the very groups – the Privy Council, the parliament in Scotland and the crown – who so strongly advocated their work. Failure to complete and make public their geography should not detract from our understanding of the social basis to construction of that knowledge.

It is now possible to understand what I have here called the constitution of geographical knowledge in more detail, both epistemologically and in terms of its sites and social spaces. In the field and in dealing with correspondence from distant others, the making of geographical knowledge had to grapple with questions shared by anyone then investigating nature: the credibility of

[133] R. Sorrenson, 'The ship as a scientific instrument in the eighteenth century', *Osiris*, 11 (1996), 221.
[134] Emerson, 'Sir Robert Sibbald, Kt, the Royal Society of Scotland', 62.
[135] Emerson, 'Sir Robert Sibbald, Kt, the Royal Society of Scotland', 66–67.

practitioners and of their results; acceptance of their findings; how to trust the word of others, gentlemen and 'the humble' alike. The making of reliable geographical knowledge depended, for Sibbald and Wodrow especially, upon networks of reliable informants, men who could be presumed upon to provide facts of benefit to the overall geographical description of the nation. They were called upon either by virtue of their local knowledge, because they were themselves geographically engaged, or, most often, because they were, simply, understood to be credible. At one level, such networks were local, with Sibbald's informants drawing, in turn, upon further local informants. At another, they were national: as answers to Sibbald's queries returned to his Canongate study, for example, or as Martin Martin met with others to report upon his travels or transmitted accounts about unknown parts of Britain to the Royal Society in London. To a lesser extent, they were international: Adair recruiting engravers and buying equipment in Holland, for example. Such networks were sustained by the travel of people themselves, exploring in the field the nature of the nation, and by the movement of knowledge itself through correspondence, the map and the printed word. The utilitarian geographical enquiries of Sibbald, Martin, Adair and Wodrow were not supported by any formal learned body, except by the Royal Society; it is impossible to know what effect Sibbald's plans for a Royal Society of Scotland would have had upon the conduct and nature of geography had he been successful.

The fact, however, that other, different, elements of geographical knowledge can be discerned makes it possible to map in outline for Scotland a 'situated geography of geographical knowledge' by the late seventeenth century. The Baconian and officially-sanctioned geographies of Sibbald, intended as useful and textual, depended upon his providing a metropolitan centre of calculation and upon networks of respondents. Adair's geographical knowledge, and Martin's, depended upon *their* being mobile and upon the reliability, respectively, of instruments and informants. Formal classes in geography in the universities of Edinburgh, St Andrews and in Aberdeen engaged practically and theoretically with Cartesian and, latterly, with Newtonian natural philosophy, but did so at different times. In the public sphere of the early eighteenth century, awareness of geography as a particular form of intellectual enterprise – earth description – was evident by 1702. Understanding of geography as a matter of national identity was perhaps heightened in the public's increasing recognition, between 1699 and 1707, of Scotland as a geographical entity separate from but dependent upon England. Geographical understanding was made, too, for practitioners and public alike, by itineraries of movement between informal information exchanges such as the printers and bookshops, where Martin's 1699 book or Darien pamphlets were sold, and the coffee houses where they were discussed.

There is no sense, however, in which we should consider geography a public science in the ways that Rupp has shown, for example, of the anatomy theatre in Holland at this time (with the selling of tickets for public dissections, the attendant strict policing of behaviour and discourse and the public constitution of natural knowledge that such events promoted).[136] Neither should we characterise the sites of geography's making as closed and private in the ways that others have documented for experimental natural philosophy.[137] Simply, different sorts of geographical knowledge were taking place in different sites in different ways and were being engaged with by different 'publics'. By 1707 there was, in combination with the clearly discernible practices of a few men with official warrant and the social networks upon which their geographical knowledge relied, a relatively strong public understanding of Scotland's geographical identity, mainly manifest in popular opposition to England and a nascent public awareness concerning geography that was to become more evident in the eighteenth century.

[136] Jan C. C. Rupp, 'The new science and the public sphere in the premodern era', *Science in Context*, 8 (1995), 487–507. [137] Shapin, 'The house of experiment'; 'A scholar and a gentleman'.

# 4

# Geography, Enlightenment and the public sphere, 1707–*c.*1830

A range of recent work has examined what may be thought of as the 'historical geography' of the eighteenth century. Amongst historians, John Brewer has paid attention to the sites and spaces of cultural production in Georgian England.[1] Others have discussed the social spaces of polite sociability, consumer culture and the urban geographies of the public sphere, in France, Germany and Britain, and the geographical and scientific making of the British empire.[2] As Miles Ogborn has observed in reviewing such work, 'These different attempts to understand new publics – in terms of popular politics, consumption, nationalism and the arts – have led to a concern for a certain sort of geography . . . a geography of public and private which acknowledges how the publics that are discussed were constituted in certain spaces'. Further, such 'political and cultural questions about these "public spheres" . . . can become geographical questions of the production of space and the meaning of place'.[3]

Historians of eighteenth-century science have explored the sites and spaces of science's making. Writing in 1980 of Enlightenment England, Roy Porter hinted at the different geographies and different publics for science: 'In some places, scientific popularization, societies and even research throve; in some, it had a halting life; elsewhere, hardly any at all. Why the differentials? And who were the audiences, the enthusiasts for science?'[4] Such questions have been considered by others. For Stafford, much public science throughout Europe

[1] J. Brewer, *The pleasures of the imagination: English culture in the eighteenth century* (London, 1997); see also D.Wahrman, 'National society, communal culture: an argument about the recent historiography of eighteenth-century Britain', *Social History,* 17 (1992), 43–72.

[2] J. Black (ed.), *Culture and society in Britain, 1660–1800* (Manchester, 1997); M. Cohen (ed.), *Fashioning masculinity: national identity and language in the eighteenth century* (London, 1996); D. Solkin, *Visual art and the public sphere in eighteenth-century England* (New Haven, 1992); K. Wilson, *The sense of the people: politics, culture and imperialism in England, 1715–85* (Cambridge, 1995); R. Draycott, 'Knowledge and empire' in P. Marshall (ed.), *The Oxford history of the British empire: the eighteenth century* (Oxford, 1998), 231–52.

[3] M. Ogborn, 'Georgian geographies', *Journal of Historical Geography,* 24 (1998), 220.

[4] R. Porter, 'Science, provincial culture and public opinion in Enlightenment England', *British Journal for Eighteenth-Century Studies,* 3 (1980), 22–46.

was a visual education affording amusement and instruction in spaces which were both public institutions and sites of private sociability.[5] Stewart's study of the rise of public science in eighteenth-century Britain focused on lecturers in London and elsewhere, and, like Schaffer, upon sites of experimental display.[6] Golinski has documented the practice and public consumption of chemistry and reviewed its place in Scotland.[7] Particular attention has been paid by scholars to the Royal Society in this period.[8] Sutton has shown how the promotion of science in France depended in large measure upon polite persuasion as to the utility of natural knowledge for the elite in society, a polite usefulness demonstrated through participatory experiments in the *salons* and other sites where 'publics' were constituted by the coming together of private individuals.[9] For much of western Europe, then, science was widely practised as a form of public culture and private sociability in the eighteenth century. Science, indeed, had in several senses, a geography: of *sites* of production, consumption and debate; of a presence in some towns but not others; of *private spaces* of polite conversation about experimental knowledge; of the *circulation of knowledge* between practitioner, author and audience; and of *practitioners* themselves on the move.

Such interest in the geography of ideas and in the importance of geography is apparent in attention by geographers to the historical geographies of modernity and to landscape and national identity in the eighteenth century,[10] and in suggestions from some Enlightenment historians that the Enlightenment should not be seen as a uniform intellectual phenomenon but as geographically variable.[11] The connections between the making of science and the Enlightenment have also been addressed in ways sensitive to 'the sites where particular knowledges were made, and the linkages that allowed them to be distributed, shared, and challenged'.[12]

[5] B. Stafford, *Artful science: enlightenment entertainment and the eclipse of visual education* (Cambridge, 1994).

[6] L. Stewart, *The rise of public science: rhetoric, technology, and natural philosophy in Newtonian Britain, 1660–1750* (Cambridge, 1992).

[7] J. Golinski, *Science as public culture: chemistry and enlightenment in Britain 1760–1820* (Cambridge, 1992).

[8] On this, see the theme issue 'Did the Royal Society matter in the eighteenth century', *British Journal for the History of Science,* 32 (1999), 130–223. Individual papers are cited below.

[9] G. Sutton, *Science for a polite society: gender, culture, and the demonstration of enlightenment* (Boulder, 1996); G. Barker-Benfield, *The culture of sensibility: sex and society in eighteenth-century Britain* (Chicago Ill., 1992).

[10] M. Ogborn, *Spaces of modernity: London's geographies 1680–1780* (New York, 1998); S. Daniels, 'The political iconography of woodland in later Georgian England', in D. Cosgrove and S. Daniels (eds.), *The iconography of landscape* (Cambridge, 1984), 43–84; *Fields of vision: landscape imagery & national identity in England & the United States* (Cambridge, 1993); R. Mayhew, 'Landscape, religion and knowledge in eighteenth-century England', *Ecumene,* 3 (1996), 454–71.

[11] D. Outram, *The enlightenment* (Cambridge, 1995); J. Bender, 'A new history of the Enlightenment?', *Eighteenth-Century Life,* 16 (1992), 1–20; R. Darnton, 'In search of the Enlightenment: recent attempts to create a social history of ideas', *Journal of Modern History,* 63 (1971), 113–32; R. Porter, *The enlightenment* (London, 1990); *Enlightenment: Britain and the creation of the modern world* (London, 2000); R. Porter and M. Teich (eds.), *The enlightenment in national context* (Cambridge, 1981); C. Withers and D. Livingstone, 'Introduction: on geography and enlightenment', in D. Livingstone and C. Withers (eds.), *Geography and enlightenment* (Chicago, 1999), 1–28.

[12] W. Clark, J. Golinski and S. Schaffer (eds.), *The sciences in enlightened Europe* (Chicago, 1999), x.

It is with these issues – with sites of knowledge, ideas of the public and the social spaces of its making – that this chapter is concerned. I consider here three related questions: what was the *nature* of geography and geographical knowledge in eighteenth- and early nineteenth-century Scotland? What were the *sites and social spaces* of its making and consumption? How far is it possible to think of the Enlightenment and the public sphere in eighteenth-century Scotland as matters of geography? The first two are concerned with the evidence for geography *in* Enlightenment Scotland. If, as I have argued, there was, by the early eighteenth century, a weak yet discernible public interest in geography, is there evidence for such public interest later in the century? Indeed, how far is it possible to determine a historical geography of public interest in geography? The third question has to do with a historical geography *of* the Enlightenment and *of* the public sphere. This is not to see the spaces of Enlightenment knowledge as, simply, matters of geographical location. It is to suggest that the Enlightenment in Scotland was, in several ways, geographically *constituted.* I mean by this the extent to which the Enlightenment varied *between* different towns in Scotland, that *different institutions* were involved with *different audiences,* and, I suggest, that understanding the nature of Enlightenment might be enhanced by our recognising that certain ideas took shape *there* and not somewhere else. In being concerned with 'the recovery of geography' in eighteenth- and early nineteenth-century Scotland, what follows examines six main themes. Consideration is given to evidence for geography teaching in the public sphere, principally to public geography classes and to the writing and publication in Scotland of works of geography. I discuss geography teaching in schools. I examine the idea of geographical survey and geography's epistemological connections with natural history. Attention is paid to the ways in which mapping and landscape painting 'visualised' the nation. The concluding section discusses the Enlightenment as a geographical question. I begin by considering something of what geography was held to be in this period.

## Geography in the Enlightenment: a partial survey

For some scholars, the origins of modern geography lie in the Enlightenment. David Stoddart has argued, for example, that modern geography's decisive moment of formation was Cook's entry into the Pacific in 1769.[13] In contrast, other work on geography in the age of Enlightenment, largely conflated with the 'long eighteenth century', disengaged geographical knowledge from its context. For Bowen, eighteenth-century geography, certainly in Britain, 'remained almost entirely in the hands of textbook writers who produced compilations of regional descriptions and showed little concern with the

[13] D. Stoddart, *On geography* (Oxford, 1986).

theory of the subject or its relation to significant issues in the scientific thought of the time'.[14]

More recent work has taken issue with these claims.[15] The Enlightenment was characterised by the production of multi-volume geographies whose form and purpose mirrored that of other contemporary encyclopedic projects designed to order the world.[16] In those terms, geography as metaphor is central to the Enlightenment: the description by Diderot and d'Alembert of the *Encyclopédie* as a map of the world, of Edmund Burke's reference to stage-by-stage theories of social advance as 'the Great Map of Mankind at once unroll'd', and earlier ideas of the encyclopedia as a means to cultivate and order fields of learning from the wilderness of ignorance all suggest as much.[17] Geography was also a matter of practice. Geography's enduring fascination with the far away, with mapping the world, with exhibiting and classifying global knowledge and with imposing European ways of thinking on encountered 'others' are all recognisably Enlightenment preoccupations, apparent in debates over the age of the Earth, the cartographic and visual representations of foreign places, and in travel accounts. Geography was used for military campaigning and knowledge of the State in France,[18] in mining and survey work more generally,[19] was discussed in the *Encyclopédie*,[20] and was taught by Kant as part of his natural science enquiries.[21] The Enlightenment idea of human progress was itself intrinsically geographical, both in the sites of its conception and in relation to the places used as models within the universal theories of, for example, Dugald Stewart and, in France, Turgot.[22] Edward Gibbon understood 'rational geography' to be fundamental to his notion of history and of historical change.[23] Samuel Johnson had a clear

[14] M. Bowen, *Empiricism and geographical thought: from Francis Bacon to Alexander von Humboldt* (Cambridge, 1981), 125.

[15] C. Withers and D. Livingstone, 'Introduction', in Livingstone and Withers (eds.), *Geography and Enlightenment*, 1–28.

[16] A. Downes, 'The bibliographic dinosaurs of Georgian geography (1714–1830)', *Geographical Journal*, 137 (1971), 379–87.

[17] C. Withers, 'Encyclopaedism, modernism and the classification of geographical knowledge', *Transactions of the Institute of British Geographers*, 21 (1996), 363–98.

[18] M. Staum, 'Human geography in the French Institute: new discipline or missed opportunity?', *Journal of the History of the Behavioural Sciences*, 23 (1987), 332–40.

[19] J. Dorflinger, *Der Géographie in der 'Encyclopédie' ein wissenschaftsgeschichlische studie* (Wien, 1976).

[20] F. Kafker and S. Kafker, *The Encyclopedists as individuals: a biographic dictionary of the authors of the Encyclopédie. Studies on Voltaire and the eighteenth century* (Oxford, 1988), 257; S. Moravia, 'Philosophie et géographie à la fin du XVIIIe siécle', *Studies on Voltaire and the eighteenth century*, 57 (1967), 937–1071; C. Withers, 'Geography in its time: geography and historical geography in Diderot and d'Alembert's *Encyclopédie*', *Journal of Historical Geography*, 19 (1993), 255–64.

[21] P. Richards, 'Kant's geography and mental maps', *Transactions of the Institute of British Geographers*, 61 (1974), 1–16.

[22] M. Heffernan, 'On geography and progress: Turgot's *Plan d'un ouvrage sur la géographie politique* (1751) and the origins of modern progressive thought', *Political Geography*, 13 (1994), 328–43; S. Moravia, 'The enlightenment and the sciences of man', *History of Science*, 18 (1980), 247–68.

[23] G. Abbatista, 'Establishing "the Order of Time and Place": "Rational Geography", French erudition and the emplacement of history in Gibbon's mind', in D. Womersley (ed.), *Edward Gibbon: Bicentenary essays. Studies on Voltaire and the eighteenth century* (Oxford, 1997), 45–72.

sense of geography as regional description, and we have seen something of the character of eighteenth-century English geography and its place, as part of natural history and mixed mathematics, in the Royal Society's interests in empire and mercantile commerce between 1720 and 1779.[24]

Such evidence suggests not just that different sorts of geography were being employed by different people in different ways, but that geographical knowledge in the eighteenth century itself had a geography: institutionally, in terms of the sites of its textual production and of its audiences, and in relation to which parts of the world figured in Enlightenment theories about the human condition. It also signals the need to know more about geography's practices and reception and about what kinds of public were engaged where and with what sort of geographical knowledge.

The idea of the public sphere and the public production and consumption of knowledge is outlined in Habermas' influential *The structural transformation of the public sphere,* published first in German in 1962. Habermas explained the rise of democratic polity within eighteenth-century Western society by reference to the place of a literary and political bourgeois culture evident in the production and purchase of printed books by a literate middling rank, by the rise of institutions and literary journals, and, above all, by the public promotion of reasoned argument as a means to the enlightenment of civil society. The 'principle' of the public sphere was critical public discourse in which the bourgeois public consisted of private individuals who joined together to form a public and to debate issues bearing upon state authority. The public sphere in the political realm and in the world of letters was connected with the private spaces of the bourgeois family and its capacity to purchase and debate the products of public culture. This new rational discourse of critical sociability was situated – in coffee houses, in lecture theatres and in new literary outlets – and was dependent upon the territorial and political power intrinsic to the rise of early commercial capitalism. For Habermas, the public sphere was essentially an urban bourgeois phenomenon, which was initially established in the world of letters before moving into the political realm. If at first apparent in Britain before becoming evident in France, Germany and America, it was everywhere integral to the birth of urbane civil society and to Western 'modernity' itself.

For some commentators, Habermas' account is flawed for its emphasis upon a totalising version of historical change and for not considering the possibility of there being more than one public sphere, with such other spheres being, at least in part, relational and dependent upon connections with other

[24] R. Mayhew, 'Was William Shakespeare an eighteenth-century geographer?: constructing histories of geographical knowledge', *Transactions of the Institute of British Geographers,* 23 (1998), 21–38; character of English geography, *c.*1660–1800: a textual approach', *Journal of Historical Geography,* 24 (1998), 385–412; R. Sorrenson, 'Towards a history of the Royal Society in the eighteenth century', *Notes and Records of the Royal Society of London,* 50 (1996), 29–46.

discourses, notably plebian cultures.[25] Others have criticised the lack of attention to gender since, it is now clear, women were actively involved in the consumption and production of knowledge-making.[26] Criticism has been levelled at Habermas' relative inattention to the forms taken by the public sphere in later periods: from the mid-nineteenth century, for example, and in the twentieth century with the rise of the mass media.[27]

He has also been criticised for his relative neglect of science in the public sphere. Yet 'science can readily be fitted into the new public culture that Habermas sees as forged in such proto-political institutions as coffee houses and salons, and transmitted through the medium of newspapers and widely read literature'. Further, 'the *publicization* of knowledge that he formulates must become an essential part of any explanation of the constitution of modernity where science is at the centre'.[28] Science was a situated public enterprise. University professors, teachers and itinerant lecturers marketed natural knowledge in ways which made science a public commodity.[29] Their audiences'

[25] C. Calhoun (ed.), *Habermas and the public sphere* (Cambridge, Mass., 1992); T. Broman, 'The Habermasian public sphere and "science *in* the enlightenment"', *History of Science,* 36 (1998), 124–49; P. Wood, 'Science, the universities and the public sphere in eighteenth-century Scotland', *History of the Universities,* 13 (1994), 99–135; H. Mah, 'Phantasies of the public sphere: rethinking the Habermas of historians', *Journal of Modern History,* 72 (2000), 153–82.

[26] E. J. Clery, 'Women, publicity and the coffee-house myth', *Women: a cultural review,* 2 (1991), 168–77; D. Goodman, 'Public sphere and private life: towards a synthesis of current historiographical approaches to the old regime', *History and Theory,* 31 (1992), 1–20; E. Klein, 'Gender, conversation and the public sphere in early eighteenth-century England', in J. Still and M. Worton (eds.), *Textuality and sexuality: readings theories and practices* (Manchester, 1993), 99–114; 'Gender and the public/private distinction in the eighteenth century: some questions about evidence and analytical procedure', *Eighteenth-Century Studies,* 29 (1995), 97–109; P. Phillips, *The scientific lady: a social history of women's scientific interest, 1520–1918* (London, 1990); M. Benjamin (ed.), *Science and sensibility: gender and scientific enquiry, 1780–1945* (Oxford, 1991); *A question of identity: women, science and literature* (New Brunswick, 1993); A. Shteir, *Cultivating women, cultivating science: flora's daughters and botany in England 1760–1860* (Baltimore, Mad., 1996); P. Findlen, 'Translating the new science: women and the circulation of knowledge in enlightenment Italy', *Configurations,* 2 (1995), 167–206; S. Tomaselli, 'The Enlightenment debate on women', *History Workshop Journal,* 20 (1985), 101–24.

[27] P. Hohendahl, 'The public sphere: models and boundaries', in Calhoun (ed.), *Habermas and the public sphere,* 99–108; G. Ely, 'Nations, publics, and political cultures: placing Habermas in the nineteenth century', in Calhoun (ed.), *Habermas and the public sphere,* 189–240.

[28] R. Cooter and S. Pumfrey, 'Separate spheres and public spaces: reflections on the history of science popularization and science in popular culture', *History of Science,* 32 (1994), 237–67; J. Golinski, 'Science *in* the Enlightenment', *History of Science,* 24 (1986), 411–24.

[29] J. Millburn, *Benjamin Martin: author, instrument maker and 'country showman'* (Leyden, 1976); 'The London evening courses of Benjamin Martin and James Ferguson, eighteenth-century lecturers on experimental philosophy', *Annals of Science,* 40 (1983), 437–55; S. Schaffer, 'Natural philosophy and public spectacle in the eighteenth century', *History of Science,* 21 (1983), 1–43; 'Machine philosophy: demonstration devices in Georgian mechanics', *Osiris,* 9 (1994), 157–82; I. Inkster, 'The public lecture as an instrument of science education for adults – the case of Great Britain *c.*1750–1850', *Paedogogica Historica: International Journal for the History of Education,* 20 (1980), 80–107; L. Stewart, 'Public lectures and private patronage in Newtonian England', *Isis,* 75 (1986), 47–58; 'Other centres of calculation, or, where the Royal Society didn't count: commerce, coffee-houses and natural philosophy in early modern London', *British Journal for the History of Science,* 32 (1999), 133–53; J. Golinski, 'Peter Shaw: chemistry and communication in Augustan England', *Ambix,* 30 (1983), 19–29; J. Secord, 'Newton in the nursery: Tom Telescope and the philosophy of tops and balls, 1761–1838', *History of Science,* 23 (1985), 127–51; S. Pumfrey, 'Who did the work?: experimental philosophers in Augustan England', *The British Journal for the History of Science,* 28 (1995), 131–56. For a review of science lecturing in the eighteenth century, see the special issue of *The British Journal for the History of Science,* 28 (1995).

reactions may have differed, of course, but teachers and audiences alike were geographically situated and mobile within the urban public sphere even if, as Barbara Stafford has it, the mathematical recreationists, empirical artists and popular science demonstrators of this period inhabited a 'shadowy world'.[30]

This term is apt of geography's public practitioners and audiences. Formal institutional records are sparse. Itinerant lecturers leave few traces beyond newspaper advertisements; even where we can identify the sites in which geography as a public discourse was given, and know the syllabus, it is difficult to know to whom they spoke. Fleeting glimpses are given: of the French geographer and physician Edmé Guyot's public entertainments, including geographical material, in his *Nouvelles recreations* (1772); of public lectures in geography (and history) in Paris in the 1730s by Noël-Antoine La Pluche, author of the *Spectacle de la Nature*, published in nine illustrated volumes between 1732 and 1750.[31] John Senex, the map- and globe-maker, gave public geography lessons in Georgian London, and Benjamin Martin, the instrument retailer and itinerant lecturer, promoted geography through 'the Knowledge of the use of the *Globes, Sphere* and *Orrery*'. In his *Young gentleman and lady's philosophy,* Martin advocated geography as a subject of polite learning.[32] In Charleston, South Carolina in May 1733, geography was taught alongside other geometrical sciences, 'At the House of Mrs Delamere in Broad Street', for example, as, from 1739 in the same town, a Mr Anderson gave lectures on the 'Science of Geography . . . to any of the Gentlemen Subscribers to the Philosophical Lecture that shall please to attend'.[33]

Eighteenth- and early nineteenth-century Scotland offers rich potential for exploring geography in the public sphere. Scots in this period had considerable interest in the problems of progress and virtue in a commercialising society and in civic enlightenment. Enlightenment Scotland had a cultural and intellectual environment 'within which science could establish its civic credentials as public culture in close conjunction with the beginnings of its academic and disciplinary structure'.[34] There was, indeed, a largely metropolitan geography of science lecturing: in agriculture, natural history and practical chemistry.[35] Yet little is known of geography's situation.

[30] Stafford, *Artful science,* 190.

[31] Stafford, *Artful science,* 56, 233; A. Morton and J. Wess, *Public and private science: the King George III collection* (Oxford, 1993).

[32] Millburn, *Benjamin Martin,* 41; B. Martin, *The young gentleman and lady's philosophy, in a continual survey of the works of nature and art; by way of dialogue,* (Second edition, London, 1772), Vol. II, 127.

[33] G. Dunbar, 'Geographic education in early Charleston', *Journal of Geography,* 69 (1970), 348–50.

[34] J. Money, 'From Leviathan's air-pump to Britannia's voltaic pile: science, public life and the forging of Britain, 1660–1820', *Canadian Journal of History/ Annals Canadiennes d'histoire,* 28 (1993), 531.

[35] C. Withers, 'William Cullen's agricultural lectures and writings and the development of agricultural science in eighteenth-century Scotland', *Agricultural History Review,* 37 (1989), 144–56; 'Improvement and enlightenment: agriculture and natural history in the work of the Rev. Dr. John Walker (1731–1803)', in P. Jones (ed.), *Philosophy and science in the Scottish enlightenment* (Edinburgh, 1988), 102–16; 'On georgics and geology: James Hutton's '*Elements of Agriculture*' and agricultural science in eighteenth-century Scotland', *Agricultural History Review,* 42 (1994), 138–49.

## Geography's public, 1708–1830

### *Public lectures in geography in Scotland, 1708–c.1830*

Examination of Scottish newspapers reveals 64 different persons giving public lectures or private classes in geography in the period 1708–*c.*1830: the same sources document some people whose lecturing continued beyond 1830 and other public lecturers from 1830 to 1861 (see pp. 160–3). This total is derived from an assessment of advertisements for people who specifically advertised 'geography': it omits any persons whose public teaching in survey, navigation and astronomy *may* have included geography but for which there is no firm evidence.[36]

This evidence allows an understanding of the historical geography of the public's engagement with geography in several ways. Most public geography classes were concentrated in Edinburgh – where the earliest class, the Rev William Smart's, was begun in 1708 – and in Glasgow, with a scattering of courses, generally beginning in the late eighteenth century or early nineteenth century, in smaller towns (Figure 4.1). The emphasis on these two larger towns is understandable. The evidence for Dumfries, however, where there were five separate public geography teachers at work offering sixteen different classes at one time or another between 1777 and 1825, is surprising for such a small centre of population. Within this picture, it is possible to address four further issues: who were geography's public teachers? what was the nature and cognitive content of this public geography? who were the audiences?; what were the local sites and social spaces of its production and consumption?

Most persons delivering this public geography were self-styled as 'teacher of geography' or 'lecturer in geography'. One or two, like Smart, were parish ministers. Smart also gave private classes in mathematics, navigation, surveying and the 'use of the globes celestial and terrestrial'.[37] Others, such as Thomas Blackwell and Robert Hamilton in Aberdeen, were university professors giving private classes and public lectures in geography, not, like Thomas Reid, as formal parts of a curriculum, but as a means to augment their salaries.[38] In Glasgow in the late 1820s, Robert Wallace and Alexander Watt combined a professorial role at John Anderson's Institution (the forerunner to the University of Strathclyde) with public geography classes at home.

John MackGregory, self-styled 'Professor of Universal History and Geography', who was active in Edinburgh (and possibly in London) between 1707–8 and 1715, appears to have used his title to substantiate his credibility.

[36] C. Withers, 'Towards a history of geography in the public sphere', *History of Science*, 37 (1999), 45–78; the full listings are taken from D. Gavine, 'Astronomy in Scotland 1745–1900' (Ph.D. thesis, Open University, 1992), Vol. 2. [37] *Edinburgh Courant*, 31 March–2 April 1708.

[38] J. Morrell, 'The University of Edinburgh in the late eighteenth century: its scientific eminence and academic structure', *Isis*, 62 (1971), 158–71.

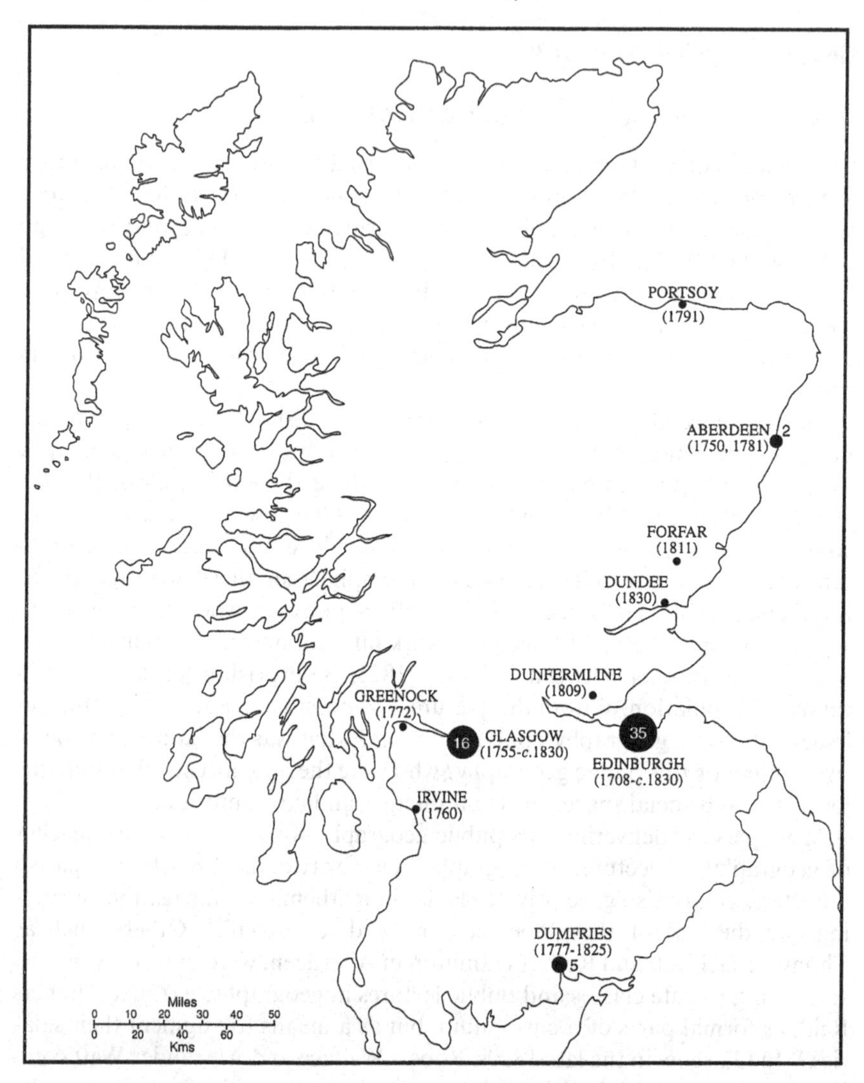


4.1 The location and numbers of persons known to be engaged in the teaching of geography in Scotland, 1708–*c.*1830.

MackGregory advertised his lectures in printed flyers and news-sheets which were available in the sites of public discourse: 'This Advertisement [that of 1713 in the *Edinburgh Courant*] is to be seen in all the Coffee-Houses in Town, and Copies on't are to be had from the Author at S[t]. James, the Grecian, and Garraway's Coffee-Houses'. A 1715 version announced that the advertisement could also be had from the author at the 'Exchange and Caledonia

Coffee-Houses'.[39] Little is known of the shadowy MackGregory. His advertisements indicate that he held a doctorate in law from Angers and that he wrote four books: *The geography and history of Lile* [Lille] in 1708, *The geography and history of Tournay (1709),* a similarly-titled work on *Mons* (1709) and a work on sepulchres, each dedicated to Prince Eugene of Savoy, all published in Edinburgh and 'to be Sold at all the Coffee-Houses in Town'. His proficiency in geography came, he tells us, from having 'Travel'd over all Europe; . . . having Liv'd at most of the Courts of Europe' and by 'having been Imployed in the Publick Business' (as a diplomat in Switzerland). His four-page '*ADVERTISEMENT to Gentlemen and Ladies*' makes clear that, having now 'come Home hither to his own Country, [he] does make Profession of Serving Gentleman and Ladies by Teaching 'Em MODERN GEOGRAPHY and UNIVERSAL HISTORY'. His overall purpose was 'to make One Understand the Descriptions and Accounts of what is to be Seen and Heard of in Traveling through the World'. Geography, understood by MackGregory as facts about foreign countries, customs and descriptions as a didactic and utilitarian practice, was taught by him 'so as to Make up the Want of Traveling to Those who have not Travel'd and to Supply them with the Next Best'.[40]

As these examples demonstrate, the diversity amongst Scotland's public geographers cautions against simple categorisation. In work on the public promotion of early modern physics, Heilbron distinguished a hierarchy of four types of lecturer: at the top, public lecturers associated with learned societies; a second group of members of learned societies who set up independently of their institutions; a third of 'unaffiliated entrepreneurs, who taught in rented rooms, and the itinerant lecturers, who taught in public houses'; and, last, 'hawkers of curiousities, the street entertainers'.[41] Adopting a similar system for those persons promoting geography in Scotland's public sphere would place most in the second and third groups with a few, like MackGregory, as entrepreneurs on the make. Any such classification should not, however, simply distinguish between public producers and private consumers, since, as I show below, much of this geography as *public knowledge* was given in *private spaces.*

Geography was taught in connection with other subjects. Most persons

[39] *Edinburgh Courant,* 3–6 September 1707; 4–6 October 1708; *Mr MackGregory's Advertisement to Gentleman and Ladies* (London, 1713 and 1715); *The Postman,* 20 October 1715.

[40] *MackGregory's ADVERTISEMENT to Gentlemen and Ladies* (London, 1715). In a letter of 12 March 1722 to Sir Hans Sloane, MackGregory seeks Sloane's patronage: 'being now returned to London in a very poor . . . way begs . . . to apply to your generosity that you may please to give him what you think proper to [support?] him in his present necessity': British Library, Sloane MS 4046, f. 213. A later letter (of 23 October 1722) uses almost the same phrasing in seeking support from a patron in Cambridge: BL, MSS Add 22,911, f. 256. It would appear that the only certain thing we can say of MackGregory is that his geography did not reward him.

[41] J. Heilbron, *Electricity in the 17th and 18th centuries: a study of early modern physics* (Berkeley, Calif., 1979), 158.

taught geography alongside astronomy, navigation, geometry, mathematics and arithmetic, with a few teaching French or Italian. Astronomy and geography had close associations for several reasons: because contemporaries saw them as subjects suitable for a liberal and polite education; because of that more learned interest, apparent in the Royal Society, for example, which saw them as crucial to maritime commerce and colonial expansion; because of their shared use of terrestrial and celestial globes and the language of mathematics; and, for a few university graduates perhaps, because of the long-standing links between geography, cosmology and astronomy they encountered in the university curriculum. Such issues are clear from Colin Maclaurin's 'Memorial' of 1741 for the establishment of an observatory in Edinburgh:

> The usefullness of astronomical learning, especially to those nations whose prosperity depends on commerce and navigation is so well known that it is unnecessary to insist upon it. And if it is considered as a part of education only, it deserves encouragement, as it fills the minds of youth with great and noble views, and has a direct tendency to instill the best sentiments. . . .
>
> An Observatory here could not but be of service for ascertaining the geography of this Country even of the distant parts, by the opportunity gentlemen would have to learn the manner of making accurate observations, and the correspondence of this kind over the Country, that would be the natural consequence of having one here.[42]

These discursive connections are important for what they suggest about the moral and intellectual imperatives Scots placed upon geography as a form of knowledge and as a means to national identity. For some, it is clear that geography was understood as a form of and a means to measurement, a type of mathematical and utilitarian practice quite in keeping with contemporary interests in estate surveying, mapping and navigation. Modern scholars have seen these practices as similar to that engagement with 'mathematical cosmography' which underlay both state mapping and private interests in geography and its allied practices of geometry and map-making.[43] Robert Darling, for example, 'private teacher of geography' and 'teacher of mathematics and geography' in Ramsay's Land in Edinburgh in 1776 and again in 1793–4, declared that he 'Plans Noblemen and Gentlemens estates in the most neat and accurate manner', and 'teacheth Youth Writing, Book-keeping, Mathematics, and geography, and Gentlemen to Measure and Plan their own estates'.[44]

It is possible to suggest, then, that the teaching of geography in the public sphere both directed and reflected the search for rational order and the reasoned improvement of individual and nation that underscored the place of the

[42] City of Edinburgh District Council Archives, Edinburgh Town Council Minutes, 10 June 1741; D. Bryden, 'The Edinburgh observatory 1736–1811: a story of failure', *Annals of Science*, 47 (1990), 445–74.

[43] E. Forbes, 'Mathematical cosmography', in R. Porter and G. Rousseau (eds.), *The ferment of knowledge* (Cambridge, 1980), 417–48; M. Edney, 'Mathematical cosmography and the social ideology of British cartography, 1780–1820', *Imago Mundi*, 46 (1994), 101–16.

[44] *Caledonian Mercury*, 1 June 1776, 1a.

articulate public in the Enlightenment. Darling was not the only such example. We might suppose that men like Robert Nichol in Greenock in 1772, Alexander Ingram in Constitution Street in Leith between 1815 and 1828, and, perhaps James Hall in the fishing village of Portsoy in 1791 with his classes on geography and navigation, were, in such ports, promoting geography as a navigational and locational discourse, a particular form of utilitarian knowledge rooted in local circumstances. Gavine suggests as much for the local teaching of navigation: *The complete navigator* (1804), written by an Aberdonian, Alexander Mackay, was particularly popular in north-east Scotland.[45] Alexander Ingram was the author of a version of Euclid's *Elements,* intended for use in schools, and of *Principles of geography, containing the use of globes, and a description of the different countries which are known to us* (1799), which went into a third edition by 1807. To judge from its contents (and assuming Ingram used his own book in teaching), his classes were about earth description and use of the globes, and may not have been directed at audiences' local needs. Yet it is not unreasonable to suppose that some public courses in geography were shaped to the needs of local circumstance – that the geography *of* a place helped constitute the geography taught *in* that place – whilst recognising, too, that there was a widely-held public regard for geography understood as discourses of mathematical utility and textual description.

For others, geography was a means to polite education, a way of promoting sociability through description of foreign countries, informed use of the globes and measured discussion of maps in ways we have noted of science and sociability elsewhere in eighteenth-century Europe. There is little evidence, however, as to *how* it was taught. There is reference to the use of globes and some mention of other apparatus. George Paterson used globes to solve 'curious Geographical and Astronomical Problems' in his classes, which ran for over twenty years from about 1747, in his house near Horse Wynd in Edinburgh.[46] Robert Lothian, chaplain to Glasgow's Trades House in the late 1780s, taught his classes in geography, military mathematics and astronomy with a set of 'improved machinery' including a planetarium, an orrery and an armillary sphere, all made by himself.[47] Robert Scott, self-proclaimed 'Mathematician and geographer', made globes for the Edinburgh map-maker James Kirkwood in 1804 and used them in his Commercial and Mathematical Academy as a means to promote navigation and surveying as useful subjects for 'young gentlemen who are soon to enter into active life'.[48] The use of globes, maps and other apparatus in public geography classes was almost certainly not

[45] D. Gavine, 'Navigation and astronomy teachers in Scotland outside the universities', *Marriner's Mirror,* 76 (1990), 6. [46] *Edinburgh Evening Courant,* 1747, 1751. [47] *Glasgow Directory,* 1790–1801.

[48] G. Clifton, *Directory of British scientific instrument makers 1550–1851* (London, 1995); A. Simpson, 'Globe production in Scotland in the period 1770–1830', *Der Globusfreund,* 35–7 (1987), 21–36; M. Wood (ed.), *Extracts from the record of the burgh of Edinburgh* (London, 1940), 93.

the exciting experimental theatre that was the public lecture on electricity or medicine, where lecturers would sometimes use their own bodies to demonstrate natural phenomena.[49] Given the fact, however, that apparatus was used at all and that some teachers made their own means we should not think of geography's place in the public sphere as just a rhetorical and textual exercise.[50]

Whilst we may infer audiences' intentions from the utilitarian and educational emphases to many courses, and from the instrumentally-mediated understanding of the globe, it is not possible to know who attended, why and in what numbers. Some classes were attended by men seeking to enter university: Robert Wallace's lectures in Glasgow in 1828 treated geography as part-preparation for natural philosophy at the University.[51] Where they can be determined, entrance fees were at a level which may have deterred the less well off. The emphasis upon practical utility suggests that surveyors, merchants, ships' captains or those intending such employment were amongst the audience. Men predominated, but six public geography classes – four in Edinburgh, two in Glasgow – either included women or were specifically designed for women in this period. Alexander Ewing ran his 40-lecture public classes in geography in Edinburgh on Mondays, Wednesdays and Fridays for men, and on Tuesdays, Thursdays and Saturdays for women, at a fee of £1 1s. for the course. He appears to have done so first in 1756, then annually for at least fifteen years from 1768 and again in 1790 and between 1793–4; his son was later to teach geography and astronomy between 1837 and 1861 at 47 George Street 'and also pupils at their own homes'.[52]

No single pattern predominates in regard to where this public geography was taught. As noted, MackGregory advertised his courses in Edinburgh's coffee-houses, yet he taught at his own lodgings and in customers' homes. Many 'public' teachers taught at home. That this is so raises questions about the production of geography in the public sphere as a matter of separate private and public spaces, and whether we should conceive of geography's situated public production as blurring distinctions between private and public *space*, and between domestic space and the *sites* of public learning. Like Alice Walters, I want to suggest 'that polite science should be situated not just in the formal public sphere of the lecture, but also in the comparatively informal domestic

[49] See Schaffer, 'Natural philosophy and public spectacle'; 'Machine philosophy'; Secord, 'Newton in the nursery'; R. Porter, 'Medical lecturing in Georgian London', *British Journal for the History of Science,* 28 (1995), 91–100.

[50] This evidence also denies the claims that popular science lecturing begins in Scotland in the 1740s with itinerant lecturers from England: J. Cable, 'The early history of Scottish popular science', *Studies in Adult Education,* 4 (1972), 34–45; J. Reid, 'Late eighteenth-century adult education in the sciences at Aberdeen: the natural philosophy classes of Professor Patrick Copland', in J. Carter and J. Pittock (eds.), *Aberdeen and the enlightenment* (Aberdeen, 1987), 168–79. More generally, see R. Whitley, 'Knowledge producers and knowledge acquirers: popularisation as a relation between scientific fields and their publics', in T. Shinn and R. Whitley (eds.), *Expository science: forms and functions of popularisation* (Dordrecht and Boston, 1985), 3–28.

[51] *Glasgow Herald,* 1832.

[52] A. Law, *Teachers in Edinburgh in the eighteenth century* (London, 1965), 172; *Caledonian Mercury,* 15 October 1759, 17 December 1768; *Edinburgh Evening Courant,* 3 May 1783.

scene of the home'. For her, polite science in the domestic sphere manifested specific characteristics: an association between the social character of polite science and its topical content; the encouragement of women as active participants; and connections between the books and instruments of learning and their display as objects of consumer culture. 'Polite science aligned the acquisition of socially appropriate kinds of scientific knowledge with the acquisition of material goods illustrative and symbolic of that knowledge'.[53]

The remarks on geography and astronomy by the anatomist Alexander Monro in his 'The Professor's Daughter' (1739), subtitled 'An Essay on female Conduct contained In Letters from a Father to his Daughter', illustrate something of these points and are of interest, too, for his closing strictures to his daughter about her making public her private learning:

> Whoever intends to read any History even the common news Papers ought to know the Situation of the different Countries, the Nature of their Climates, the Distances of the most remarkable Places from each other and the other particulars which are to [be] learned from Globes and Maps. This knowledge is what I call geography. Without it one can not understand the different Claims and Interests of Potentates, the Causes of their different Complexions, of the gaining or losing of Battles, the difficulties or advantages of Armies or people in their Marches &c. A Months Application in this Study will make you very sensible how much People ignorant of it are blundering every day in common Conversation. I design to steal as much time from my other Business as to instruct you in this necessary piece of Knowledge by which I shall regain what I have forgot of it. I hope you will be as much entertain'd hereafter with the Globes and Maps as you was the other Day when I shewed you the more general things upon your small Set being brought home.
>
> I must think you ought to know as much of Astronomy as explains the common System and Motion of the Planets, if it was only to shun the Extravagancies and fears which many of your Sex so frequently express upon seeing an Eclipse or some such natural Appearance. Learn but so much as to read Fontenelle's Plurality of Worlds with pleasure. I engage to instruct you this far in five or six Lessons, but must give you the Caution never to discover this part of your Knowledge to your female acquaintances or the ignorant foplings of my Sex, for they will fix the name of Virtuosi, Pedant, and I don't know what on you.[54]

Public institutions were also, therefore, sites of private sociability as individuals came together to discuss matters of geography. At the same time, private spaces were used to debate a geography that was seen as publicly useful, a geography that demanded knowledge of maps, of the globes and, even, of one's local world (see Figures 4.2 and 4.3). Because so much geography in the

[53] A. Walters, 'Conversation pieces: science and politeness in eighteenth-century England', *History of Science*, 35 (1997), 122.

[54] P. Monro, 'Introduction, and the Professor's Daughter', *Proceedings of the Royal College of Physicians of Edinburgh*, 26 (1996), 1–189 (p. 19). In a note in the original manuscript, Alexander Monro records that on Friday 16 January 1741, John Monro, Margaret's eldest brother, 'gave an oration on Geography to the youthful "Latin Society" which met each evening in the Professor's room: the written account was corrected by his father': (Monro, p. 185, n. 37).

4.2 'Group representing the children of Henry (Dundas), First Viscount Melville', painted by David Allen *c.*1785. The activities shown here illustrate the place of geographical knowledge in the polite education of the upper classes in the eighteenth century: one girl is standing by a globe demonstrating measurement to her sister, who is holding a map of Europe in her hand. A third girl is sketching. The globe represented is still in the possession of the family.

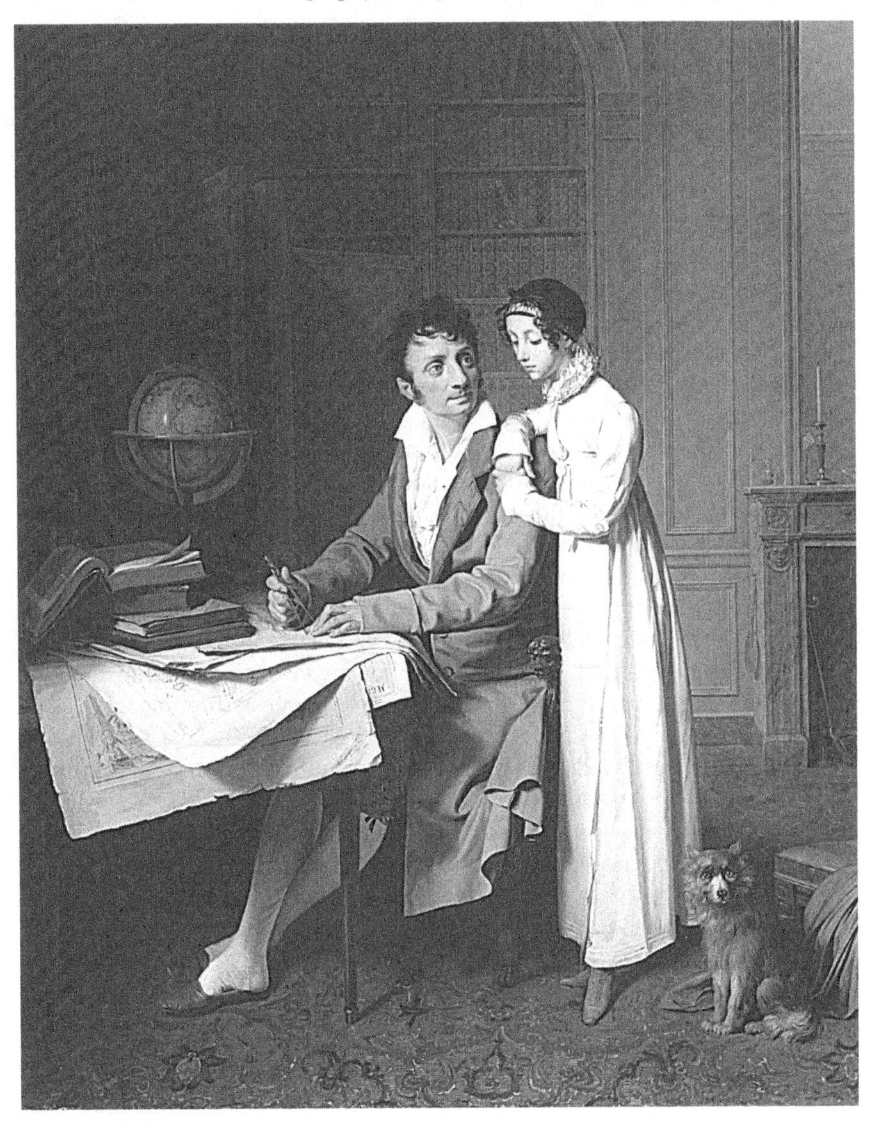

4.3 'The Geography Lesson', by the French artist, Louis-Leopold Boilly (1812). The painting beautifully illustrates the use of domestic space as geographical teaching space.

public sphere involved globes and other instruments, we perhaps ought not to see sharp distinctions between the home and the laboratory, or, at least, between domestic space and sites of scientific display. To judge from those persons who used apparatus in their classes, at least part of their domestic space was given over for display, experiment or to lecture. This was not true for all: Thomas Longstaffe lectured in geography and astronomy at Glasgow's Theatre Royal in 1825, for example, with an orrery 10 feet in diameter, accompanied by music; others took their classes to the local observatory.[55] For the Rev Smart, making geography public knowledge was certainly a domestic affair: 'They may be instructed in the most useful problemes of geographie' [his 1708 advert read], 'at their own house or at his own house in the head of the Canongate, . . . or any other convenient place as the party shall direct'.[56]

Several geography teachers in Edinburgh and in Glasgow continued to teach the subject after they had relocated within the city, but it is impossible to know if their audiences moved with them. James Dinwiddie moved between towns. In November 1777, aged 29, he advertised a course of lectures on 'the principles of Geography and Astronomy, to be explained "on an elegant eighteen inch *Globe,* of a new construction, and the ORRERY" in Dumfries, at which time he also signalled an intention to give classes in French, mathematics and experimental philosophy.[57] The following year, and again in 1779, he was advertising natural philosophy, geography and astronomy classes in Edinburgh. He graduated MA from Edinburgh in 1778 and LLD there in 1792, the year he travelled to China as astronomer to Lord Macartney's British diplomatic delegation. From 1794 to 1805 (the year he returned to Scotland), Dinwiddie was Professor of Mathematics and Natural Philosophy (and teacher of geography) at the College of Fort William, Calcutta.[58]

Such evidence hints at considerable complexity in terms of the local spaces and sites in which geography was being made as polite learning in eighteenth-century Scotland. It suggests private spaces used for instrumental displays; lecturers moving to their audiences and *vice versa*; classes and teachers meeting in certain specialised sites, scientific or popular; and, even, trips to the shops to purchase texts and instruments.

### *Geography's instruments: the production of geography books, globes and maps in Scotland, 1708–1830*

On 20 January 1783, Edinburgh's *Evening Courant* carried news that 'Martyne's New System of Geography', just published, would be delivered to

[55] *Glasgow Herald,* 1 April and 5 August 1825. [56] *Edinburgh Evening Courant,* 31 March to 2 April 1708.
[57] *Dumfries Weekly Journal,* 11 November 1777, 4A.
[58] *Caledonian Mercury,* 13 June 1778, 2 January 1779; A. Law, *Teachers in Edinburgh in the eighteenth century* (Edinburgh, 1965), 162–4; W. Proudfoot, *Biographical memoir of James Dinwiddie* (Liverpool, 1868), 115.

subscribers next week. The announcement also informed readers that globes useful in understanding the book were on display in a local shop and that 'those who choose to advance the subscription-price, may have the globes as soon as they can be brought from London'.[59]

Sitwell's analysis of special geography books (books which purported to describe the whole world) reveals a steady level of publication from the second decade of the eighteenth century at about five or six adult special geography books per decade, with a 'peak' to an average of twelve books per decade from about 1791, before a decline from 1830. The supply of special geography books aimed at the young, always more numerous than books for adults, rose steadily until the early 1860s.[60] Sitwell's work makes it possible to chart where such books were printed. I am not suggesting that, alone, the location of the publication of geography books is a meaningful measure of the public's engagement with geography. But, even as a single distributional measure – the geography of the production of books of geography – it is part of that historical geography of geographical knowledge I am trying to recover.

Most special geography books were published in Britain and, within Britain, in London. Beginning with Drummond's *Treatise,* twenty-five special geography books were published in Scotland between 1708 and 1830, all except three in Edinburgh. Law lists twenty-six geography books and textbooks published in Scotland between 1708 and the publication in 1804 of Thomas Ross' *A compendious system of geography, as connected with astronomy, and illustrated by the use of the globes.*[61] That total includes, however, MackGregory's books, and the Scottish editions of three well-known geography texts produced in England: Thomas Salmon's *A new geographical and historical grammar,* first published in London in 1749, and printed in Edinburgh in 1767 for James Meuros, a Kilmarnock bookseller (there were Edinburgh editions produced in 1771, 1778, 1780 and 1782); Salmon's *The modern universal gazeteer,* first published as *The modern gazeteer* in 1746, in Edinburgh in 1777, 1781, in Glasgow in 1785 and in numerous London editions; and the work of the Brechin-born William Guthrie, *A new geographical, historical and commercial grammar,* which was published first in London in 1770, and which had Edinburgh editions (1790, 1799), and three editions produced in Montrose (in 1807, 1808 and 1810).

The chronology of publication in Scotland is distinguished by three features: a gap between 1708 and the publication in Edinburgh in 1762 of John Mair's *A brief survey of the terraqueous globe* (see this chapter, p. 137); infrequent publication between 1762 and 1788 when only three books were published; and the period 1788–1830, in which eighteen works appeared. Publication of maps of Scotland, produced within Scotland and beyond,

[59] *Edinburgh Evening Courant,* 20 January 1783.
[60] Figures derived from Sitwell, *Four centuries of special geography* (Vancouver, 1993).
[61] A. Law, *Schoolbooks and textbooks in Scotland in the eighteenth century* (Edinburgh, 1989), [no pagination].

shows a similar trend. Relatively few were published between Herman Moll's 1714 map and 1750, with a marked increase between 1750–1807 and again between 1807 and 1830.[62] The effective origin of globe production in Scotland was mid-1776, when the Jedburgh-born surveyor John Ainslie and John Miller, previously assistant to George Adams the London-based instrument maker, announced in the Scottish press plans for a pair of 12-inch table globes. Their proposal was initially a failure and Miller did not succeed in producing a globe until 1793. The first substantial Scottish globe was made by James Kirkwood in 1804. By 1830, there was a thriving market, based in Edinburgh, served by several manufacturers.[63]

Taken together, such evidence suggests that the production in Scotland of instruments and geographical texts was a feature of the later eighteenth century and after, and that it was concentrated in Edinburgh, with Glasgow and smaller towns producing only later a handful of books by the early nineteenth century. It is difficult to know if the authors of special geography texts and makers of globes in Scotland responded directly to the geographical discoveries of the period. We can only infer that public geography classes drew upon such information, perhaps especially so from the rise in numbers of classes from 1790 (see Figure 4.4). The growth in periodical literature from the later eighteenth century also meant that the public's engagement with geography's books and with explorers' exploits was mediated through other literary works. Several examples are illustrative here. James Bruce, the Scots-born discoverer of the source of the Nile in 1774, had his spoken reports of his journeys to audiences in Rome, Paris and London received with incredulity and disbelief, particularly his accounts of dietary practices amongst the Abyssinians. Bruce's rather muddled accounts of his travels, published in 1790, only reinforced doubts about his credibility and the geography of the area he had visited. Not until the early nineteenth century was Bruce found to have been telling the truth: an account of the means by which his credibility was established was carried in the *Scots Magazine* in 1819, over forty years after the events he recounted had occurred. What for Bruce was a matter of personal trust and public credibility may for his audiences have been a prompt to buy geography books and to discuss, as a public, the nature of geographical 'truth'.[64] John Pinkerton's *Modern geography,* published in two volumes in London in 1802, is also instructive here. Pinkerton, born in Edinburgh in 1758 and trained as a lawyer there, had earlier come to public attention through his ill-informed *Dissertation on the origin and progress of the Scythians or Goths* (1786), in which he had argued for the innate racial inferiority of the Celts, seeing in the Irish, the Welsh and the Scottish Highlanders a 'fatal moral and

[62] D. Moir (ed.), *The early maps of Scotland to 1850* (Edinburgh, 1973), Vol. I, 180–225.

[63] Simpson, 'Globe production in Scotland', 21–31.

[64] C. Withers, 'Travel and credibility: towards a geography of trust', *Géographie et Culture,* 33 (2000), 3–17.

intellectual weakness, rendering them incapable of susceptibility to the higher influences of civilisation'.[65] His 1797 *History of Scotland* was attacked by critics, but his 1802 *Modern geography* was well received. The *Edinburgh Review* ran a lengthy review of Pinkerton's *Geography* in October 1803, with the reviewer, William Stevenson, beginning his review with the words 'There is no science so attractive as geography'.[66] A greatly-enlarged edition (2,800 pages in three volumes) also attracted extensive, if less favourable, attention.[67] Members of the Scottish public disinclined to read Pinkerton's geographical work or prejudiced against him after his 1786 book had, in Stevenson's review, an accessible digest. As the supporters of the never-realised *Scottish Chronicle* wrote in 1788, the papers attached great importance to 'information from every quarter of the globe' in promoting 'the happiness and intelligence' of its local audiences.[68]

The public's encounter with geography's instruments was influenced, no doubt, by cost. The six volumes of James Playfair's *System of geography, ancient and modern,* published in Edinburgh 1808–14, and the accompanying *General atlas* (1814) 'Drawn and Engraved for Dr Playfair's Geography', would have been beyond all but the wealthy and serious-minded.[69] There is evidence, moreover, of differences within geography's public and of a recognition of such differences by publishers of books and atlases. Pinkerton's *Modern geography* was among several books with an edition produced 'for the Use of Schools and Young Persons'. In the Introduction to his 1815 *Modern atlas,* Pinkerton recognised – in condescending tones – that women were part of geography's public audiences:

> Geography is a study so universally instructive and pleasing, that it has, for nearly a century, been taught even to females, whose pursuits are foreign from serious researches. In the trivial conversations of the social circle, in the daily avidity of the occurrences of the times, pregnant indeed above all others with rapid and important changes, that affect the very existence of States and Empires, geography has become an habitual resource to the elegant female, as well as the profound philosopher.[70]

John Thomson's 1832 *Atlas of Scotland* in contrast was prefaced with a 'short history of the progressive Geography of Scotland'[71] in order that readers could appreciate the role of Scottish geographers in the past. It is also clear that some of those private individuals making up the geographically-minded Scottish public had some formal education in geography.

[65] John Pinkerton, *Dissertation on the origin and progress of the Scythians or Goths, being an introduction to the ancient and modern history of Europe* (London, 1786), 17. See also F. Sitwell, 'John Pinkerton: an armchair geographer of the early nineteenth century', *Geographical Journal,* 138 (1972), 470–9.

[66] *Edinburgh Review,* 3 (1803), 67: the whole review is 67–81.

[67] *Edinburgh Review,* 10 (1807), 154–71.

[68] W. Couper, *The Edinburgh periodical press* (Stirling, 1908), Vol. II, 177.

[69] NLS, MS Acc.11592, Letter of 1810 from James Playfair to Mr Hood, London, on the sale of Playfair's *System of geography.*

[70] J. Pinkerton's 'Memoir on the recent progress, and present state, of geography', in J. Pinkerton *Modern geography* (1807 edition), xxix.

[71] J. Thomson, *Atlas of Scotland* (Edinburgh, 1820), ii–vi.

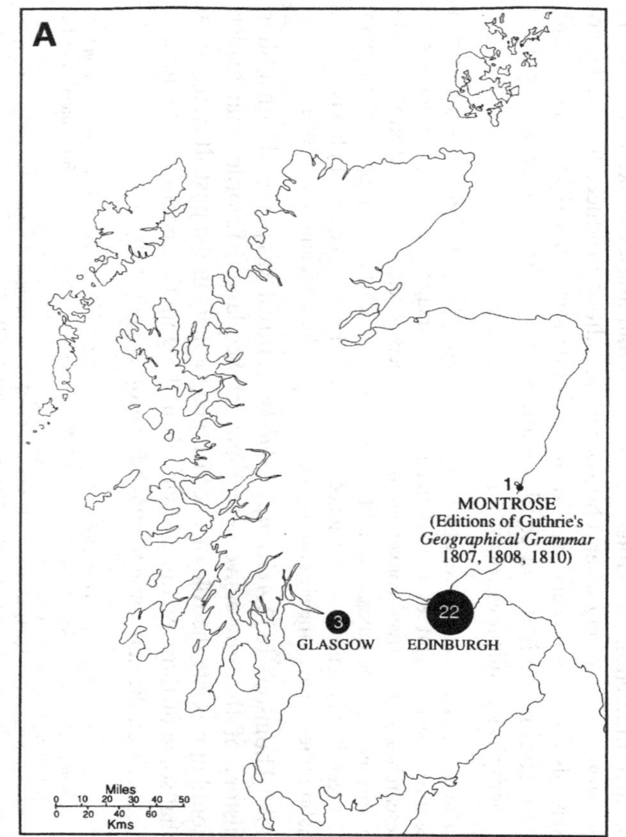

A
1
MONTROSE
(Editions of Guthrie's
*Geographical Grammar*
1807, 1808, 1810)
3
GLASGOW
22
EDINBURGH
Miles
0 10 20 30 40 50
0 20 40 60
Kms


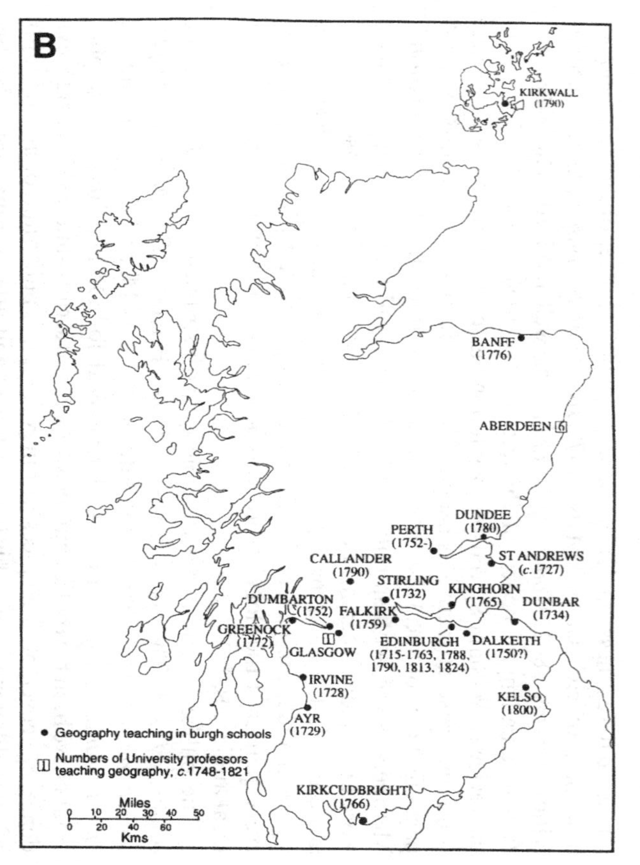

B
KIRKWALL
(1790)
BANFF
(1776)
ABERDEEN 6
DUNDEE
(1780)
PERTH
(1752-)
ST ANDREWS
(*c.*1727)
CALLANDER
(1790)
STIRLING
(1732)
KINGHORN
(1765)
DUNBAR
(1734)
DUMBARTON
(1752)
FALKIRK
(1759)
GREENOCK
(1772)
1
GLASGOW
EDINBURGH
(1715-1763, 1788,
1790, 1813, 1824)
DALKEITH
(1750?)
IRVINE
(1728)
KELSO
(1800)
AYR
(1729)
Geography teaching in burgh schools
Numbers of University professors
teaching geography, *c.*1748-1821
KIRKCUDBRIGHT
(1766)
Miles
0 10 20 30 40 50
0 20 40 60
Kms

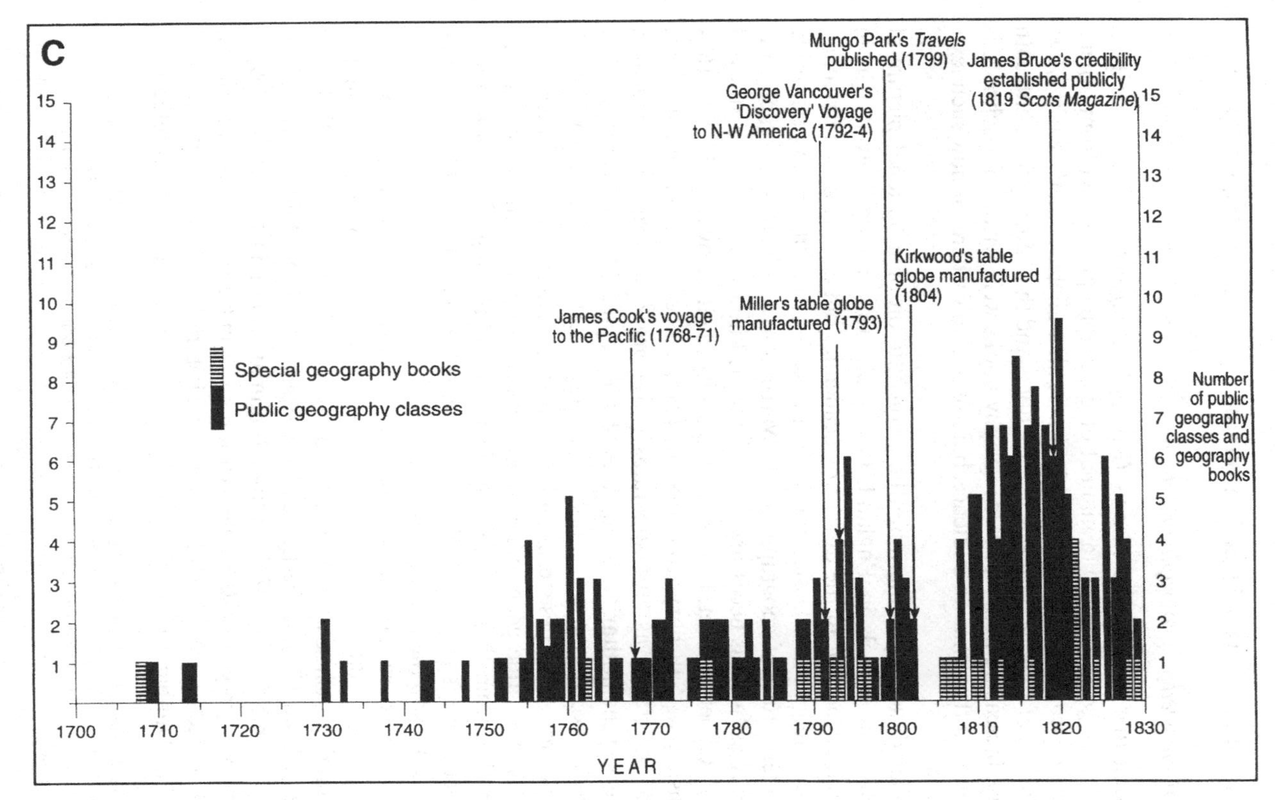


4.4 The geography of geographical activity in Scotland, 1700–1830. Figure 4.4A shows the location of publishers of geography books with the numbers of different titles; 4.4B shows geography's teaching sites in burgh schools and in the universities; and 4.4C charts the number of special geography books and of public teaching in geography in relation to significant moments of geographical discovery and the publication of geography books.

## Teaching places: geography in Scottish schools and universities, 1727–1821

### *Sites of knowledge: geography in schools*

Knowing that Matthias Symson's *Geography compendiz'd* (1702), his 1704 *Encheiridion geographicum* (a manual of geography plagiarised from others' work) and Drummond's *Treatise* (1708) were all available from the early 1700s, that the last of these was produced for schools and that, during the eighteenth century, increasing numbers of geography books were written for schools, is to know that there was an audience, but not to know where, or how such texts were used.

Similar difficulties undermine the reconstitution of a national picture of geography in schools in England. Geography was not formally taught in grammar schools or in the dissenting academies but as part of a curriculum which centred on the classics. For Joseph Priestley – credited with pioneering the modern study of geography at the Warrington Academy – geography was seen only as 'what is necessary or useful to be known previous to a study of history'. Gilbert Burnet, Professor of Divinity in Glasgow from 1669 to 1674 and Bishop of Salisbury from 1689 to 1715, echoed these sentiments in his view of geography as 'ane apparatus to history'.[72] Commercial academies emphasised the utilitarian advantages of geographical knowledge, a sense of geography's worth that was supported by public lecturers and the work of the Royal Society. If, in these terms, it is the case that 'Three types of reader, then, could use the same geographical compendia, but would cull very different information from them because of their sites of reading',[73] any attempt to chart geography's use in schools must be sensitive to the nature of the institution as well as to the nature of the geography.

There are records of geography being taught in burgh schools in Edinburgh in 1755, 1763, 1788, 1790, in the Edinburgh Sessional School in 1813 and in Dr Young's parish school in 1824; it was also taught in Irvine in 1728, in Ayr Academy from 1729, in Perth Academy from 1752, in St Andrews *c.*1727, in Stirling High School in 1732, in 'the Grammar school' in Falkirk in 1759, in Dundee Academy in 1790, in the parish schools in Callander and in Kirkwall in 1790, in Kelso in 1800, and, at sometime before 1779, in the 'mathematical school' at Dunbar[74] (Figure 4.4b). This evidence must be interpreted cau-

[72] J. Clarke, *Bishop Gilbert Burnet as educationalist being his thoughts on education with notes and life of the author* (Aberdeen, 1914), 45–6.

[73] Mayhew, 'The character of English geography', 399.

[74] Law, *Teachers in Edinburgh in the eighteenth century*; *Education in Edinburgh in the eighteenth century* (1965); *Edinburgh schools of the nineteenth century* (Edinburgh, 1996); W. Boyd, *Education in Ayrshire through seven centuries* (London, 1961), 70–7; J. Strawhorn, *750 years of a Scottish school: Ayr Academy 1233–1983* (Ayr, 1983), 20, 24, 34, 41–4; Perth Archives, MS B 59/24 *et seq.*; E. Smart, *History of Perth Academy* (Perth, 1932), 66, 80–5, 120, 158; Anon., *Memorabilia of the City of Perth* (Perth, 1806), 341–9; A. Hutchinson, *History of the High School of Stirling* (Stirling, 1934), 92, 106–7, 145–6; J. Love, *The schools and schoolmasters of Falkirk from the earliest times* (Falkirk, 1898), 53–7, 108, 183–9; J. Miller, *History of Dunbar* (Dunbar, 1859), 220–4; J. Muir, *Story of my boyhood and youth* (Edinburgh, 1985 edition), 18; J. Smith, *History of Kelso Grammar school* (Kelso, 1909), 76–7.

tiously: school records have not survived equally well. These dates note the first explicit mention of 'geography', but it is clear that navigation and mixed mathematics, with which geography was closely associated, were taught before then, if not widely. Some of those persons who advertised geography classes in the public sphere were also school teachers. James Barclay, for example, who advertised classes in geography in the *Edinburgh Evening Courant* in 1742, gave them up on being appointed to the city's High School in June 1742. Barclay was, from 1750, Rector of Dalkeith High School and, in 1743, author of *A treatise on education.* Barclay's *Treatise* gives a good idea of what geography was understood to be and how it was taught. Geography was useful in understanding other subjects and itself extensive: 'poetry, either ancient and modern, can no more be read without the Heathen fables and ancient geography, than modern history without some antecedent knowlege of modern geography': 'The study of geography . . . is no less a review of the habitable world.' Geography was taught in order to understand history, ancient and modern, and contemporary world affairs and used maps and globes. Travel reports 'of such as have narrowly examined particular parts of the globe . . . are very agreeable to children, and of great use in improving their geography'. Further, noted Barclay, 'I shall only observe, that it increases attention, if before boys saw either globes or maps, the master would explain *viva voce* the figure of the earth, her annual and diurnal motion, all the greater and lesser circles, zones, climates, *&c*'. Attention to the globe then followed, then by attention to others' texts and, lastly, to maps. For want of such geographical understanding, Barclay argued, 'a great many often mistake their true interest, and, by the false representations of ambitious men, blindly engage in schemes destructive of all trade and liberty'.[75]

Barclay's summary of the nature of geography, and its methods of teaching, was repeated by many others in this period, in Scotland and beyond. In some schools, geography was taught as an 'additional subject', that is, additional to the normal curriculum (of mathematics, reading, writing, and, for some, Latin and Greek) for which masters would get paid more. There is no evidence that financial advantage for teachers promoted attention to geography. Maps and globes were used where they could be afforded. In Stirling High School, funds were secured in 1732 for two 'geograficall maps to be putt up', and, in 1755, Mr John Livie, who taught 'the voluntary subjects – French and geography', received a 'sett of proper maps' to use with the school's globes. A later master, John Weir, included map drawing in geographical teaching from 1814.[76]

The clearest pictures we have of how geography was taught and of contemporaries' views of it in relation to ideas of enlightened citizenship and social utility come from Ayr Academy and Perth Academy. In Ayr, John Mair

[75] J. Barclay, *A treatise on education* (Edinburgh, 1743), 196–201.
[76] Hutchinson, *History of the High School of Stirling*, 92, 106–7, 145–6.

was appointed as 'doctor' or second master in 1727. He taught navigation from that year and, in 1728, submitted an application to the burgh for maps and globes in which he indicated the advantages to be derived from their use: 'Being convinced that the Council with the wiser part of mankind look upon the flourishing of the liberal arts and sciences as a common interest, [he] prays their honours to procure a set of maps and globes, the knowledge of which as their honours may very well know is highly necessary for forming the man of business'.[77] By 1735, geography was an accepted part of the school course, was regarded as a necessary supplement to study of grammar, and was taught by Mair alongside arithmetic, navigation, surveying and geometry. 'As the world now goes', noted Mair in 1729, 'the mathematical part of learning is a principal part of a gentlemen's education'.[78] In 1773, for three weeks only, John Murdoch, the English master at Ayr, enrolled the young Robert Burns in order 'to perfect his grammar'.[79] Whether Burns studied geography directly is uncertain, although in a letter of August 1787, Burns noted 'I spent my seventeenth summer . . . a good distance from home at a noted school, to learn Mensuration, Surveying, Dialling, &c. in which I made a pretty good progress', and that 'My knowledge of ancient story was gathered from Salmon's and Guthrie's geographical grammars'. In a letter of 1791, Burns again referred to 'the latest edition of Guthrie's Geographical grammar' as among 'the valuable books' he was to be sent.[80]

Geography was particularly promoted at Ayr from 1826 by the rector, Dr John Memes. In addition to geography, he taught mathematics, natural philosophy, mathematics, history, botany and English. He prepared his own wall maps for his geography classes and painted two large terrestrial spheres on the ceiling of the school hall. From 1837, he introduced geographical excursions into the local countryside, and, to extend the study of geography and natural history, he acquired botanical specimens and created a botanical garden at the school. Pupils recalled his extraordinary geographical energies: '[Memes] sketch[ed] large maps for his Geographical classes till eight at night [from 6 pm] . . . Next morning, at four or five o'clock, he was to be found actively engaged with his Mathematics class, making plans of the Town Harbour, to be exhibited at the annual examination . . . to the Ayr Mechanics' Institute he gave gratuitous instruction . . . With a party of senior pupils he made a survey and prepared a report on the feasibility of bringing piped water from Carrick Hill into Ayr – for which he was rewarded with a public dinner in his honour'. Fortified as he sometimes was in his senior geography class by a glass of wine and a biscuit, Memes would, unsurprisingly perhaps, often fall asleep in his own classroom.[81]

[77] Boyd, *Education in Ayrshire*, 75; Strawhorn, *Ayr Academy*, 20. [78] Strawhorn, *Ayr Academy*, 19–20.
[79] Boyd, *Education in Ayrshire*, 76.
[80] G. Ross Roy (ed.), *The letters of Robert Burns* (Oxford, 1985 edition), Vol. I, 138, 140; Vol. II, 66.
[81] Strawhorn, *Ayr Academy*, 43–4.

By 1761, John Mair had moved to become rector of Perth Academy. Geography had been earlier taught there: Mr Andrew Cornfute had petitioned the town council in 1752 for funds to purchase maps and globes to accompany his geometry and geography classes.[82] Cornfute's vision for the curriculum was that it should closely relate to 'the business and experience of life'. It was for that reason that geography, along with French and book-keeping, was added from 1752–3. Mair further promoted geography from 1760. His geographical teaching at Perth provides the first clear evidence in Scotland of a formative relationship between geographical texts and the taught geographical curriculum. The structure of Mair's 1762 *Brief survey of the terraqueous globe* is exactly that laid down in the prospectus of the Academy from 1760. The first-year course included:

> I. A short view of natural history in its different parts, viz. The constitution of the material world, the nature and property of the elements, and vegetable, mineral, and animal economy, as a proper introduction is well calculated to fix the attention and awaken the curiosity of young people, being all illustrated by experiments . . .
> VI. The general principles and most useful problems in astronomy.

The second-year course included:

> III. . . . a practical course of geography, as an introduction to civil history, which should then follow . . .
> IV. The history of commerce, and a short view of the present state in the different, nations, particularly in Britain.[83]

This scheme had marked similarities with those of the dissenting academies in England, but was revolutionary in Scotland.[84]

Mair died in 1769 and was succeeded by Robert Hamilton, who developed the experimental aspects of natural philosophy and geography in his purchase of air pumps and globes before, in 1779, being appointed to the Chair of Natural Philosophy at Marischal College, Aberdeen.[85] His successor, Alexander Gibson, previously a teacher of mathematics at Dunbar for thirteen years, also had large sums of money spent on scientific apparatus and globes.[86] Gibson's appointment was supported by Professors John Playfair and Dugald Stewart as well as members of the gentry like Robert Ainslie, who noted how Gibson had 'put the school [at Dunbar] on a more regular footing' by getting local worthies to contribute 'a compleat set of instruments for mathematics and natural philosophy'.[87] Gibson left letters by which we can see him doing the same things in Perth. In a letter of 29 October 1794 to James Ramsay, Perth's lord provost, Gibson emphasised the poor condition of the

[82] Perth Town Council Minutes, 10 August 1752, 11. [83] Smart, *History of Perth Academy*, 66.
[84] Law, *Schoolbooks and textbooks in Scotland in the eighteenth century*, n.p.
[85] Smart, *History of Perth Academy*, 84–5.
[86] Perth Town Council Minutes, 5 February 1770, 17; Smart, *History of Perth Academy*, 85–6.
[87] Perth Archives, MS B59/24/6/42, Robert Ainslie to George Sandyman, 3 May 1779.

Academy's globes – 'thro' the lapse of time and the frequent use of the Globes in the Academy the figures and marks are so obliterated, that it is with great difficulty any question can be answered by them' – and stressed the importance of new ones with up-to-date information. Gibson clearly got the town council's approval: by 26 November 1794, he was in communication with George Adams in London over the purchase of a good pair of used table globes.[88]

Generalising from such limited, if individually rich data, is dangerous. Yet it is clear not only that geography was taught in some Scottish schools from at least the mid-eighteenth century, but that such teaching involved the negotiation by pupils of the evidence of the globe, maps and others' accounts. Geography was a necessary part of a useful polite education, a means to understand, and to get on in, the world. Some pupils and masters may have used books written for general adult audiences, just as a recognition amongst geographical authors of a developing school market may have prompted them, as it did Mair, to write school books. Like their counterparts in the public sphere, some teachers moved to their audiences. Mary Somerville, later to receive formal recognition for her geographical texts, recalls how the family had two small globes which were used by Mr Reed, the village schoolmaster in Burntisland, 'who came to teach me for a few weeks in the winter evenings'.[89] Other teachers wrote books for their audiences. Thus, David Foggo, private teacher of English and geography in Edinburgh in 1824, who wrote *Elements of modern geography,* (subtitled *with a great variety of problems, on the terrestrial and celestial globes, for the use of schools*), noted that 'the primary object of the compiler of this compendium was to furnish for his own Pupils, at a moderate price, a comprehensive outline of Modern Geography'.[90] A namesake, John Foggo, had taught geography in Kelso in 1800, giving different instructions to boys and girls using 'all manners of instruments'.[91]

It was just such interests that prompted the provision of geography books by the Scottish School-Book Association, begun in 1818. In 1823, they published *A short system of geography upon an improved plan,* compiled by a committee of schoolmasters in Scotland, although the work also drew from the geographical articles in the *Edinburgh Review* and the *Quarterly Review.*[92] The fact that those and other journals – such as the *Edinburgh Monthly Review,* which discussed the 1819 *First book of geography, for the use of schools and*

[88] Perth Archives, MS B59/24/6/112, Alexander Gibson to James Ramsay, 29 October 1794; MS B59/24/6/114, George Adams to William Wallace [Assistant master at Perth Academy], 26 November 1794.

[89] M. Somerville, *Personal recollections, from early life to old age of Mary Somerville with selections from her correspondence* (London, 1873), 29–30.

[90] David Foggo, *Elements of modern geography* (Edinburgh, 1822), 2.

[91] J. Smith, *History of Kelso Grammar School* (Kelso, 1909), 76.

[92] A. Belford, *Centenary handbook of the Educational Institute of Scotland* (Edinburgh, 1946), 15–20. The claim about the derivation of the works from periodical literature is made by an anonymous reviewer of the *A first book of geography, for the use of schools and private teachers* (Edinburgh, 1819) which appears in *The Edinburgh Monthly Review,* 2 (1819), 333–8.

*private teachers* – were prepared to debate geography's educational worth says much about public recognition of its value. The fact, too, that Robert Hamilton had taught geography in Perth Academy before taking up a chair, and that David Jack, teacher of geography in Irvine in 1728, had studied under Robert Simson in Glasgow, highlights the connections between geography's teaching places.

*Spaces of practical reason: geography in the universities*

Geography was taught by six different professors in Aberdeen during the eighteenth century. Matthew Mackaile, Professor of Medicine at Marischal 1717–33, may have taught it as part of his natural philosophy classes.[93] Mackaile's father, also Matthew, had corresponded with Sibbald over his geographical enquiries forty years earlier (see Appendix, no. 1, p. 256). The Newtonian natural philosopher and Professor of Mathematics in Marischal College, John Stewart, certainly taught geography to his students from at least 1748: he had been appointed in 1727, but there is no certain evidence of his including geography before 1748. In those 'mathematics texts' referred to by Stewart in 1748 are books of geometry, surveying, navigation, fortification, architecture and geography, which last category included Gordon's *Geography anatomiz'd,* and a 1733 edition of Varenius' *Compleat system of general geography, . . . since improved and illustrated by Sir Isaac Newton and Dr. Jurin.* To judge from notes of 1753, Stewart included geography in his first-year practical mathematics classes. The College's minute book also records that 'the Semi Year, or Second of the Course shall be spent in the most usefull parts of Natural History, in geography, and the Elements of Civil History'.[94] Thomas Blackwell Junior, Professor of Greek in Marischal from 1723, gave 'a private class, for ancient *History, Geography* and *Chronology*' in the 1750s.[95] Later in the century at Marischal, James Beattie, Professor of Natural History, lectured in geography, chronology and civil history and, to judge from notes of 1799 taken by William Knight from Beattie's lectures, geography was intended to provide, with civil history, a historical background to the study of classical literature.[96] Robert Hamilton, Professor of Natural Philosophy and Mathematics from 1779 to 1829, taught his first class the elements of arithmetic, algebra, the 'principles of geography, and use of globes'. Navigation was taught in the second class and advanced astronomy in the third class: 'In every part of the course' [noted a student], 'the application of the principles to the practical parts of life is pointed out, and illustrated with examples'.[97]

[93] P. Wood, *The Aberdeen enlightenment: the arts curriculum in the eighteenth century* (Aberdeen, 1993), 23.

[94] Wood, *Aberdeen enlightenment,* 63; AUL, MS M41, f. 42r. Wood has Stewart's list of texts for his classes in 1748 (pages 229–30); see also AUL, MS CT Box 43, J. Lorimer to Roderick MacLeod, 25 June 1794.

[95] Wood, *Aberdeen enlightenment,* 60; SRO, GD/18/50536/5,9,10,18.

[96] Wood, *Aberdeen enlightenment,* 147; AUL, MS 189, ff. 9–11v.

[97] Wood, *Aberdeen enlightenment,* 88.

Hamilton also gave evening classes, advertising for two years in the *Aberdeen Journal* from 1781; his classes aimed at 'those Subjects which are useful in different Departments of Life: Geography, Navigation, Fortification, Perspective, Astronomy and the Like'.[98] Geography was also taught in Robert Gordon's Hospital in 1781.

In King's College, Aberdeen, geography was a formal part of the second-year curriculum in the mid-eighteenth century. The outline of the course for the years 1752–3, given by the philosopher Thomas Reid, shows 'The Elements of Geography so far as they can be delivered in a Historical or Narrative Way', broken down into four main parts: a general regional geography of the globe; the main lands and seas named; 'A very few of the Most remarkable Things belonging to each Country or City'; and accounts of the principal trade winds. For Reid, geography was an essential part of that philosophy 'which may qualify Men for the more useful and important Offices of Society', rather than, noted Reid, merely making men 'subtle Disputants, a Profession justly of less value in the present Age'.[99] Thomas Gordon, the Humanity Professor, taught 'the Geography of the Antients' as part of his classics lectures in 1761.[100]

In Glasgow University, James Millar, Professor of Mathematics there from 1796 to 1832, delivered a lecture programme in geography between 1802 and 1821.[101] His course outline for the 1802–3 session shows a 43-lecture course under three headings: the Earth as connected with the heavenly bodies; the Earth 'in relation to its own Properties'; and the Earth in connection with its inhabitants. The first part should be thought of as cosmography and astronomy. The second was concerned with physical features and problems of measurement. Millar here exposed his students to contemporary disputes in the earth sciences, notably between the Neptunists and the Vulcanists and, thus, whether water or heat was responsible as a major formative force in Earth history. Hutton's 1785 'Theory of the Earth', with its emphasis upon heat, volcanic activity and 'deep time', had only recently been made more accessible by John Playfair in his 1802 *Illustrations of the Huttonian theory of the earth.* 'Dr Hutton' is mentioned by Millar, but not Playfair's work. Even so, Millar's lectures kept students abreast of current theoretical speculation as well as

[98] *Aberdeen Journal,* 15 October 1781, 4; 7 October 1782, 4; 6 October 1783, 3; Reid, 'Late eighteenth-century adult education in the sciences at Aberdeen', in Carter and Pittock, *Aberdeen and the enlightenment,* 168–79.

[99] AUL, MS 2131/8/V/I, Thomas Reid's Course Outline of 1752: Wood, in his *Aberdeen enlightenment,* notes that 'Thomas Reid's course outline of 1752 contains the only evidence we have about the teaching of geography at King's in the period 1753–1800' (p. 222, n.185); *Abstract of some statutes and orders of King's College in old Aberdeen. M. DCC. LIII. With Additions M. DCC. LIV* (Aberdeen, 1754), Article XXVII, p.13: [see also AUL, MS M41, f. 42r, Minute Book of Marischal College, 1729–90], 11 January 1753.

[100] J. Pickard, *A history of King's College library, Aberdeen, until 1860* (Aberdeen, 1982), Part 3 (1700–99), 211; AUL, MS K 45 [Minute Book of King's College, Vol. X, 1761–65], 63–4.

[101] NLS, MS 2801 [this is the complete set of a student's notes of Millar's 1802–3 course: v + 312ff]; there are some other extant but partial student notes of Millar's lectures on geography: GUL, MS. Gen. 1355/105 [notes dated 4 December 1821]; MS Gen. 1355/106/1 [notes on 11 lectures, undated]; MS Gen. 1355/106/2 [ff.1v-15v.: notes on geography lectures 1821–2].

teaching them practical interpretation. The third part of Millar's course, as with many others, sought to promote commercial understanding through regional description by placing emphasis upon the 'advancement of Commerce in modern Nations' consequent upon modern voyages and discoveries.[102]

As these examples show, geography's discursive space was principally shared with mathematics and history. In the first sense – mathematics – it was allied with geometry and astronomy. In the second sense – history – it was associated with civil history and as a means to understand the classics. Geography was seen in Enlightenment Scotland as a means to useful knowledge and enlightened thinking about the nature of the world. Geography and its practices were part of the world of reason. So keen, indeed, were Mackenzie of Delvine's two sons to learn geography and the use of the globes in their time at St Andrews between 1711–14 that they taught themselves in the evenings. The boys' guardian implored their father to send on new globes recently purchased: 'none are allowed the use of the publick Globes out of the Library, + you may mind the old one we had at Delvin was so abused before we got it, that we could do little with it, therefore if your new Globes could be got transported safely to this place, it would be a great help to my Twins'.[103]

Success by some in universities' teaching spaces even effected geography's public expression, as the case of Ebenezer MacFait shows. MacFait taught geography, mathematics and natural philosophy in Edinburgh – 'at the east turnpike of Mealmarket Stairs' – in 1754 and in 1760, and at five other addresses in the city's Old Town between 1746 and 1785. Walter Scott attended MacFait's class. Much of MacFait's teaching material, was included in his 1780 *A new system of general geography.*[104] MacFait's letter of 24 November 1782 to an Edinburgh MP illuminates something of the circumstances of one of geography's public teachers: 'I am now seventy one years of age + feel myself turning every year more infirm; and tho' I think myself as fit for teaching as ever, yet partly from ungenerous prejudices against old age, partly from the abilities of Mr Stewart the professor of mathematics who is a young man of good address and easy elocution, the run for him is so great that I have very few scholars at present'.[105]

Further, the geography of geography's teaching places may now be seen not just to be about *sites of knowledge*, public or private, but also to be a question of *movement* – of geography's instruments, audiences and teachers and of ideas themselves – between such sites. Yet geography's making was not just about the situated production of geography in public classes, private drawing rooms, school rooms and texts. Seeing geography as useful was intimately

[102] NLS, MS 2801 [Lectures 31–3].
[103] W. Dickinson, *Two students at St Andrews, 1711–1716* (Edinburgh, 1952), 5, 6–7, 21, 55–9.
[104] Law, *Schoolbooks and textbooks,* n.p.
[105] NLS, MS 5391, f.53–53v, Ebenezer MacFait to A.[lexander] Stuart, 24 November 1782. A note in the *Edinburgh Advertiser* for 1 December 1786 records MacFait's death 'on Saturday last', and speaks of him as an eminent mathematician: NLS, MS 8250, f. 279.

connected with other ways of seeing. To illustrate this point, I want to turn away from the previous discussion of particular sites and social spaces to examine the different ways in which Scotland was 'visualised' in the eighteenth and early nineteenth centuries.

## 'Visualising the nation': survey, geographical knowledge and national identity

By 'visualising', I mean here both the ways in which Scotland was *actually realised* and *figuratively imagined* – in mapping practices, in landscape paintings and in literary geographies – and the ways in which geographical knowledge *of* Scotland was undertaken in order to understand its present constitution and to provide a vision of the nation in the future. In these terms, of course, Scotland provides a particular expression of the ways in which geographical knowledge was central to contemporaries' understanding of the world in the Enlightenment. Mapping and survey were Enlightenment practices of nation building.[106] The idea of survey as a question of relative social advance was central to the ways in which Europeans viewed the rest of the world.[107] Those illustrators who accompanied geographers on their travels represented 'new worlds' to the polite audiences of Europe's public sphere, as, at the same time, emergent traditions of landscape painting established enduring symbolic importance in their portrayal of certain parts of Britain and Europe as icons of national identity. Vision and visualisation in such ways has, simply, been central to the processes by which geographers have determined the 'shape' of the world *and* to the constitution of geography itself. It is for just such reasons of different meanings in practice that I want to move away from situated textual definitions of geography to consider geography at work. As the authors of the Glasgow-based 1805 *System of geography* recognised, 'It may not be improper . . . to remark, that the term Geography has been used in different acceptations. . . . Geography comprehends many discussions, moral, political, and historical'.[108]

Two themes in particular are addressed to make sense of the 'visualising' of Scotland. I first explore the idea of survey as a geographical practice and consider the attempts made during the eighteenth century to provide a national geographical survey or, as some contemporaries later put it, a 'statistical account', of Scotland. Second, I consider something of Scotland's visualisation in maps and paintings and its literary representation by looking at the work of the Military Survey from 1747 and at the connections between landscape painting and national identity before about 1830.

[106] A. Godlewska, 'Map, text and image. The mentality of enlightened conquerors: a new look at the *Description de l'Egypte*', *Transactions of the Institute of British Geographers*, 20 (1995), 5–28; M. Edney, *Mapping an empire: the geographical construction of British India, 1765–1843* (Chicago, 1997), 39–120.

[107] N. Dirks (ed.), *Colonialism and culture* (Ann Arbor, 1992); B. McGrane, *Beyond anthropology: society and the other* (New York, 1989).

[108] Anon., *A system of geography by a literary society* (Glasgow, 1805), i–ii. This was almost certainly produced by members of the Glasgow Philosophical Society.

*Geographical survey, natural history and national knowledge, 1704–1799*

Eighteenth-century practices of national survey have, as we have seen, intellectual origins in ideas of political and natural arithmetic, and, in Scotland, earlier expression in the work of Sibbald. Probably the first attempt to undertake a national survey for Scotland (other than Sibbald's) was that of Andrew Symson (sometimes 'Simpson'), the printer and former minister, himself a contributor to Sibbald's 'Queries' (see Appendix, no. 35), whose son, Matthias, was the geographical author. In a letter of 1708 to Edward Lhuyd, Andrew Symson made reference to his 'Villare Scoticum, wherein I intend to give an account of all the parishes in Scotland'. To judge from Bishop Nicholson, Mathias brought a copy of the 'Villare Scoticum' (presumably in manuscript) to Nicholson in Carlisle.[109] No trace has been found of this work. At much the same time, in a letter of 1708 by two men involved with the failed attempt to establish a Royal Society for Scotland, William Carstares wrote to George Ridpath, noting that 'the Geography and Natural History of our Country stand in need of being done as well as our political and Church History, with submission I should think it proper that it were recommended to ye Minr. of every parish to do it for their parish and Communicate it to ye presbytery in order to be sent to ye Commission'.[110] This, too, came to naught.

In 1748–9, the antiquarian Walter Macfarlane brought together and transcribed in manuscript 'Geographical collections relating to Scotland', amongst which were papers dating from 1721–44 concerning an intended 'Geographical Description of Scotland'. This material was published in 1906: as Mitchell, the editor, noted then, 'It is rare to find Accounts of localities which are made so much as these are from a geographer's point of view'.[111] Emery's examination of the Macfarlane collections has established the chronology of their production – most (59) of the 75 extant parish accounts date between 1722 and 1726 – and shown that the accounts resulted from the Church of Scotland's concern to map Scotland and provide an accompanying description. The 1720s survey is important less as a forebear of Sinclair's project and more because, like Sinclair's, the collection of geographical information was easily facilitated through the distinctive hierarchical structure of the Church of Scotland. The contents vary from statements of distances and accounts of the history and antiquities of the parish to summaries of economic activities and natural products. They were based on returns to a single sheet of 'directions and rules' to ministers on 'the geography of the parish'. Emery has also shown that the remaining 26 accounts were compiled for the

[109] Bodleian Library, Ashmole MS 1817a, f.499; T. Gray and E. Birley (eds.), 'Bishop Nicholson's diary (1703–4): Part II', *Transactions of the Cumberland and Westmorland Antiquarian and Archaeological Society*, 50 (1951), 124.

[110] EUL, MS DK.1.1.2; R. Emerson, 'Natural philosophy and the problem of the Scottish Enlightenment', *Studies on Voltaire and the Eighteenth Century*, 242 (1986), 275.

[111] A. Mitchell (ed.), *Geographical collections relating to Scotland made by Walter Macfarlane* (Edinburgh, 1906), Vol. II, v.

churchman and historian William Maitland from returns from parish ministers via 'printed queries concerning their respective parishes' for his projected 'History of Scotland'; only one volume, *The history and antiquities of Scotland,* was published, in 1757, the year he died.[112] The geographical coverage of both surveys is uneven – only the north-east Lowlands, southern Perthshire and northern Sutherland were surveyed – but their existence testifies, I suggest, to the ways in which geographical knowledge was seen: by Maitland, as central to a historical understanding of the nation, and, by the Church, to management of the nation in the present.

The sense in which geography was understood as an inductive practice and as a means to useful national knowledge is apparent in the 1781 proposal of David Erskine, earl of Buchan, that the Society of Antiquaries in Scotland should conduct 'a general parochial survey' of the nation in order 'to excite a taste for Natural History among our countrymen' and to further 'national improvement'. Buchan founded the Society in 1780. Part of its objectives was the comparative examination of 'the ancient state of Scotland' with the modern condition of the nation. Attention to the 'Geography, Hydrography and Topography of the Country' was central to this concern. Buchan's intended national survey was structured in seven sections: section I to examine the situation, boundaries and landownership of each parish 'geographically and topographically described'; II to consider 'from actual survey' the agricultural economy and demographic circumstances of each parish; III and IV to cover roads, communications and mineral workings; V to examine trade and manufactures; VI to consider the 'antiquities of the parish'; and a final section for miscellaneous observations. This was an exercise in knowing one's national geography in order to improve it: 'Accounts of the parishes in Scotland . . . would exhibit a noble and complete survey of this part of the united kingdoms . . . and . . . be a most interesting and useful national attainment'.[113]

For reasons to do with Buchan's own inabilities to focus the proposal, lack of money, and conflict between cultural institutions, this survey was never realised.[114] Only a handful of parishes was covered, reports on them appearing in the delayed first volume of the Society's *Transactions* in 1792. Membership of the Society included men whose professions would have suited them to the intended task: John Ainslie, the surveyor and globe-maker (described in the Society's membership lists as 'geographer'[115]); John Bain, topographer; John Clark, land surveyor; John Williams the mineral surveyor;

[112] F. Emery, 'A "geographical description" of Scotland prior to the statistical accounts', *Scottish Studies,* 3 (1959), 1–16; C. Withers, 'How Scotland came to know itself: geography, national identity and the making of a nation, 1680–1790', *Journal of Historical Geography,* 21 (1995), 371–97.

[113] William Smellie, *Account of the institution and progress of the society of the antiquaries of Scotland* (Edinburgh, 1782), 23.

[114] S. Shapin, 'Property, patronage and the politics of science: the founding of the Royal Society of Edinburgh', *British Journal for the History of Science,* 7 (1974), 1–41; C. Withers, 'Natural knowledge as cultural property: disputes over the "ownership" of natural history in late eighteenth-century Edinburgh', *Archives of Natural History,* 19 (1992), 289–303.

[115] *Transactions of the Society of the Antiquaries of Scotland,* I (1792), xxxii.

and the painters Alexander Runciman, Gavin Hamilton and John Knox. John Knox spoke of his fellow members as 'real Patriots the true friends of a much injured Country', and informed Buchan that he had 'a warm Zeal for the preservation of antient remains; the cultivation of Science and the improvement of my native Countrey'.[116] It was precisely that remit, however, that was objected to by others: the review of the Society's *Transactions* carried in the *Critical Review* in 1792 remarked that 'the idle titles of seal-engraver, geographer, topographer, etc, etc, to the Society are ostentatious and unbecoming the modesty of a literary body'.[117] By the time of its limited publication, Buchan's plan was being overtaken by Sinclair's *Statistical account.*

Sir John Sinclair, MP for Caithness, lay member of the General Assembly of the Church of Scotland and known as 'Agricultural Sir John' for his interests in estate improvement and work for the Board of Agriculture, first articulated his proposals for a detailed survey of Scotland in May 1790. He knew the intellectual tradition in which he stood, citing Sibbald, Robert Gordon of Straloch and Macfarlane as well as something of the history of statistical enquiries elsewhere in Europe.[118] Sinclair, who wanted to elucidate 'the Natural History and Political State of Scotland', knew of Buchan's plans (he had been elected a member of the Antiquaries in 1787), and may have known of another failed call to national knowledge, the proposals of Sir James Steuart in his 1767 *Enquiry into the principles of oeconomy,* which recommended the examination of Scotland 'parish by parish', since, in Steuart's words, 'every other plan for benefiting a nation has only been a vain and deceitful expedient for walking in the dark'.[119]

Sinclair's plan for Scotland's geographical enlightenment, like that of the General Assembly seventy years before, was to use parish ministers working to an agreed scheme through whose local knowledge Scotland would be geographically constituted. He understood that science, in the form of geographical survey, was vital to national identity. His plans demonstrate methodological links with the empirical inductivism of natural scientists examining 'nature's economy' from the ground up:

> The superiority, which the philosophy of modern times has attained over the ancient, is justly attributed to that anxious attention to facts, by which it is so peculiarly distinguished. Resting not on visionary theory, but on the sure basis of investigation and experiment, it has risen to a degree of certainty and pre-eminence, of which it was supposed incapable. It is by pursuing the same method, in regard to political disquisitions, by analysing the real state of mankind, and examining, with anatomical accuracy and minuteness, *the internal structure of society,* that the science of government can alone be brought to the same height of perfections.[120]

116 Society of Antiquaries of Scotland, Letters Books (1781–2), Letter from John Knox to Buchan, 2 August 1781. 117 *Critical Review,* 5 (1792), 405.

118 J. Sinclair, *Analysis of the statistical account of Scotland* (London, 1826), 64–9.

119 J. Steuart, *Enquiry into the principles of oeconomy* (Edinburgh, 1767), 127.

120 D. Withrington and I. Grant (eds.), *The statistical account of Scotland* (Edinburgh, 1983), Vol. I, xv. This statement is part of a French advertisement for the project, widely distributed throughout Europe in 1790 and 1791.

In those terms, his intentions echo Sibbald's utilitarian views on the medical-topography of Scotland, find a parallel expression in work on human physiology and anatomy in the Scottish Enlightenment and mirror that commonly-understood contemporary distinction, voiced by David Hume and others, between the anatomist and the painter.[121] Via survey, the political and natural body of Scotland was to be dissected and inductively reconstituted, parish by parish.

Others have detailed the nature and chronology of Sinclair's enterprise.[122] In summary, a set of 160 questions was asked in four sections: 1–40 on the geography and topography of the parish, climate, natural resources and natural history; 41–110 on population and related matters; 111–16 on parishes' 'agricultural and industrial production'; and a final section on miscellaneous matters. A further six questions were asked as an appendix to his circular of May 1790 and five more in a circular of May 1791. Sinclair's letters to recalcitrant ministers get shorter and more intemperate: by June 1796, he was resolving 'to send *Statistical Missionaries*' to different parts of the country to hurry ministerial replies. Yet, by 3 June 1799, the project was complete, and he laid before the General Assembly 'a unique survey of the state of the whole country, locality by locality'. Scotland's ministers represented local credible informants. This was knowledge gained from people 'in the field' as it were. What he represented in his 1825 summary as 'the Pyramid of Statistical Enquiry' (Figure 4.5) symbolised what he had intended in the advertisement giving notice of the project over thirty-five years before: 'the completest survey of a kingdom, of which we have any knowledge'.[123]

Sinclair understood 'statistics' not just as knowledge of the state (the sense in which it was then understood and practised in Germany, which he had toured extensively in 1786–87), but, rather, as 'an inquiry into the state of a country, *for the purpose of ascertaining the quantum of happiness enjoyed by its inhabitants, and the means of its future improvement*'.[124] The *Statistical account* was widely and well received. The political statistician and zoo-geographer, Eberhardt Zimmerman, author of *A political survey of the present state of Europe* (1787) – which influenced Sinclair from the outset – included notice of the *Account* in his geographical and statistical journal. Geographical authors in St Petersburg and in Lisbon, natural historians and political economists across Europe, and John Adams, then president of the United States of America, praised the work highly as, variously, a statistical enquiry, a work of geography and a means to political knowledge.

[121] C. Lawrence, 'The nervous system and society in the Scottish enlightenment', in B. Barnes and S. Shapin (eds.), *Natural order–historical studies of scientific culture* (London, 1979), 19–40; 'Ornate physicians and learned artisans: Edinburgh medical men, 1726–1776', in W. Bynum and R. Porter (eds.), *William Hunter and the eighteenth-century medical world* (London, 1985), 153–76.

[122] Sinclair does so, in self-laudatory tones, in his *Analysis of the statistical account*; see also Withrington and Grant (eds.), *The statistical account.*

[123] Sinclair, *Analysis of the statistical account*, iv.

[124] Sinclair, *Analysis of the statistical account*, 63–4.

THE

PYRAMID

OF

STATISTICAL INQUIRY.

4.5 'The Pyramid of Statistical Enquiry', from Sir John Sinclair's *Analysis of the Statistical Account of Scotland* (1825). Note the iconic thistle to the left of the pyramid.)

This was, I submit, geographical knowledge of a very different sort from those persons teaching earth description in Edinburgh's Mealmarket Close, navigation in Portsoy or geography's connection with civil history in Marischal College. The difference rests in contemporaries' understanding of Sinclair's survey (and of Buchan's failed attempts) as a *scientific* nationally-focused project. Yet, if the methods were different, the intended end was the same. As Sinclair noted, 'The plan of that work is considered to be the best hitherto adopted, for acquiring a full and accurate knowledge of the internal state of a country: and fortunately, it combines, with the pleasures of novelty and variety, the solid satisfaction of authentic information, and public utility'. These benefits stemmed from the inductive and empirical method employed.

For Sinclair, 'Founding systems of political economy, on minute and extensive investigations of local facts, is following the example of Bacon, who tested the basis of natural philosophy, on minute inquiries, accurate experiments, and inferences deduced from them. No system can be better entitled to adoption'.[125] It is this methodological practice, as well as its political utility, that gave this form of geographical knowledge authority, gave it – in terms I have outlined above – epistemic credibility *and* a particular social space.

Calling such geographical work 'natural history', as so many did, was not accidental. In their attention to place, to the facts of nature, to the authority of observation and to survey as a means of natural knowledge, the discursive practices of geography and natural history were closely connected.[126] What natural historians like the Rev. Dr John Walker, Professor of Natural History at Edinburgh from 1779–1803, did in his geographical natural histories of Scotland was repeated by Banks and Solander in the Pacific and in northern Europe by Linnaeus and Pallas.[127] In Scotland, men like James Robertson, John Lightfoot, author in 1777 of *Flora Scotica,* (a regional geography of Scotland's plants organised on Linnaean principles), and Thomas Pennant with his use of distributed 'local queries' were all geographical historians of what Linnaeus termed the 'economy of Nature'.[128] Their enquiries depended upon fieldwork, classification, and documenting via circulated queries in order to produce a taxonomy of national natural knowledge. It is for those reasons that Buchan thought of his survey as a 'natural history', that we should see Sinclair's emphasis upon 'statistical philosophy' as a matter of natural knowledge and that we should understand the plaudits of contemporaries like Zimmerman. Knowing Scotland geographically involved direct encounters with Scotland's landscape; as Lightfoot put it in his *Flora Scotica,* 'scaling the highest mountains, climbing the most rugged rocks, penetrating the thickest woods, treading the fallacious bogs, winding upon the shores of seas and lakes'.[129] It also involved the representation of that landscape.

[125] Sinclair, *Analysis of the statistical account, 60.*

[126] W. Ashworth, 'Natural history and the emblematic world view', in D. Lindberg and R. Westman (eds.), *Reappraisal of the scientific revolution* (Cambridge, 1990), 303–32; J. Larson, 'Not without a plan: geography and natural history in the late eighteenth century', *Journal of the History of Biology,* 19 (1986), 447–88; P. Sloan, 'Natural history, 1670–1802', in R. Olby, G. Cantor, J. Christie and M. Hodge (eds.), *Companion to the history of modern science* (London, 1990), 304–19; 'The gaze of natural history', in C. Fox, R. Porter and R. Wokler (eds.), *Inventing human science: eighteenth-century domains* (Berkeley, 1995), 112–51; D. Outram, 'New spaces in natural history', in N. Jardine, J. Secord and E. Spary (eds.), *Cultures of natural history* (Cambridge, 1996), 249–65; C. Withers, 'Geography, natural history and the eighteenth-century enlightenment: putting the world in place', *History Workshop Journal,* 39 (1995), 136–63; R. Porter, 'Measure of ideas, role of language: mathematics and language in the eighteenth century', in T. Frangsmyr, J. Heilbron and R. Rider (eds.), *The quantifying spirit in the eighteenth century* (Berkeley, 1990), 113–40.

[127] L. Koerner, 'Linnaeus' floral transplants', *Representations,* 47 (1994), 144–69.

[128] Withers, 'Geography, natural history and the eighteenth-century enlightenment'; D. Henderson and J. Dickson, *A naturalist in the Highlands. James Robertson, his life and travels, 1767–1771* (Edinburgh, 1994); D. Miller and P. Reill (eds.), *Visions of empire: voyages, botany, and representations of nature* (Cambridge, 1996).

[129] J. Lightfoot, *Flora Scotica: or, a systematic arrangement, in the Linnaean method of the native plants of Scotland and the hebrides* (London, 1777), 21.

*Portraying Scotland's landscape: mapping, the Military Survey and landscape painting, 1747–1830*

That the process of mapping is also a matter of survey, dependent upon ideas of measurement and the power of vision, was certainly apparent to George Mark, a schoolmaster and surveyor (he taught at the mathematics school in Dunbar 1737–40 and was a correspondent of Colin Maclaurin, Professor of Mathematics at Edinburgh, who himself taught geography to his mathematics students in 1741), in his 1728 'Proposal for publishing, by subscription, An Accurate Map or Geometrical Survey of Lothian, Tweddale and Clydesdale.' Mark had earlier produced a map of the Roman walls and antiquities in northern England and Scotland: it was this 'which first gave him Occasion to take Notice of the horrid and unaccountable Blunders and Mistakes committed in the Maps of his own Country'. For Mark, all extant map representations of Scotland were flawed, a fact which reflected badly, he argued, upon the leading social groups in Scotland. His work had found

> Rivers sometimes made to run over the Tops of the highest Mountains, and on the wrong side of Towns and Villages, and sometimes Rivers are represented in Places where there are none, and as often omitted where they should be represented: And in a Word, they are no more than a Parcel of crooked wavey Lines drawn at random, and not only blundered in their several Turnings and Windings, but even in their general Tendencies and Bearings. And for the Distances and Positions of Places, there are scarcely to [be] found Three agreeable to Truth: Those Villages which are not in Fact above half a Mile distant one from another, are often set at more than two or three Miles Distance; and in one Word, the whole is no more than one continued Blunder. 'Tis truely strange why our Scotish [*sic*] Nobility and Gentry, who are so universally esteemed for their Learning, Curiosity and Affection for their Country, should suffer an Omission of this Nature! Can they, whom their Neighbours as it were by way of Distinction call National, contentedly see themselves outdone by those Neighbours, in what so much concerns the Honour of their Nation?[130]

Mark's rhetoric is typical of many geographers and others seeking personal advancement and patronage. But it is illustrative, too, of a more important general point.

Despite earlier efforts by Pont and Adair and others, what we might think of as the 'interior geography' of the nation was still not known in detail by about 1730, and, as we have seen, the 'exterior shape' was certainly not agreed upon (see Figure 1.1). Such circumstances prompted some Scots to undertake local map-making surveys in order to give shape to their nation: the survey of north-east Scotland by the Rev. Alexander Bryce of Thurso in 1741–2, made at the behest of the Philosophical Society in Edinburgh in order to correct the

[130] G. Mark, *Proposals For publishing by subscription, an accurate map or geometrical survey of the shires, of Lothian, Tweddale* [sic], *and Clydsdale* (Edinburgh, 1728). (This is a rare two-page pamphlet: I have consulted that copy held by the NLS.)

Gordon–Blaeu outlines, and the survey of the Orkneys in the 1740s by Murdoch Mackenzie, who later (1776) provided the first accurate outlines of the Western Isles, are the most significant here.[131] Estate maps from the 1740s likewise illustrate the ways in which the land was, cartographically, 'put to order' in Enlightenment Scotland.[132] The first national map survey of Scotland was produced, however, as a political instrument.

The Military Survey of Scotland was begun in 1747, under Colonel William Roy, in order to provide a picture of northern Scotland for military purposes following the 1745–6 Jacobite rebellion. The Duke of Cumberland understood the importance of knowing the lie of the land in stressing to King George the 'Inconvenience and Want of such a Survey'.[133] Roy reflected in 1785 upon the 'infinite importance' it was to the State of having unknown Britain brought to order 'by establishing military posts in its inmost recesses, and carrying roads of communication to its remotest parts'.[134] One modern authority has even claimed that 'but for the Young Pretender [Prince Charles Edward Stuart], we would have had no survey of Scotland – the first comprehensive, large scale national survey . . . in the British Isles'.[135]

The essential features of the Military Survey are well understood: the use of a standard scale and agreed practices of survey amongst organised teams; the attention to northern Scotland before turning, from 1751, to southern Scotland; the emphasis upon speed which, in turn, determined omission of detail but attention to roads, bridges and the houses of the 'great families'; differences between the 'Protracted' and the 'Fair' copies; the relative lack of attention paid to it by contemporaries given war with the French from 1754; and the prompt it later gave to the foundation of the Ordnance Survey. In these terms, the 'Roy Map' was, for its originator, 'rather . . . a magnificent military sketch, than a very accurate map of a country'. For modern scholars, it represents 'one of the most intriguing and at the same time infuriating documents available to the researcher into Scotland's past landscapes'.[136]

The Military Survey is not the product of institutionalised 'native'

[131] Moir, *Early maps of Scotland*, 86; D. Smith, 'The progress of the *Orcades* survey, with biographical notes on Murdoch Mackenzie senior (1712–1797)', *Annals of Science*, 44 (1987), 277–88.

[132] I. Adams (ed.), *Descriptive list of plans in the Scottish Record Office* (four volumes, Edinburgh, 1966–88).

[133] SRO, RH1/2/511.

[134] W. Roy, 'An account of the measurement of a base on Hounslow Heath', *Philosophical Transactions*, 75 (1785), 385–6.

[135] Y. Hodson, 'The Highland survey 1741–1755 and the Scottish school of cartography' (mimeographed conference paper given to 'Opening up the Highlands: Highland history and archives' Conference of the Scottish Records Association, 17 November 1991, Edinburgh), 2. See also Y. O'Donoghue, *William Roy 1726–1790: pioneer of the Ordnance Survey* (London, 1977); R. Skelton, 'The Military Survey of Scotland 1747–1755', *Scottish Geographical Magazine*, 83 (1977), 5–16; G. Whittington and A. Gibson, *The Military Survey of Scotland 1747–1755: a critique* (Lancaster 1986). Roy was also concerned to understand the *antiquarian* record in the landscape: see G. Macdonald, 'General William Roy and his "Military Antiquities of the Romans in North Britain"', *Archaeologia*, 68 (1917), 161–228; R. Gardiner, 'William Roy, surveyor and antiquary', *Geographical Journal*, 143 (1977), 439–50.

[136] Roy, 'An account of the measurement', 385; Whittington and Gibson, *The Military Survey of Scotland*, 61.

geographical knowledge. I do not mean the principal surveyors – many were Scots. I mean by this that, despite there being private teachers of geography, geography and surveying classes in Marischal College and bodies in Edinburgh, like the Philosophical Society, which promoted local surveys, there was no single institution that could be called upon to undertake such geographical work other than the state. Further, its accuracy depended, in an important sense, upon *not* being in the field. The results of fieldwork were transferred in the winter months to the Ordnance Room in Edinburgh Castle and to London for inspection. Shaping the nation depended upon measurement in the field but also upon its calibration elsewhere, just as Banks and others managed the *field-based collection* of geographical and natural historical facts and the *metropolitan calculation* of their significance.[137] Moreover, the Military Survey, in contrast with other forms of survey, was not simply a means to statistical anatomy, but rather, in the work of Paul Sandby, its chief draughtsman, was one expression of Scotland's 'portraiture'.

Sandby's representation of the survey as a matter of landscape aesthetics (Figure 4.6) reflects not just his own engagement with the portrayal of Scotland – he produced several topographic scenes between 1746 and 1752 – but a developing sensitivity to landscape and to upland scenery by Scottish and British painters in the later eighteenth century.[138] The 'taste for mountains' in art both directed and reflected a shift in attitudes towards nature that moved away from the idea of barren landscapes as moral reproof and towards a cultural awareness about 'the sublime', the delights of rugged nature and the supposed 'primitivism' of inhabitants of mountainous regions.[139] Cultural attitudes to nature changed: Sandby's portrayal of Strath Tay in 1747, for example, was, by 1780, considered 'not sufficiently Highland' and was amended by making the landscape appear more dramatic, in line with contemporary fashion.[140] The Military Survey itself did not prompt such shifts in cultural appreciation. But the survey should be seen as a key moment in the mapped and painterly visualisation of Scotland's landscape. Sandby's work on behalf of the survey helped make visual those connections between landscape, poetic sensitivity and historical association that were later apparent in the work of Scottish landscape painters such as James Norie, Jacob More,

[137] D. Miller, 'Joseph Banks, empire, and "centers of calculation" in late Hanoverian London', in Miller and Reill (eds.), *Visions of empire*, 21–37; 'The usefulness of natural philosophy: the Royal Society and the culture of practical utility in the later eighteenth century', *British Journal for the History of Science*, 32 (1999), 185–201; J. Gascoigne, 'The Royal Society and the emergence of science as an instrument of state policy', *British Journal for the History of Science*, 32 (1999), 171–84.

[138] J. Christian, 'Paul Sandby and the Military Survey of Scotland', in N. Alfrey and S. Daniels (eds.), *Mapping the landscape: essays on art and cartography* (Nottingham, 1990), 18–22; J. Holloway and L. Errington, *The discovery of Scotland: the appreciation of Scottish scenery through two centuries of painting* (Edinburgh, 1978); P. Howard, *Landscapes – the artists' vision* (London, 1991).

[139] H. Andrews, *The search for the picturesque: landscape aesthetics and tourism in Britain, 1760–1800* (Aldershot, 1989).

[140] Holloway and Errington, *The discovery of Scotland*, 37.

4.6 'A surveying party at work in the Highlands', by Paul Sandby, *c.*1748.

Alexander Runciman and Alexander Nasmyth, who portrayed the Scottish landscape in what Howard has called the 'Romantic Period' (1791–1830).[141]

There is not the space here to document the chronology of Scottish landscape painting, but several points may be made in support of my contentions. Jacob More's paintings of the three Falls of Clyde (1771–3) represent what MacMillan terms not just 'a natural wonder on one of the great rivers of Scotland' – in those terms alone a *geographical feature* to be marvelled at – but also its place 'as a kind of natural national monument', given associations with William Wallace, earlier paintings of the falls by Sandby and the poetry of James Thomson and others.[142] The connections between upland landscape, national identity and the discovery of 'wilderness Britain' in Highland Scotland – just that region whose 'inmost recesses' needed bringing to light and order through the power of the map – were particularly apparent in the Ossianic cult of the late eighteenth century. In 1760, James Macpherson published a work entitled *Fragments of ancient poetry, collected in the Highlands of Scotland, and translated from the Galic or Erse Language.* In 1762, he published *Fingal, an ancient epic poem in six books,* and, in 1763, *Temora, an ancient epic poem in eight books.* Macpherson claimed these texts as, in translation, the works of Ossian, an ancient Caledonian bard. In context with those other trends making the Highlands culturally fashionable – landscape painting, literary representations, the beginnings of scenic tourism – the Europe-wide impact of Ossian reinforced that part of Scotland as the desolate refuge of a primitive people in a sublime landscape and helped establish new 'ways of seeing' the Scottish landscape.[143] As one commentator on 'Ossianic geographies' has shown, the myth is, geographically, both everywhere and nowhere in the Highlands: the region is saturated with 'Fingalian sites', yet one cannot precisely locate the tradition in any place.[144]

## Conclusion

This chapter has shown that the weak public interest in geography apparent in Scotland by 1707 had developed by the 1760s, in the form of public classes in geography at least, and in terms of the number of such classes was stronger by the 1790s and stronger still by the 1820s. The production within Scotland of books of geography that effectively began with Drummond's *Treatise* was not further realised, however, until Mair (1762), and was not strongly evident thereafter. Whilst we may speculate that the timing, in 1708, of Smart's class and Drummond's text reflects a heightened geographical

[141] Howard, *Landscapes – the artists' vision*, 56–79.

[142] D. Macmillan, *Painting in Scotland: the golden age* (Oxford, 1986), 139.

[143] C. Withers, 'The historical creation of the Scottish Highlands', in I. Donnachie and C. Whatley (eds.), *The manufacture of Scottish history* (Edinburgh, 1992), 143–56.

[144] P. Baines, 'Ossianic geographies: Fingalian figures on the Scottish tour, 1760–1830', *Scotlands,* 4 (1997), 44–61.

consciousness in the wake of the Act of Union, such an explanation does not satisfactorily account for the relative lack of public classes until mid-century.

The geography of public interest in geography in Scotland was, principally, centred upon Edinburgh and Glasgow. In the former, geography's public teachers lived and delivered their classes within the 'Old Town'. Only from the second decade of the nineteenth century do 'New Town' addresses appear. Such evidence must be treated with some caution, however, since classes were given within teachers' homes, teachers moved to their audiences and other particular sites – coffeehouses, theatres and observatories – provided spaces for the making of geography's public audiences. If, for our purposes, the historical geography of public interest in geography in Enlightenment Scotland should be seen as a largely metropolitan and chiefly late eighteenth-century affair, the actual making of geography as a form of polite public discourse amongst enlightening Scots was, for some, a matter of using private domestic space for instrumental display, reading reviews of geography books within the periodical literature, or, for Professor Monro and others, instructing their children. Bruckner's argument that maps and geography texts in America at this time helped promote a national 'geo-literacy', as regional identities were replaced by a sense of nationhood, does not have a parallel in Scotland.[145] But eighteenth-century Scots did use geography to improve themselves, to attain an education and to know about world affairs. As the example of the Scottish linen map sampler from *c.*1800 suggests, doing geography was not just a textual exercise.[146]

Outside the home and other sites of the urban public sphere, geography was taught in a number of burgh schools and, within the universities, was a major element in the curriculum in King's and Marischal Colleges in Aberdeen from about 1748. In geography's teaching places, the subject was taught with an emphasis on practical utility and in association with geometry, astronomy and civil history. Globes were used to teach mathematical measurement, the historical evolution of civil society and the current political divisions of the world. If the evidence of James Barclay's 1743 *Treatise* may be more widely applied and inferences drawn from the many instances of globes being purchased within schools, globes and maps were more important than books as instruments for geographical understanding. Not until the early nineteenth century – in Meme's classes in Ayr and in Millar's classes in Glasgow University – was geography taught through local fieldwork or, directly, in association with contemporary theoretical speculation about the Earth's physical processes.

Wherever situated, geography's audiences, public teachers and authors were overwhelmingly male. This is not to say that geography was not deemed

[145] M. Bruckner, 'Lessons in geography: maps, spellers, and other grammars of nationalism in the early republic', *American Quarterly*, 51 (1999), 311–43.

[146] On this linen map sampler, see *The Map Collector*, 56 (1991), 1.

suitable for women: both Monro and his daughter thought it so, and John Pinkerton was commenting upon a growing trend in his comments of 1815. This is to recognise that what was considered useful in one context – for men, geography as a source of and a means to practical reason about the shape and nature of the Earth – was, in another and for women especially, a means to informed understanding of contemporary political affairs.

Geographical knowledge about Scotland was also being made through survey (and failing to be made, since part of the story I have recovered here is of failure): both in the sense of inductive enquiries at the parochial level designed to understand the 'anatomy' of Scottish society and in the work of those who sought to represent, in painting and in maps, 'the face' of the land. What connected the two was 'vision': visualising as a means to portray the topography of the present, and visualising about Scotland's future geographical condition. Maclaurin's never-realised vision for an observatory in Edinburgh was for that site to be a centre of geographical vision. Such knowledge was institutionally sited – in the Society of Antiquaries, for example, with its membership, in several senses, of 'surveyors' – and, notably, in the Church of Scotland in its partially-successful attempts at a 'Description of Scotland' in the 1720s, and in the activities of its ministers in constituting Sinclair's 'statistical philosophy' in the 1790s. Only from the Military Survey of 1747 can we think of geographical survey in map form as a national state enterprise.

Given such evidence, geographical knowledge about and knowledge of geography within Scotland should, I argue, now figure much more prominently in understanding the nature of Enlightenment knowledge. Since it has also proved possible to identify the sites and spaces in which the ideas of reason and utility were situated and to know something of the movement over space of ideas and of people, it may also be possible, if only in preliminary ways, to begin to address questions about the geographies *of* the Enlightenment and *of* the public sphere. Robertson has noted how 'a great deal of scholarly achievement over the past twenty years has . . . been stimulated by an insistence upon the national particularity of the Scottish Enlightenment'. As he further notes, such attention reflects 'a tendency to treat national Enlightenments in isolation, losing sight of the extent to which the Scots' intellectual interests, their concepts, methods and subjects of enquiry, were commonplace to contemporaries across Europe'.[147] Geography's making as a form of Enlightenment knowledge in the ways I have here shown for Scotland clearly had parallels elsewhere. In those terms, the Enlightenment might be seen as an *international* matter and our task should be to understand better its *geographical diversity*, not to essentialise national distinctions. It is also clear, as Mayhew has shown of William Guthrie's

[147] J. Robertson, 'The Enlightenment above national context: political economy in eighteenth-century Scotland and Naples', *The Historical Journal*, 40 (1997), 670.

*Geographical Grammar* (1770), that geographical works were drawn upon by other subjects as, in turn, geography incorporated the stadial theories of the conjectural historians.[148]

Such a geographical conception of Enlightenment is not at all incompatible with thinking of the Enlightenment as a matter of *local knowledges*: in terms of differences between towns or, even, between different institutions. The earlier view that the Enlightenment in Scotland was the Enlightenment in Edinburgh has been replaced by recognition of differences between, for example, the commercially-minded nature of the Enlightenment in Glasgow, which looked to America for its markets and to the mercantile classes for its audiences, and Aberdeen, with its European outlook and interest in the Common Sense philosophy of Thomas Reid and his circle.[149] It is clear, too, that the geography of Enlightenment was about relationships between towns and their local areas, and, in general terms, about the philosophical expression of ideas in polite urban settings and their practical implementation in rural ones. James Hutton, for example, sought in his (unpublished) 'Elements of Agriculture', 'to make philosophers of husbandmen and husbandmen of philosophers'.[150] It is not unreasonable to suppose that those men constituting the audiences of Robert Darling's public geography classes, or Reid's class in 1753, or Millar's in Glasgow in 1803 took back to their estates and studies ideas of civic advance, national utility and social propriety. Enlightenment theories were, in this sense, practically situated and, literally, 'grounded' in Scotland's soil.

Finally, it is possible to conceive of *individual geographies* of Enlightenment. I mean not just that individual institutions clearly had a geography, but also that Enlightenment knowledge was made, disseminated, debated and engaged with by a public made up of private individuals. These individuals' lives came together at certain times, and in *particular spaces*: philosophers to debate and dine, students to classrooms, audiences to listen. Others have charted William Cullen's 'personal geography' in these terms, as he engaged in the medical world of the Enlightenment and moved, as he did so, within the local spaces bound by his home off Edinburgh's Cowgate, the University, the Infirmary and Nicholson's Tavern.[151] A more complete under-

[148] R. Mayhew, 'William Guthrie's *Geographical Grammar*, the Scottish enlightenment and the politics of British geography', *Scottish Geographical Journal*, 115 (1999), 19–34.

[149] The essentialist place-specific claim that the Enlightenment in Scotland was the Enlightenment in Edinburgh is particularly associated with Nicholas Phillipson: see N. Phillipson, 'The Scottish Enlightenment', in Porter and Teich, *The Enlightenment in national context*, 19–40; 'Culture and society in the eighteenth-century province: the case of Edinburgh and the Scottish Enlightenment', in L. Stone (ed.), *The university in society* (Princeton, 1974), Vol. II, 407–20. Partial counters are offered by the essays in Carter and Pittock (eds.), *Aberdeen and the enlightenment*, and in A. Hook and R. Sher (eds.), *The Glasgow enlightenment* (East Linton, 1997).

[150] J. Jones, 'James Hutton's agricultural research and his life as a farmer', *Annals of Science*, 42 (1985), 573–601; Withers, 'On georgics and geology'.

[151] A. Doig, J. Ferguson, I. Milne and R. Passmore (eds.), *William Cullen and the eighteenth-century medical world* (Edinburgh, 1993), show (pp. 62–3) a map of mid-eighteenth-century Edinburgh with the locations charted of the sites of Cullen's professional and social activities.

standing of the historical geography of Enlightenment and of the public sphere might depend, then, upon our being able to site the places in which knowledge was made and contested and upon recovering the connections between them.[152] The following chapter considers these issues in exploring geography's role as a subject of civic endeavour after 1830.

[152] I have discussed these ideas further in C. Withers, 'Toward a historical geography of enlightenment in Scotland', in P. Wood (ed.), *The Scottish enlightenment: essays in re-interpretation* (Rochester, 2000), 63–97.

# 5

# National identity, geographical knowledge and civic enterprise, *c.* 1830–1884

In 1834, Roderick Murchison, the leading geologist, co-founder in 1830 of the Royal Geographical Society and, from 1854, its president, noted that Scotland 'stood almost alone in Europe as a Kingdom without a map', having earlier stated that Scotland 'was in a disgraceful condition in respect to geography'.[1] Murchison's claims should not be taken literally. They were directed at the lack of an accurate geological map of Scotland and at the slow progress of the Ordnance Survey's mapping of Scotland – its 'trigonometrical survey' – begun in 1809.[2] What is of interest is that Murchison, together with others such as the Royal Society of Edinburgh and the provost and council of the City of Edinburgh, could speak to a matter of national geographical importance from a position of institutional authority. Robert Jameson, for example, Professor of Natural History at the University of Edinburgh, represented the views of the Wernerian Natural History Society (of which he was president) in its memorial to the government, in May 1837, about the delayed state of the survey's mapping of Scotland:

> Your Memorialists do not consider it necessary to enter into any detailed observations on this occasion on the numerous and important advantages which must result to navigation, commerce and agriculture, or the scientific interest which would arise from the completion of the trigonometrical survey of Scotland, . . . they have the fullest confidence in the desire of His Majesty's Government to extend the benefits of accurate geographical knowledge to all parts of His Majesty's dominions.[3]

Such evidence represents in one form the institutionalisation of scientific knowledge – a distinguishing feature of the social authority of science and of

[1] British Parliamentary Papers [BPP], 1854, XLI, 76; 'Trigonometrical Survey, Great Britain', BPP, 1837, XLVII, 89.

[2] R. Boud, 'The Highland and Agricultural Society of Scotland and the Ordnance Survey of Scotland, 1837–1875', *The Cartographic Journal,* 23 (1986), 3–26.

[3] 'Trigonometrical Survey, Great Britain' [Copies of the memorials addressed to the Government], BPP, 1837, Vol. XLVII, 2.

specialisation in science in this period.[4] It also illustrates in a particular sense what I consider here as the recognition by contemporaries of geographical knowledge as a form of civic enterprise, and of the institutionalised expression of science as part of civil society in Scotland.

Several studies of national identity in nineteenth-century Scotland have examined the country's 'civic nationalism', evident in its distinctive church, legal and educational systems, and in Scotland's role in the making of the British empire.[5] Morton has emphasised the importance of Scottish civil society, understood as that institutionalised social structure existing between the household and the state, and has suggested that national identity 'maintained through the institutions and the civic culture of civil society' was different from 'the unthinking patriotism of the British state'.[6] In those terms, my concern here is more to understand the place and nature of geographical knowledge in Scottish civil society than it is to chart the connections between Scottish national identity, imperialism and the making of Britishness. These are discussed in the following chapter.

I will examine three themes. First, I will consider geography's place in the public sphere from *c.*1830 to 1860, thus extending the earlier exploration of the public engagement with geography. The accompanying examination of the *Edinburgh Journal of Natural and Geographical Science* demonstrates differences – in social class and in intellectual purpose – within geography's public audiences by about 1830, and allows us to consider the nature of the connections between geography's texts and audiences. Discussing further the public place of geography allows us to know *how* geographical understanding was used to deal with the problems facing Scottish civil society by the

[4] S. Cannon, *Science in culture: the early Victorian period* (New York, 1978) and the discussion of what she calls 'Humboldtian science' in the early nineteenth century. See also R. Home, 'Humboldtian science revisited: an Australian case study', *History of Science,* 33 (1995), 1–22; M. Dettelbach, 'Humboldtian science', in N. Jardine, J. Secord and E. Spary (eds.), *Cultures of natural history* (Cambridge, 1996), 287–304; T. Heyck, *The transformation of intellectual life in Victorian England* (London, 1982).

[5] H. Hanham, 'Mid-nineteenth century Scottish nationalism: romantic and radical', in R. Robson (ed.), *Ideas and Institutions of Victorian Britain* (London, 1967), 143–79; Rosalind Mitchison, 'Nineteenth century Scottish nationalism: the cultural background', in R. Mitchison (ed.), *The roots of nationalism: studies in northern Europe* (Edinburgh, 1980), 131–42; R. Morris and G. Morton, 'Where was nineteenth-century Scotland?', *Scottish Historical Review,* 73 (1994), 89–99; G. Morton, 'What if . . .? the significance of Scotland's "missing" nationalism in the nineteenth century', in D. Broun, R. Finlay and M. Lynch (eds.), *Image and identity: the making and remaking of Scotland through the ages* (Edinburgh, 1998), 157–76; *Unionist-nationalism: the governing of urban Scotland, 1830–1860* (East Linton, 1998); 'Civil society, municipal government and the state: enshrinement, empowerment and legitimacy – Scotland, 1800–1929', *Urban history,* 25 (1998), 348–67. On imperialism and national identity, see D. Forsyth, 'Empire and union: imperial and national identity in nineteenth century Scotland', *Scottish Geographical Magazine,* 113 (1997), 6–12; R. Finlay, 'Controlling the past: Scottish historiography and Scottish identity in the 19th and 20th centuries', *Scottish Affairs,* 9 (1994), 127–43; 'The rise and fall of popular imperialism in Scotland, 1850–1950', *Scottish Geographical Magazine,* 113 (1997), 13–21; 'Imperial Scotland: the British empire and Scottish national identity, c.1850–1914', in J. Mackenzie (ed.), *Scotland and the British empire: studies in imperialism* (Manchester, forthcoming); J. Mackenzie, 'Essay and reflection: Scotland and the British empire', *International History Review,* 4 (1993), 714–39.

[6] Morton, 'What if . . . ?', 169.

1840s.[7] In that regard, I briefly discuss the attempt made by the Rev. Thomas Chalmers to ameliorate urban deprivation through attention to 'the principle of locality', by which he meant dividing cities into districts small enough for social reformers to know an area's problems. Chalmers also included geography in the curriculum of the voluntary school he established in Edinburgh's Cowgate. For Chalmers and others, geographical knowledge was both a means for *and* an end in the moral improvement of civil society. Second, I examine the returns in 1841 to a parliamentary enquiry into the state of education in Scotland. From such evidence, it is possible to construct a historical geography of geographical education in Scotland for the late 1830s, and to know something about the texts used to promote geography then and in later periods. Third, I explore the situated nature of scientific knowledge in Scotland before about 1880 by considering the historical geography of local scientific bodies, particularly of natural history societies, and the ways in which they promoted local scientific knowledges about different parts of Scotland.

What connects these themes is a concern to understand further the sites of geography's making and reception, the different purposes to which geographical knowledge was put, and the different scales at which we can begin to think of the historical geography of science, including geography. In arguing that these matters are questions to do with geography as a form of civic enterprise, this chapter is also concerned to place geography and science within those debates on the social constitution of national identity in nineteenth-century Scotland from which they have hitherto been almost entirely absent.

## Geography and the public sphere, *c.*1830–1860: polite learning and civic improvement

### *Public lecturing in geography after 1830: a matter of social class?*

It is possible to identify thirteen persons giving public lectures in geography in Scotland between 1830 and 1860: six persons in Edinburgh, six in Glasgow and one in the small town of Newton Stewart.[8] Several individuals had given classes before 1830: men like Adam White, for example, who taught history, geography and astronomy in Edinburgh between 1812 and 1850 at 79 South Bridge, then at 10 Nicholson Street and finally at 3 St David's Street. Some, like Alexander Ewing who taught geography and astronomy from 1837–1861 at 47 George Street in Edinburgh, continued a tradition begun by his father's 40-lecture course in geography held at various times between 1756 and 1790

[7] These included rapid urban growth; the failure of the educational system to cope with urban expansion and factory labour; squalid sanitary management and, not least, the 1843 schism in the Church of Scotland: Mitchison, 'Nineteenth century Scottish nationalism', 133–34.

[8] C. Withers, 'Towards a history of geography in the public sphere', *History of Science,* 37 (1999), 45–78.

(see p. 124). As for earlier periods, geography in public classes after 1830 was chiefly a metropolitan affair and was commonly taught with astronomy and mathematics. Four of the thirteen lecturers intimated use of the globes and two specified 'geography for ladies'.

The number of classes in geography after 1830 was fewer than the 'high-point' of such public engagement between about 1808 and 1820 (Figure 4.4c). It is difficult to explain this. It is possible that those local institutions for the study of natural history, geology and botany considered here offered an outlet for interest in the natural world in ways that public classes in geography did not. As several Scottish newspapers for this period show, many towns had visiting lecturers in geology, astronomy, or others presenting their work as missionaries or explorers. Bodies such as the Society of Arts held lectures in its 1836–7 session on Scotland's mapping, for example.[9]

What is apparent for this period is that not all public classes in geography were aimed at the bourgeois public sphere. David Mackie, for example, taught geography and the globes at five different addresses in Glasgow between 1820 and 1835, including 9 St George's Place, the Mechanics' Institution Hall. In 1835, John Gullan gave a course of twenty-five lectures on geography, geology and astronomy to the Gorbals Popular Institution for the Diffusion of Science, which had opened in 1833 to provide popular and useful lectures and a library for the labouring classes.[10]

The Mechanics' Institute movement began in Scotland with the establishment of the Edinburgh School of Art (1821) and the Glasgow Mechanics' Institution (1823).[11] The movement was not just a metropolitan enterprise. In Dunbar, the third session of the Dunbar Mechanics' Institution began in October 1827 with the opening of a class on elementary geometry and algebra, 'and another for Geography with the use of the globes'. The latter was more popular: 'Owing to an erroneous dread of the insurmountable difficulties to be encountered, the Mathematical Class, we regret to say, mustered very poorly. That in Geography was making excellent progress'.[12] In 1836, the Brechin Mechanics' Institution, begun the year before, established a library in which geography books were numerous, together with works on history, statistics and science.[13] In contrast, the directors of the Edinburgh School of Art made it plain that they would 'confine themselves to subjects *which will be*

[9] Edinburgh University Library Special Collections, MS Gen 422D [Society of Arts Programme, Session 1836–37], Item No. 5. Notice is given in *The Stirling Journal and Advertiser* (for 5 December 1873) of a 'Geographical Society' in Stirling but no other records exist of this body.

[10] Withers, 'Towards a history of geography in the public sphere', 56.

[11] The establishment in Scotland of Mechanics' Institutes has been dated to the early 1820s by W. Marwick, 'Mechanics' Institutes in Scotland', *Journal of Adult Education,* 6 (1932–34), 292–309. A contrary opinion – arguing for the early nineteenth century – has been advanced by J. Jessop, *Education in Angus* (London, 1931), 296–7. The evidence favours Marwick's view: see also W. Marwick, 'Early adult education in Edinburgh', *Journal of Adult Education,* 5 (1930–2), 389–404.

[12] *Third Report of the Dunbar Mechanics' Institution* (Haddington, 1828), 4.

[13] Marwick, 'Mechanics' Institutes in Scotland', 301.

*directly useful to mechanics in the exercise of their trade'* [original emphasis] and, accordingly, declined offers of lectures on geography and astronomy (though from whom it is not known). In the Edinburgh Philosophical Institution, however, geography was included in their evening classes established upon foundation in 1846.[14]

In Glasgow, from the mid-1820s, several men actively promoted geography as useful popular knowledge for the working classes through the Mechanics' Institution, in associated publications, and through public lectures and private teaching. The first editor of *The Scots Mechanics' Magazine,* begun in Glasgow in 1825, was Robert Wallace, who, whilst Professor of Mathematics at Anderson's Institution, gave private lecture courses in geography and related subjects between 1828 and 1832 at his home in South Frederick Street and later in St Vincent Street.[15] This publication contained several papers by one James Bell on physical geography and on applied mathematics understood as geographical problems.[16] James Bell was a failed cotton manufacturer from Jedburgh in the Scottish Borders who had turned to geographical writing to pay off his debtors. He was the author in 1824 of *Critical researches in philology and geography* and editor of the six-volume *A system of geography, popular and scientific, or a physical, political and statistical account of the world and its various divisions* (1829–32), which drew upon the Danish geographer Conrad Malte-Brun and his 1822 *Universal system of geography.* It is likely that Bell was the moving force behind *The Glasgow geography, containing a physical, political and statistical view of the various empires, kingdoms, states in the known world,* published in 1822 in five volumes with an accompanying atlas. He was also probably the author, in 1812, of the *System of geography* and, almost certainly, author of a similarly-titled work in 1805 (see p. 142). All these works were published in Glasgow.

Sitwell has hinted at this genealogy for these works and has noted of the 1829–32 *System* that it is 'the last British example of a massive description of all the countries in the world written by a single author'.[17] This judgment should be tempered perhaps, given the work's similarities to others' and the fact that the title pages of several of Bell's works acknowledge the 'work of several literary gentlemen', and, for the 1805 work, 'a literary society'. Bell's works are better seen as compilations of extant geographical knowledge used, in Glasgow and elsewhere, to promote geography to mechanics and others in the urban public sphere. For Mayhew, Bell's 1832 *System* 'betokens the break-down of

[14] *First Annual Report of the Edinburgh School of Art* (Edinburgh, 1822), 5; Marwick, 'Early adult education in Edinburgh', 400. [15] Withers, 'Towards a history of geography in the public sphere', 56.

[16] Bell had three articles in *The Scots Mechanics' Magazine,* 1 (1825): 'On the magnitude and velocity of the great rivers of the globe, and the quantity of water discharged by them into the ocean' (258–64, 305–10); 'Table of the most remarkable points on the surface of the globe, with the limit of the visible horizon' (299–304); and 'Remarks on the table on rivers given in the article Physical Geography, of the Supplement to the Encyclopaedia Britannica' (409–13).

[17] O. F. G. Sitwell, *Four centuries of special geography* (Vancouver, 1993), 95.

the humanist conception of geography' at a time of reform within British politics and within the new age of statistical enquiry.[18]

Geography as a form of popular interest and intellectual enquiry in these terms was not just centred in Scotland's large towns. What little is known of one Nathaniel McCleary in 1834 illustrates well the connections between subjects, the way even a self-taught artisan promoted geographical knowledge, and the social distinctions within geography's public sphere.

> The inhabitants of Newton Stewart and its vicinity, have lately been highly entertained by a course of lectures on Astronomy and Geography, delivered by Mr Nathaniel McCleary, a self-taught Astronomer, and a native of the banks of the Cree. Mr McC has for many years past assiduously applied himself to the study of the above sciences, in the course of which time, although labouring under great disadvantages – he is of the number of those who earn their bread by the sweat of their brow – he notwithstanding has made himself acquainted with the theories of all the ancient, as well as the principles of the most eminent modern Astronomers. His first lecture, on Friday evening, the 28th March, was on the Solar System, in the course of which he exhibited a mechanical figure, the workmanship of his knife, which showed the situation of all the primary planets from the sun; as also the satellites from their central bodies. The second night he lectured on the fixed stars; the third and last night on Geography. In his lecture on the latter science, he exhibited an astonishing strength of memory.
>
> Mr McC. intends in the course of the present summer, visiting the principal towns in the Stewurtry and Shire of Galloway. We wish him success, as he seems to deserve it; and also as a wife and family are dependent on his exertions.[19]

Geography's public lecturers did not all work, then, in the centres of population, use manufactured instruments or treat lecturing as a means to the polite education of the middle and leisured classes. McCleary's straitened circumstances, the audience attending Gullan's lectures in Glasgow's Gorbals in 1835, and the type of knowledge being transmitted in such sites were very different from the social make-up and intellectual interests of the Royal Society of Edinburgh in the same period, for example, with its high proportions of landowners and professors in its audiences for earth science lectures.

### *Publishing for the public sphere: the Edinburgh Journal of Natural and Geographical Science*

This journal, published in Edinburgh and distributed there and in London and Dublin, appeared in 21 monthly parts between October 1829 and June 1831, under the editorship of William Francis Ainsworth, president of the Royal Physical Society of Edinburgh and former president of the Edinburgh Plinian Society, and Henry Cheek, formally involved with the city's Royal Medical Society amongst others. Prior to editing the journal, Ainsworth had

[18] R. Mayhew, *Enlightenment geography: the political languages of British geography, 1650–1850* (London, 2000), 207–8. [19] *Dumfries Telegraph*, 16 April 1834.

travelled in France in search of data to support Hutton's theorisations about Plutonism: he, like Cheek, was profoundly anti-Wernerian, a view apparent in their dealings with Robert Jameson. Ainsworth, a founder member of the Royal Geographical Society in London in 1830, was physician-naturalist on Chesney's 1835 Euphrates expedition, and widely travelled in Asia Minor.[20]

What is important here, however, is not the credibility of the editors, but the journal's purpose. As the initial Preface stressed, the *Edinburgh Journal* was begun precisely because of contemporary interests in natural science and geographical knowledge. New knowledge in those areas should be available to all:

> The Edinburgh Journal of Natural and Geographical Science was instituted with the view of supplying a deficiency long contemplated with regret by all men of science and information. That no periodical, devoted to the prosecutions of geographical enquiry, and the careful collection of important facts which every month brings forth, was to be found in this country, seemed to argue a degree of supineness very inconsistent with the character of the nation: that natural science should be the exclusive property of those only who could afford to purchase the expensive periodicals of the day, appeared to be an injustice to the public, and a drawback on the progress of knowledge, which ought no longer to exist; and that the cumbersome quarterly publications should occupy the whole field . . . was evidently incompatible with that anxious desire for information, which is now felt by all ranks and classes in this country.
>
> This Journal, was, therefore, established for the purpose of affording to the public, with the requisite rapidity, in a condensed form, and at a cheap rate, those discoveries and observations, which could hitherto only be arrived at, by a slow process, at a high price, and in a form the principal merit of which seems to be the respectability of its bulk: and we invite the public to open this volume, and judge how far we have executed our design.[21]

In advertisements accompanying the journal's launch, the public promotion of natural and geographical science was again stressed:

> A leading consideration which actuates the editors of this Journal is, that Science ought to be no longer hidden under the covers of expensive volumes, and prostituted to the unworthy elegances of mere gain. They wish their numbers, uniting cheapness with elegance, at the same time to be found in the mechanics' cottage and on the table of the scientific soiree. Therefore, and at a moderate price and in a condensed form, they propose to publish the novelties of the day.[22]

This commitment might have placed the *Journal* both in commercial and in intellectual opposition to other Edinburgh-based scientific periodicals such as the *Edinburgh Review*, the *Transactions* of the city's Royal Society and *Blackwood's Magazine* given the contest between these periodicals for the pro-

[20] L. Sanders (ed.), *Celebrities of the century* (London, 1890), 45.

[21] *Edinburgh Journal of Natural and Geographical Science*, 1 (1829), Preface [hereafter *EJNGS*].

[22] This phrasing does not appear in the printed Preface to the first number of the *EJNGS*, but in a separately printed scarce Prospectus advertising the *EJNGS* bound in with the first full volume. The quote is from p. 4 of this Prospectus, contained in the copy of the *EJNGS* held in the Library of the Royal Botanic Gardens in Edinburgh.

motion of science in the capital. In fact, most periodicals in Edinburgh welcomed it: in March 1830, one rival in the city's literary public sphere spoke of the *Journal* as 'a star of the first magnitude in the often turbid heaven of Athenian [Edinburgh-based] science'.[23]

The *Journal* organised its contents around three main headings: original articles; collections of facts; and analyses of new books and papers. The collections were further divided into six main listings: geographical, geological, anatomical, botanical, mineralogical and chemical. A further section, on the proceedings of scientific institutions, ensured that public audiences in London, Edinburgh and Dublin knew what was being debated in scientific circles there and in European cities and provincial bodies throughout Britain. Attention was paid, for example, to the business of the Berlin and Paris Geographical Societies, and, from 1830, to the formation of the Royal Geographical Society in London. 'We did not anticipate when we commenced our labours . . . that we should so soon have the pleasure of announcing the intended formation of a Geographical Society in London', noted the editors in May 1830, 'and we can only say for ourselves, that our voices and pens will be devoted to its advancement'. A later part noted: 'Indeed, we shall make it a duty to keep our eye constantly fixed on the progress of this Society, which is so intimately connected with the objects of our periodical; and we feel assured that its directors will take every opportunity of enabling us to extend the knowledge of their invaluable undertaking. We would wish to identify ourselves, in a measure, with the Geographical Society of London, which may thus stretch out its own arm to the remotest corners of the land'.[24]

For historians of geography, this highlights a context to the origins of the RGS and to public interest in geography not before identified. It is interesting that the journal did not result from the foundation of a formal institution for geography or science, as was the case in London, but from the coming together of 'an association of naturalists' for whom geography was a 'regenerated science' given the wealth of new facts, yet also a science without a periodical devoted to it (Figure 5.1).

In those terms, we should think of this pioneering journal as more than just a reflection of science in Edinburgh, and as more than just a geography journal. The *Edinburgh Journal* was part of the networks of scientific knowledge operating nationally and internationally as well as locally. It carried, for example, extracts from Cuvier's lectures on the history of the natural sciences to the Parisian Académie des Sciences. It carried information about local public lectures and demonstrations, notably in chemistry but also Ainsworth's

[23] *Edinburgh Observer,* October 1829; *Edinburgh Literary Journal,* 6 March 1830. The 'contest' between journals and scientific bodies is outlined in S. Shapin, '"Nibbling at the teats of science": Edinburgh and the diffusion of science in the 1830s', in I. Inkster and J. Morrell (eds.), *Metropolis and province: science in British culture, 1780–1850* (London, 1983), 151–78; see also, 'The audience for science in eighteenth-century Edinburgh', *History of Science,* 12 (1974), 95–121.

[24] *EJNGS,* August 1830, 433.

*This day is published,*

Price 2s.

BY DANIEL LIZARS, No. 5, ST. DAVID STREET, EDINBURGH;
WHITTAKER, TREACHER, AND ARNOT, LONDON;
AND WILLIAM CURRY & CO. DUBLIN,
TO BE HAD ALSO OF ALL THE BOOKSELLERS OF THE UNITED KINGDOM,

No. I.

OF THE

EDINBURGH JOURNAL

OF

NATURAL AND GEOGRAPHICAL SCIENCE.

CONDUCTED BY

AN ASSOCIATION OF NATURALISTS.

ILLUSTRATED OCCASIONALLY WITH MAPS, CHARTS, AND ENGRAVINGS.

---

*TO BE CONTINUED MONTHLY.*

---

THE advantages and necessity of periodical communication between the labourers in the fields of science, where observers are so numerous, and local discovery so great, are sufficiently attested by the variety of publications of this class, with which the press perpetually teems. And if in the progress of observation, new branches of knowledge are pressed upon our attention, the active mind looks around with eagerness for new sources of information, and additional means must be afforded of supplying the increase of demand.

GEOGRAPHY, in all its departments, is comparatively within this few years a regenerated science; but it has assumed an importance, which has not yet met, in this country, with an equivalent devotion. For, constant as have been our missions abroad, and ponderous as are the tomes which have sprung from these labours, there does not exist a single British periodical in which any leading attempt is apparent at particular exertion in this department.

But the rapid advances in maritime discovery which Europeans have lately effected, and the beautiful generalizations in Physical Geography which Genius has exposed, have stirred up a spirit of inquiry, which pervades all classes of society; and it thus becomes an object of the highest importance to provide a channel in which the thoughts and discoveries of the people may flow, with express reference to this new subject.

The EDINBURGH JOURNAL of NATURAL and GEOGRAPHICAL SCIENCE, owes its origin to these facts, and has been instituted with the hope, that the strenuous exertions of a body of men, united in one common cause, must be equal to a task of the nature of that which they have imposed upon themselves.

It would not, however, be judicious to exclude from a Journal of this kind, those branches of knowledge which are so requisite to make the value of Geography effective. NATURAL HISTORY, and the whole train of kindred sciences, are indispensable to the Physical Geographer; and the knowledge of Mathematics, Physics, and Astronomy, are equally requisite to the Traveller: accordingly, a great portion of this Publication will be devoted to these objects. To Natural History and Geology especial attention will be directed, as it is hoped that, by an energetic mode of treating these subjects, in them alone sufficient attraction will be held out to that class of readers who may not wish to embrace the whole design.

5.1 Advertisement announcing the publication in 1829 of the *Edinburgh Journal of Natural and Geographical Science.*

public lectures in geology and physical geography, in which he related fossil evidence 'to the new Theories of the Earth'.[25] Articles were included by scholars such as R. Kaye Greville and J. H. Balfour, the distinguished botanists; de la Beche, Murchison and Charles Lyell amongst the geologists; and William MacGillivray, the ornithologist; and reviews were carried of the latest publications. James Bell wrote several articles, ranging from errors in maps of the Orient to the political geography of Russia.[26]

The *Journal* also tried to coordinate a meeting (never realised) of Scottish naturalists, to be held in Edinburgh, which was intended to be the largest gathering of natural scientists ever convened. The organisers hoped to enhance Edinburgh's status as a site for the promotion of science and, in that respect, strengthen Scotland's national identity: 'Our southern friends had better be on the *qui vive,* for the metaphysical nation is becoming clear-headed, and threatens soon to take a lead in the cultivation of national science'.[27] In the event, such a function was performed by the meetings of the British Association for the Advancement of Science in Edinburgh in 1834 and in Glasgow in 1840. Geography within the BAAS – itself begun in 1831 – was largely physical geography, although early correspondence within that body shows debates about the subject's political and botanical context in ways which mirror its treatment within the *Edinburgh Journal*.[28]

What, then, are we to make of this periodical's 'enthusiasm in the cause of a noble and favourite science'? We cannot determine either circulation or subscription rates nor patterns of readership. Did private individuals buy it? Was it bought by Robert Wallace for the Glasgow Mechanics' Institute or other sites of popular learning and read aloud there, or was it privately and silently consumed in domestic spaces by those persons who either gave or attended public geography classes? It was considered important enough to be noticed in other scientific periodicals, in Edinburgh anyway, but the price of 2/- per monthly part may have meant that it only ever got as far as the scientific soirée, not to the mechanics' cottage.

What we can say is that, in the *Edinburgh Journal*, we have evidence of trained scientific men in Edinburgh and further afield seeking to promote geography, which was understood by contemporaries as a form of natural science, and doing so in order to bring knowledge of geography and related matters to as wide a public as possible through an accessible journal rather than through lecture, instrumental demonstration or private class. It reflected local public awareness about geography and about science in Edinburgh, but,

[25] See also the lectures by Thomas Kemp, lecturer in chemistry: *EJNGS,* December 1829, Article IV, 'Experiments on exciting Galvanism, with compound and simple substances, at high temperature', 183–6, and Kemp's further advertisements in *EJNGS,* 2 April 1830.

[26] *EJNGS,* November 1829, 81–91; August 1830, 259–62, 329–40, 416–24.

[27] *EJNGS,* June 1830, 234.

[28] J. Morrell and A. Thackray (eds.), *Gentlemen of science: early years of the British Association for the Advancement of Science* (Oxford, 1981), 286, 333; *Gentlemen of science: early correspondence of the British Association for the Advancement of Science* (London, 1984), 54.

in being bound up with the international circulation of geographical knowledge, it also had a wider remit. In both documenting work on natural knowledge and the geography of Scotland, and in aiming to extend scientific knowledge as a form of civic improvement, the *Edinburgh Journal* was seen as one means to reinforce national identity through science. Other means involved more direct action.

### *Social statistics, 'civic economy' and moral improvement: the geographical imagination of Thomas Chalmers*

In 1844, one year after he led the Free Church in its separation from the Established Church of Scotland, Thomas Chalmers, the churchman, social reformer and Professor of Divinity announced a 'Church Extension' campaign to create an additional sixty working-class churches in Edinburgh. These churches were to be managed by local philanthropic societies, each with responsibility for a given district of the city, divided into twenty sub-districts, or 'proportions'. Chalmers focused particularly on the West Port in Edinburgh's Cowgate, then one of the city's most impoverished districts, and plans for his 'model operation' there involved a school in which geography, together with reading, writing, natural science and Bible study was taught.[29]

Chalmers was one of many like-minded individuals concerned with the social problems of urban Britain and with the 'moral geography' of the poor. The philanthropic intervention of concerned individuals and institutions into the nature of British civil society was closely allied with the development of social statistics and with monitoring the condition of the nation.[30] Chalmers' attempt to create a 'Godly Commonwealth' of parish communities was part of the emergence of certain sorts of statistical and political knowledge about the nature of civil society, and resulted from his seeing urban social problems as not just having geographical expression but being geographically constituted. Chalmers, I contend, was thinking and acting geographically. Indeed, Chalmers' work should be understood as marking a 'geographical civic consciousness' in that he sought to understand *and* to propose solutions to contemporary social issues and because he saw geography as a means to moral improvement.

Chalmers, born in Anstruther in 1780, was minister in nearby Kilmeny from 1803 before moving to the Tron parish, Glasgow, in 1815. In 1820, he moved to the newly-created St John's parish, one of the poorest in the city. From at least 1819, Chalmers was interested in the moral value of the parish as the

[29] S. Brown, *Thomas Chalmers and the Godly commonwealth in Scotland* (Oxford, 1982), 352–5.

[30] L. Goldman, 'The origins of British social science: political economy, national science and statistics, 1830–1835', *Historical Journal*, 26 (1983), 587–616; F. Driver, 'Moral geographies: social science and the urban environment in mid-nineteenth-century England', *Transactions of the Institute of British Geographers*, 13 (1988), 275–87.

'right' size for social communities and was committed to the management of social problems through the 'principle of locality'. This meant, simply, extending the idea of the small rural parish (Chalmers drew upon his own experiences) to the social problems of urban Scotland. As he wrote to William Wilberforce in 1820, only this would 'ever bring us back again to a sound and wholesome state of the body politic'.[31] In this thinking, he was supported in Glasgow by the social statistician James Cleland, author of several books on Glasgow's civil condition.[32] It was Cleland, who, in a letter of 1819, sanctioned Chalmers' plans to divide his parish into 'such Proportions as you may judge proper':[33] Chalmers proposed areas of 2,000 persons.

Chalmers outlined his thinking most clearly in his *The christian and civic economy of large towns,* published in three volumes between 1821 and 1826. In writing 'On the influence of locality in towns', Chalmers considered that 'there is a charm in locality [felt by every individual] who has personally attached himself to a manageable portion of the civic territory', and argued that 'the established ministers of a large town, should be enabled, each to concentrate . . . on his own district and separate portion of the whole territory . . . his own geographical vineyard'.[34] Chalmers' arguments about the right scale for the constitution of civic society echo Sinclair's recognition of the parish as the basis to Scotland's statistical enumeration and to knowledge, via the 'locality', of the political anatomy of the nation (see pp. 145–8). More than that, they indicate his concern to understand Scotland geographically, to see the problems of its cities especially as rooted in distributional differences in health, wealth, access to education and spiritual supervision, and, in turn, a concern to see geography as a means to ameliorate such things. As he noted in 1839 whilst travelling through the Grampians, 'I confess myself to be nearly as much on edge after novelties for the purpose of geographical truth, as for the purpose of a spectacle'.[35]

In his 1841 *The sufficiency of the parochial system,* Chalmers again stressed the value of social statistics. He drew analogies with the work of Alexander von Humboldt and employed geographical metaphors in emphasising their value: 'And so the magnificent sketches of Humboldt could be broken down into an atlas of successive landscapes, which would present us with what may be called the statistics of scenery. Statistics in short stands in the same relation to general science that topography does to geography'.[36] His concern was less

[31] S. Mechie, *The church and Scottish social development, 1780–1870* (London, 1960), 51.

[32] On Cleland's civil statistics, see his *Enumeration of the inhabitants of Scotland* (Glasgow, 1828); *Enumeration of the inhabitants of Glasgow* (Glasgow, 1832); and *Historical accounts of Bills of Mortality of the probability of human life in Glasgow and other large towns* (Glasgow, 1836).

[33] Edinburgh University, New College Library [hereafter NCL], MS CHA. 4.10.53, James Cleland to Thomas Chalmers, 26 June 1819.

[34] T. Chalmers, *The Christian and civic economy of large towns* (Glasgow, 1821–6), I, 55.

[35] W. Hanna, *Memoirs of the life and writings of Thomas Chalmers* (Edinburgh, 1852), 4, 69.

[36] T. Chalmers, *The sufficiency of a parochial system, without a poor rate, for the right management of the poor* (Glasgow, 1841), 227.

with facts about localities than with what they might mean as a basis to social policy, a claim which is borne out in Chalmers' dealings with William Pulteney Alison, the Edinburgh Professor of Physiology from 1822 and, from 1842, of Medicine. Alison's views on the management of the Scottish Poor Law in his 1840 *Observations on the management of the poor in Scotland,* which emphasised formal assessment unlike Chalmers' preference for voluntary contributions, formed the basis to the new Poor Law in Scotland from 1845.

What for Alison was a matter of what we might think of as political physiology and, for Chalmers, was a spiritual-statistical question, was for them both a concern for the nation's civil and moral constitution. Others shared this concern: William Collins, for example, preached with Chalmers and published his works. Richard Poole, author of the *Report on examination of medical practitioners* in 1833 and, from 1820, the moving force behind plans for a new infirmary in Edinburgh prepared, in manuscript, 'Observations and Statistical Tables illustrative of Political Geography' (which he used in lectures) in which he advocated the role of 'the science of geography' in 'promoting the liberal constitution of human affairs in general'. Poole was also in correspondence with John Thomson over the latter's 1832 *Atlas.*[37] Keith Johnston, the Edinburgh geographical and atlas publisher, knew Chalmers, sought his views on his own 1844 *Physical atlas* and expressed support for 'your noble enterprise in the West Port, in the success of which I feel the highest interest'.[38] Johnston lowered the prices of several works in order that he should 'not overlook the wants of the poorer classes in their desire for instruction in geography'. This fact, noted one commentator in 1873, 'has done more for popularising geography in this country than any modern geographer'.[39] The BAAS provided a platform for discussions concerning such interventionist social science: the 1834 Edinburgh meeting heard a paper from Cleland, for example, about Chalmers' geographical principles of Poor Law management in Glasgow.[40]

Several things follow from this evidence. The first concerns the success of Chalmers' plans. For the Glasgow journalist George Troup, educating the urban poor was wasted effort: 'Reading is luxurious, writing is extravagant, arithmetic is wasteful. Geography, and all similar sciences, are most ruinous. Every locality has its own predilections, and none of those connected with the Old Wynd are literary'.[41] In contrast, Chalmers' West Port school was considered successful, even if his other plans were not.[42] Even so, I would not wish to diminish Chalmers' geographical imagination nor to underestimate the

[37] Royal College of Physicians, Edinburgh [hereafter RCPE], Richard Poole MSS, MS 17/125–19/132, No. 13, ff.1–9v; RCPE, Richard Poole MSS, MS 18 [folders 128–9].

[38] NCL, MS CHA.4.335. 36, Johnston to Chalmers, 22 January 1846.

[39] T. Johnston (ed.), *In memoriam of the late A. Keith Johnston L.L.D., Geographer to the Queen for Scotland* (Edinburgh, 1873), 16.

[40] Morrell and Thackray (eds.), *Gentlemen of science,* 293.

[41] George Troup, [Accounts of visits to the Glasgow slums], *The North British Daily Mail,* 2, 4 and 5 September 1848.

[42] Brown, *Thomas Chalmers and the godly commonwealth,* 117–21.

concerns that he and others had over using geographical thinking to address the problems of contemporary civil society.

A second outcome concerns this sense that for Chalmers and others, understanding Scotland's civil society was itself a geographical matter. I mean by this – and, to use a phrase common to their interests – that such men saw Scotland as a 'mission field'. In investigating Scotland's inner cities and in, for example, extending spiritual supervision and education to the Gaelic Highlands, Chalmers and others were at work amongst those they deemed ill-educated in just the same way as many Scots and others saw 'natives' overseas. One of the aims of the Free Church Ladies' Highland Association, for example, established in 1850, was to educate destitute Highlanders at home, including 'grounding the children well in geography' by supplying them with maps and globes.[43] One self-proclaimed 'missionary', writing of work 'reclaiming the outcastes among the Red Indians', said she had 'never seen such vice and misery as among "the blacks in the Canongate"'.[44] And one observer of Edinburgh's Cowgate in 1879 wrote of it in ways common – as Livingstone has shown[45] – to much of that racialised discourse through which 'the West' geographically imagined 'the Rest' in the nineteenth century: 'The Cowgate is the Irish portion of the city. Edinburgh leaps over it with bridges; its inhabitants are morally and geographically the lower orders'.[46] Geographers and others could use the new 'science' of photography to document the geography – physical but, in the cities, chiefly moral – of Scotland's 'lower orders' in ways which have been shown of geography, photography and the representation of empire[47] (see Figure 5.2). As McEwan has shown, the Dundee-born Mary Slessor used the lessons from her missionary work in West Africa to improve the lot of Dundee's industrial poor.[48]

In those terms, whilst acknowledging that Scotland played a part in the making of the British empire and, thus, of Britishness from the mid-nineteenth century, national identity was also being made in the geographical exploration of Scotland's own body politic for purposes of social amelioration. The culture of exploration in geography was played out in the dark 'heart of empire', Britain's teeming and ill-understood cities, just as much as it was in the

[43] J. Watson, *Pathmakers in the isles* (Edinburgh, 1956), 19–20.

[44] J. Blumenreich, *The missionary: his trials and triumphs. Being nine years' experience in the wynds and closes of Edinburgh* (Edinburgh, 1864), 183.

[45] D. Livingstone, 'The moral discourse of climate: historical considerations on race, place and virtue', *Journal of Historical Geography,* 17 (1991), 413–34; 'Climate's moral economy: science, race and place in post-Darwinian British and American geography', in A.Godlewska and N. Smith (eds.), *Geography and empire* (Oxford, 1994), 132–54; 'Tropical climate and moral hygiene: the anatomy of a Victorian debate', *British Journal for the History of Science,* 32 (1999), 93–110.

[46] Alexander Smith, *A summer in Skye* (Edinburgh, 1880), 20.

[47] J. Ryan, *Picturing empire: photography and the visualisation of the British empire* (London, 1997).

[48] C. McEwan, '"The mother of all the peoples": geographical knowledge and the empowering of Mary Slessor', in M. Bell, R. Butlin and M. Heffernan (eds.), *Geography and imperialism, 1820–1940* (Manchester, 1995), 125–50.

5.2 'The Cowgate Arch of George IV Bridge' [Edinburgh], by William Donaldson Clark, *c.*1860.

unknown spaces of 'darkest Africa'.[49] Thinking and acting geographically aimed to deal with the problem of what, for Scotland anyway in the 1830s, was 'a half educated nation'.[50]

## Historical geographies of schools and texts, 1838–1884

### *The geographical state of the nation in 1838*

Before the 1872 Education (Scotland) Act and the foundation of the Scotch Education Department, both differences between educational establishments and the fact that in many, geography was only an optional subject for which extra fees were paid, make it difficult to know what proportion of Scotland's school-going population received a geographical education, and, of those who

[49] F. Driver, *Geography militant: cultures of exploration and empire* (London, 2000), 170–98.

[50] J. Smith, 'Manners, morals and mentalities: reflections on the popular enlightenment of early nineteenth-century Scotland', in W. Humes and H. Paterson (eds.), *Scottish culture and Scottish education 1800–1980* (Edinburgh, 1983), 25–54; D. Withrington, '"Scotland a half-educated nation" in 1834? reliable critique or persuasive polemic?', in Humes and Brown (eds.), *Scottish culture and Scottish education,* 55–74; G. Morton and R. Morris, 'Civil society, governance and nation: Scotland 1830–1914', in R. Houston and W. Knox (eds.), *Penguin History of Scotland* (London, forthcoming).

did, exactly which books were used and how pupils were taught. John Gibson, appointed Her Majesty's Inspector of Schools in Scotland in 1840, reported in 1841, for example, that of 1,873 pupils in 27 parish schools he had visited, all were taking reading, 964 writing, 644 arithmetic, 341 English grammar and 331 geography. Gibson encouraged parents to pay the extra fees so that their children could be taught geography by suggesting to schoolmasters that they invite parents to watch the geography lessons, but nothing seems to have come of his proposals.[51]

Using evidence collected in 1838 in submission to the Select Committee on Education, however, it is possible to document the national state of geography.[52] The work of the Select Committee itself reflected contemporary interests in the educational state of the nation and the need for the 'moral regeneration' of Scottish civic society.[53] The committee circulated queries about the state of education to every parochial and non-parochial school throughout Scotland. Some information, such as numbers on the school roll, relates to the years 1836 and 1837, but most of the data is for 1838. In some respects it is inconsistent: the age of pupils taught was not everywhere the same; numbers on the school roll are not always given; no replies were received from several schoolmasters. From an assessment of questions concerning the 'present course of instruction', 'what books are used?' (geography was one of ten subjects listed) and teachers' qualifications, we can know not only where geography was taught, and, thus, chart for the first time a national picture of geography in education, but also know which books were used and something of how texts, maps and globes were employed.

Table 5.1 shows the county-level picture for geography teaching in Scotland in 1838. For Scotland as a whole, 82 per cent of all parochial schools were teaching geography, whilst fewer than half of all non-parochial schools (private establishments, charity foundations and so on) did so. The least well-served county was Argyllshire, with fewer than half of all parochial schools and just over one-third of all non-parochial schools teaching geography in 1838. Figure 5.3 shows that geography was taught in most parochial schools in most parishes in Scotland in 1838. Argyll's low total overall was due to poor provision in its central parishes: there, and in several parishes elsewhere, geography was taught in one only of several parochial schools. There is no explanation to account for the handful of parishes not returning answers to the 1838 queries. Table 5.2 gives a more detailed picture of the provision of geography in urban Scotland in non-parochial schools. Most such schools were located in Edinburgh and Glasgow; Greenock offered the poorest chance of an urban child being taught geography in the non-parochial school system.

[51] T. Bone, *School inspection in Scotland, 1840–1966* (London, 1968), 23, 31, 33.

[52] BPP, 'Answers made by schoolmasters in Scotland to queries circulated in 1838 by order of the Select Committee on Education in Scotland', 1841, XIX.

[53] Withrington, 'Scotland a half-educated nation'; Morton and Morris, 'Civil society, governance and nation'.

Table 5.1 *The number of schools in which geography was taught in Scotland, 1838, by county*

| County | Parochial schools | | Non-parochial schools | |
|---|---|---|---|---|
| | Number of schools teaching geography | Proportion of total number of schools (%) | Number of schools teaching geography | Proportion of total number of schools (%) |
| Aberdeen | 85 | 91 | 97 | 60 |
| Angus | 37 | 84 | 73 | 58 |
| Argyll | 33 | 49 | 47 | 36 |
| Ayr | 36 | 92 | 84 | 54 |
| Banff | 19 | 95 | 28 | 35 |
| Berwick | 27 | 87 | 24 | 51 |
| Bute | 6 | 86 | 10 | 42 |
| Caithness | 10 | 100 | 23 | 46 |
| Clackmannan | 2 | 50 | 10 | 67 |
| Dumbarton | 11 | 85 | 22 | 58 |
| Dumfries | 47 | 84 | 46 | 52 |
| East Lothian | 24 | 96 | 21 | 64 |
| Fife | 44 | 90 | 78 | 60 |
| Inverness | 24 | 83 | 26 | 38 |
| Kincardine | 17 | 89 | 17 | 26 |
| Kinross | 2 | 67 | 7 | 70 |
| Kirkcudbright | 39 | 85 | 20 | 48 |
| Lanark | 43 | 74 | 104 | 50 |
| Midlothian | 24 | 96 | 97 | 54 |
| Moray | 16 | 80 | 19 | 45 |
| Nairn | 2 | 67 | 7 | 100 |
| Orkney and Shetland | 14 | 54 | 29 | 30 |
| Peebles | 13 | 93 | 3 | 43 |
| Perth | 55 | 86 | 68 | 53 |
| Renfrew | 13 | 72 | 43 | 46 |
| Ross and Cromarty | 20 | 77 | 18 | 24 |
| Roxburgh | 38 | 88 | 25 | 74 |
| Selkirk | 5 | 83 | 6 | 75 |
| Stirling | 28 | 97 | 47 | 49 |
| Sutherland | 9 | 82 | 8 | 35 |
| West Lothian | 6 | 55 | 5 | 50 |
| Wigtown | 12 | 86 | 29 | 52 |
| [SCOTLAND | 761 | 82 | 1141 | 49] |

*Source*: *British Parliamentary Papers,* 1841, Vol. 19, *'Answers made by Schoolmasters . . .'*: (non-parochial schools), pp. 1–790; (parochial schools), pp. 1–312.

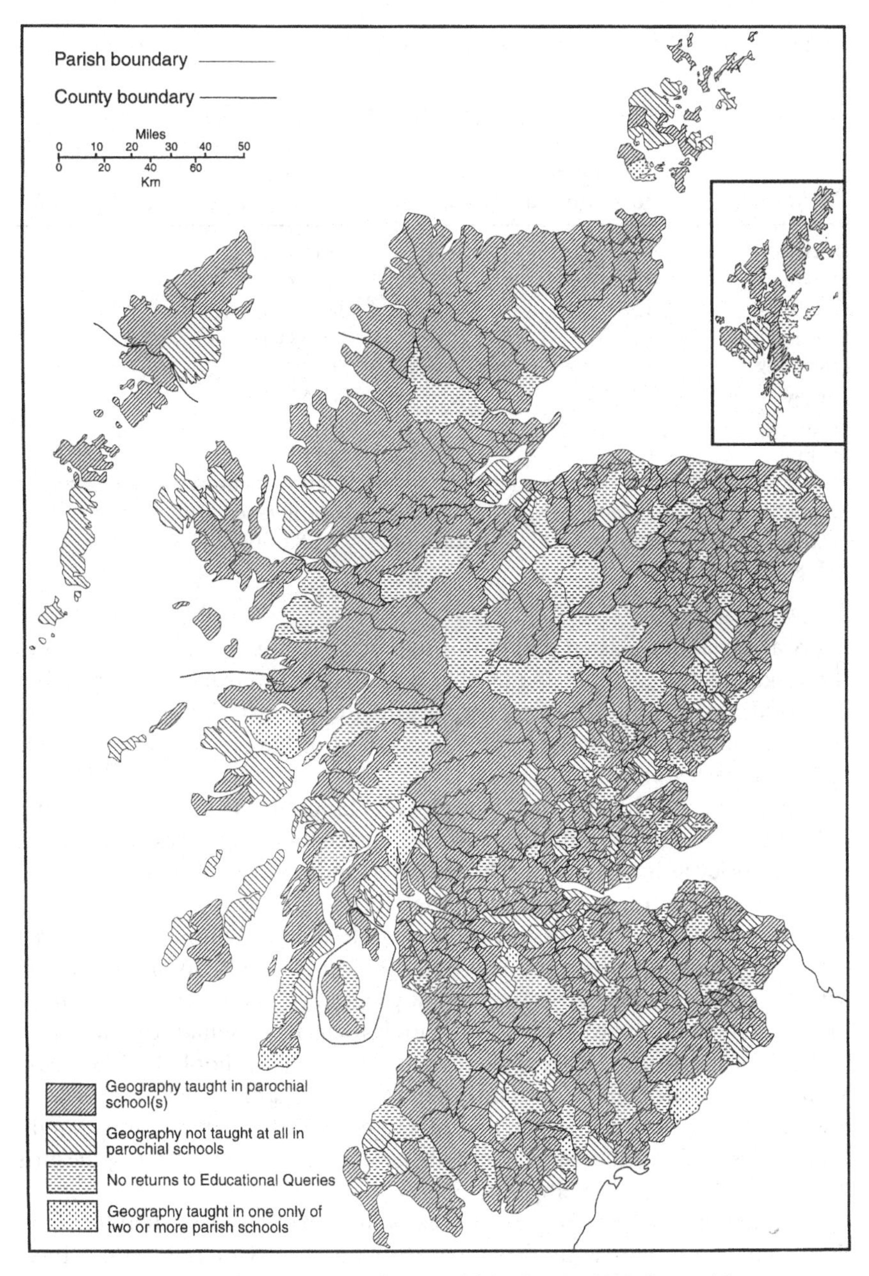


5.3 Geography teaching in Scotland's parochial schools, 1838, by parish.

Table 5.2 *The teaching of geography in non-parochial schools in urban Scotland, 1838*

| City/town | Total number of non-parochial schools | Number of non-parochial schools teaching geography (and as % of total no. of schools) |
|---|---|---|
| Aberdeen | 47 | 31 (66%) |
| Dundee | 41 | 24 (59%) |
| Edinburgh | 62 | 38 (61%) |
| Glasgow | 86 | 50 (58%) |
| Greenock | 20 | 6 (30%) |
| Paisley | 35 | 20 (57%) |
| Perth | 10 | 4 (40%) |
| Stirling | 12 | 6 (50%) |

*Source*: *British Parliamentary Papers*, 1841, Vol. 19, '*Answers made by Schoolmasters . . .*': (non-parochial schools), pp. 1–17, 283–303, 393–405, 518–47, 645–8, 655–61, 750–4.

Local variations existed within this national picture. In Cockburnspath in Berwickshire, for example, the teacher was qualified to teach geography, but did not do so. In Strachen, Kincardineshire, by contrast, 'The teacher is prevented from introducing geography on account of having no spare room for maps and he regrets this very much'.[54]

Many returns indicate that geography was taught using maps – sometimes *only* using maps as a prompt to *viva voce* instruction – and, less commonly, globes. Fee levels for geography classes, where indicated at all, ranged from 1/- to 2/6d per class. Inconsistencies in the recording of numbers of pupils on schools' rolls make it impossible to enumerate the total number of children receiving education in geography, but the schools' rolls do show that, in general, more boys than girls were taught geography. There were exceptions to this, notably in non-parochial urban schools focusing on female education. In St David's parish in Glasgow there were three such schools in 1838: Mrs Strachan's Female School, Graham's Charity School and Miss Roger's School – with 172 girls in total. In the first of these, 'sacred geography' was taught using maps, and, in other female schools, geography was used to teach the girls 'to read the Scriptures intelligently'.[55]

No consistent pattern appears from examination of teachers' qualifications. Parochial schools were administered by the Church of Scotland through presbyterial authorities, and it appears that satisfying local presbyteries of compe-

[54] BPP, 'Answers made by schoolmasters . . .', 77, 186.
[55] BPP, 'Answers made by schoolmasters . . .', 526, 527, 529.

tence to teach English reading and writing, arithmetic and, importantly, the scriptures, was considered a higher priority than capacity to teach geography. In some Highland parishes, greater emphasis appears to have been placed on capacity to teach through Gaelic, although the evidence of Figure 5.3 in respect of geography teaching throughout the Gaelic-speaking north and west Highlands suggests this was not widely adhered to. Several teachers had a university education, which raises the question of where and from whom they learned their geography.

Geography was taught in Marischal College Aberdeen in the 1830s by James Davidson, Professor of Civil and Natural History, but only in a limited way. As he told visiting Commissioners in 1837, 'I do nothing with Geography, because I consider that so extensive a subject; except in Natural History, when treating of mountains, I give a good deal of Geography there; but it is Natural Geography, and not Civil Geography'.[56] In answer to the question 'do you generally find that they are tolerably acquainted with Geography before they come to College?', Davidson noted, 'Very much the reverse; they are very much deficient in general knowledge. Students that come to my class have generally had their attention entirely directed to the study of languages; and the study of nature is so new to them'.[57]

In Glasgow James Millar, Professor of Mathematics, included lectures on astronomy and geography to his senior class at this time;[58] it is not clear when he dropped or modified his earlier lecture programme on geography (see p. 140). In St Andrews Thomas Duncan, Professor of Mathematics, taught geography to his senior class, together with the principles of navigation and fortification. In his view, geography could be taught with civil history:

> The two branches are nearly allied, and one half of the Session might be advantageously devoted to the one branch, and the other half to the other. In the Mathematical Class, I give only the Mathematical Principles of Geography, and am scarcely able to spare a fortnight of my course for these; but Geography, considered in its fullest extent, viz. in a Physical, Statistical, Civil, and Political view, is a most extensive and important subject. Without doubt, both History and Geography could be studied privately in future life, but the same thing might be said of many other branches that usually form a part of education.[59]

In Edinburgh, Robert Jameson's lectures in 1830 on 'the waters and climate of the Globe' included one class on the physical geography of Scotland.[60] Geography seems not to have been taught by Edinburgh's mathematicians. In contrast, James Pillans, Professor of Humanity there from 1820–63, paid

[56] BPP, 'Evidence, oral and documentary, taken before the Commissioners for visiting the Universities of Scotland', 1837, XXXVIII, 86. [57] BPP, 'Evidence, oral and documentary . . .', 86–87.

[58] BPP, 'Evidence, oral and documentary . . .', 145. [59] BPP, 'Evidence, oral and documentary . . .', 153.

[60] National Library of Scotland [hereafter NLS], MSS 14142–54, 'Notebooks containing essays and notes from lectures, kept by W. S. Walker as an undergraduate at Edinburgh University, 1828–1831'. For material relating to Jameson's teaching (in 1830), see MS 14148, ff.1–87v; MS Gen. 1996/9/70–88.

considerable attention to geography. Pillans had taught classical geography during his teaching in Edinburgh's High School from 1810. His university teaching included geography as an explanation to classical history. Lack of appropriate texts led Pillans to write three geography texts for his students: *Outlines of geography* (1847), *First steps in the physical and classical geography of the ancient world* (1853), which ran to thirteen editions by 1882, and *Elements of physical and classical geography* (1854).[61] More widely, several texts were commonly used.

### *Works for civic improvement? The nature of geography books used in Scotland, 1838–1884*

In Scotland's schools in 1838, six books were commonly used in the parochial and the non-parochial schools alike: Thomas Ewing's *System of geography,* published in Edinburgh in 1816 'for the use of schools and private students'; Joseph Guy's *School geography* (London 1810), Hugh Murray's *Catechism of geography* (1833), which ended with sections on globes and geographical problems; William Pinnock's 1821 *Catechism of geography*; the Rev. Alexander Stewart's *A compendium of modern geography* (1829), published, like Ewing and Murray, by Oliver and Boyd; and John White's 1822 *System of modern geography, with an outline of astronomy,* published in Edinburgh.[62] Ewing, Pinnock and Stewart were the most commonly referred to in the 1838 survey. These and other books were differently used according to age and ability: several teachers indicated that White's work was for beginners and Stewart's for advanced classes. Ewing's *System* was tremendously popular, running to at least twenty-five editions by 1875: its structure emphasised problem-solving exercises; it provided clear discussion of the historical, political, civil and natural geography of the world's countries; and its short preface outlined 'The Manner in which this Geography may be used in Teaching.'

By the 1830s the evidence that geographical authors, including editors and producers of atlases, were working to promote geography – apparent in earlier evidence (see pp. 128–32) and in Johnston's connections with Chalmers in the 1840s – is reinforced by the correspondence between Alexander Stewart (who had taught classes in geography in Edinburgh in the 1820s) and his publishers at Oliver and Boyd over his *Compendium of modern geography.* Throughout 1828, the correspondence between author and publisher documents the production of a book of 'a very decided superiority', which, for Stewart, was demanding a great deal of 'reading and turning over of a number

[61] C. Withers, 'Notes towards a historical geography of geography in early modern Scotland', *Scotlands,* 3 (1996), 111–24; Sitwell, *Four centuries of special geography . . .*, 469 makes mention only of Pillans' 1847 *Outlines of geography.*

[62] This is clear from detailed examination of teaching texts in BPP, 'Answers made by schoolmasters . . .'. For details on the texts, see Sitwell, *Four centuries of special geography . . .*, 206, 287–8, 429–30, 471, 538–40, 585.

of books'.[63] On 27 April 1828, Stewart wrote: 'My geographical labours are now very near a close. They can bring me but little fame; but they will afford me at least the gratification of thinking that they are calculated to be useful, + [I] shall be much disappointed if they be not somewhat popular. The trouble the work has cost me, you can scarcely imagine'.[64] The work, much revised, went to a thirty-sixth edition in 1889, and did much better than Stewart's 1839 *The first book of modern geography.*

Hugh Murray's correspondence with George Boyd over his 1834 *Encyclopedia of geography* – the work for which he is best known, despite his *Catechism* being widely used in teaching – also reveals what should be thought of as specialist book production for a target audience. Murray, an excise clerk, wrote numerous geographical essays for Edinburgh's learned periodicals and reviewed for the *Edinburgh Journal of Natural and Geographical Science.*[65] His *Encyclopedia* was 'assisted by the following gentlemen, in their respective departments of science': William Wallace, Professor of Mathematics at Edinburgh, who wrote on astronomical and mathematical geography; Robert Jameson who dealt with geology; William Hooker, Professor of Botany at Glasgow; and William Swainson, the zoologist.

Recognising that the teaching of geography in Scotland's schools and, less widely, in her universities was, by 1838, supported by a variety of geography books tailored to given audiences is not to claim that these were works by persons who had actively explored the countries they described. These were works of geographical synthesis aimed, as Ewing put it, at promoting geography 'as a useful science'. Utility was understood as mathematical knowledge and historical understanding. Furthermore, some of geography's texts were dependent upon contributions from specialists in other sciences. Particular needs could be accommodated: for example, James Gall's 1837 pamphlet on methods of teaching geography to the blind by maps made of pin-cushions and twine; braille maps of 1851; and the fact that, in St Mungo's Asylum for the Blind in Glasgow in 1838, geography was taught 'by globes and maps, made for the use of the Blind'; each illustrates this point for a particular constituency. Such geographical materials for the blind were on public display in Edinburgh in 1839.[66]

[63] NLS, MS Acc. 5000/194 [1828 and 1829 folders].

[64] NLS, MS Acc. 5000/194, Alexander Stewart to George Boyd, 27 April 1828.

[65] NLS, MS Acc. 5000/197. Murray's geographical essays include 'On the ancient geography of central and eastern Asia, with illustrations derived from recent discoveries in the north of India [1816]', *Transactions of the Royal Society of Edinburgh,* VIII (1818), 171–204; 'Observations on the information collected by the Ashantee Mission, respecting the course of the Niger and the interior of Africa', *Edinburgh Philosophical Journal,* 1 (1819), 163–70; 'Observations on the results of the late expedition of Captain Parry, including a view of previous discoveries made in the same direction', *Edinburgh Philosophical Journal,* 11 (1824), 225–49.

[66] James Gall, *An account of the recent discoveries which have been made for facilitating the education of the blind* (Edinburgh, 1837), 34–6; Thomas Haig, *Geography of Scotland for the use of the blind with historical notes* (Edinburgh, 1851); BPP, 'Answers made by schoolmasters . . .', 543; *Catalogue of the exhibition of arts, manufactures, and practical science in the Assembly Rooms, Edinburgh, Dec. 1839–Jan. 1840* (Edinburgh, 1840).

More generally, geography was closely associated with Christianity, in Scotland's parochial schools particularly. This is unsurprising, perhaps, given the authority of the Scottish Church over these schools before 1872, the Church's missionary work at home and abroad and the educational circumstances of its clergy. It was for such reasons and, in part, because their contents could be learnt or spoken as almost a litany, that several of the more widely-used texts were entitled 'Catechisms' of geography. It is possible, for example, that the Rev. Alex Mackay's *A rhyming geography for little boys and girls* (1873) was meant to be sung to psalm tunes while instilling geographical learning.[67]

Alwall has commented upon the religious trend in secular Scottish schoolbooks for the second half of the nineteenth century, noting a strong religious emphasis to geography books in the period 1850–61 and, relatively, less emphasis for the period 1873–82 in comparison with descriptive or commercial geography.[68] I suggest that the religious emphasis was present earlier, a claim supported both by the geography books used in Scotland's schools in 1838 and by those geography texts produced by the Scottish School Book Association. Begun in 1818 (and dissolved in 1891), this body produced texts for reading with emphases upon 'religious teaching and cheapness of price' in order to reach as large an audience as possible. Its principal geographical publications were *A short system of geography* (1823), which emphasised rote learning of geographical descriptions, *Outlines of modern geography* (1842), written for the association by William Whyte, which stressed the map as a teaching aid and which was to be read alongside the association's *Complete system of modern geography,* also published that year. By 1873, the Association had produced thirteen geography texts, twelve atlases and as many maps for use in schools. These books were much sought after by many organisations, including by the Geographical Society of Belgium.[69]

That these and other books stressed geographical understanding as a means to understanding the Christian scriptures is apparent from comments by the Rev. Hugh Martin, minister of Greyfriars in Edinburgh, who noted of his own experiences: 'With the solitary exception of arithmetic, all that we teach in our Scottish schools is taught in indissoluble connection with religious education. We teach geography religiously. I was myself taught geography at school from a little book, the preface of which began by saying that the teacher should take

[67] This suggestion is made as a footnote (p. 83) in Kenneth Maclean, 'Scottish geographical roots 3: Rev. Alexander Mackay', *Scottish Association of Geography Teachers' Journal,* 27 (1998), 71–84. Mackay begins his book with the observation that 'it is almost incredible how ill-informed most people are in everything pertaining to that most useful and even necessary branch of popular education [geography]. In our public schools, with the exception of the junior classes, very little geographical knowledge is communicated.' (Alexander Mackay, *A rhyming geography for little boys and girls* (Edinburgh, 1873), iii).

[68] E. Alwall, *The religious trend in secular Scottish school-books 1850–1861 and 1873–1882* (London, 1970), 141–2.

[69] A. Belford, *Centenary handbook of the Educational Institute of Scotland* (Edinburgh, 1946), 15–20.

every opportunity of instilling religious truths into the minds of his pupils.'[70] And the sense in which some geography texts engaged with theological interpretations of current scientific debate is apparent, for example, from the Rev. Alexander Mackay's works of geography, notably his *Manual of modern geography* (1861) which, while focusing largely on physical geography and physiography and incorporating the latest geological ideas, rejected Darwinian biology completely. To have embraced Darwinism would have compromised Mackenzie's own views, and for many in the church would have denied the Word of God.[71]

These remarks serve to highlight several issues, not least of which is how much more there is to know about the study of geography in education as part of the production, dissemination and reception of geographical knowledge. Other things follow from this. One has to do with national comparisons. In England, many of those geographical texts in use in education after 1830 stressed discourses of race and gender in relation to Britain's imperial authority.[72] In Scotland, geography books in use before the 1880s were more explicitly concerned with civic *self*-improvement than with matters of racial difference or imperial advantage. Yet we should not simply distinguish in absolute or national terms between those works advocating the British empire as the consequence of military and commercial hegemony and those which aimed at the improvement of Scottish (and British) citizens through attention to geography as a form of moral, even religious, education. Some books, like Ewing, stressed earth description and teaching method. Others, like Murray's, were more the work of contemporary scientists in other disciplines than of 'geographers' alone. Some, like Mackay in his attention to theological concerns, Geikie in geology and Mary Somerville for physical geography, placed geographical knowledge within current scientific debates as part of teleological explanations.[73]

Whilst it is clear that different sorts of geographical texts were being produced for different markets amongst those for whom geography, however differently understood, was both useful and interesting, we should not look alone at geography's *books* in seeking to understand its place in the public's imagination or, more specifically, the place of geography in the home. The sense that geographical literatures and geographical questions were taken seriously by professional scientists in other disciplines is not just evident in the case of the *Edinburgh Journal*, but in the many reviews of geographical works by Sir David

[70] K. Ross, *Church and creed in Scotland: the Free Church case 1900–1904 and its origins* (Edinburgh, 1988), 328.

[71] Maclean, 'Scottish geographical roots 3: Rev. Alexander Mackay', 79–81.

[72] A. Maddrell, 'Discourses of race and gender and the comparative method in geography school texts 1830–1918', *Environment and planning D; Society and space,* 16 (1998), 81–103; see also W. Marsden (ed.), *Historical perspectives on geographical education* (London, 1980) and '"All in a good cause": geography, history and the politicization of the curriculum in nineteenth and twentieth-century England', *Journal of Curriculum Studies,* 21 (1989), 506–26.

[73] Maclean, 'Scottish geographical roots 3: Rev. Alexander Mackay', 80–1.

Brewster in the *North British Review.*[74] That journal, like others, provided both a medium for the publication of scientific ideas and a means by which a geographically aware and scientifically curious public could be kept up to date. For example, the autumn 1863 issue of the *North British Review* carried an article entitled 'On recent geographical discovery and research' in which the explorations of Speke and Grant in Africa were discussed and illustrated with a map published by W. and A. K. Johnston.[75] For Chalmers, periodical literature provided a further outlet in which public opinion could be made aware about pauperism in urban Scotland and his locality-based plans to deal with it.[76]

Such journals provided an important outlet for discussions about the nature and value of science, in Scotland and elsewhere. The *North British Review* of February–May 1858 included a 30-page article on science in Scotland, which began: 'It cannot be doubted, that the educated mind of Scotland has been more largely devoted to the study of Theology than to that of Physical Science. Even a partial acquaintance with Scottish history, and a comparatively limited knowledge of national literature are sufficient to convince us of this'.[77] This was not strictly true, as the following section shows. But it symbolises, nevertheless, as did similar articles within the *Edinburgh Journal*, one of the many 'literary sites' in which civil society engaged with scientific understanding as a matter of national identity, both in the sense of knowing about one's nation and in regard to what one's nation was to be known for by others. Let me now turn away, however, from textual and urban matters to the making of geographical knowledge and civic identity 'in the field'.

## Historical geographies of local natural knowledge: field clubs and natural history societies, 1831–1887

David Allen has claimed that the natural history field club – that 'masterpiece of social mechanics', as he termed it – did more than anything else to promote natural knowledge in Victorian society. Allen identifies four defining features for such bodies in England: each had a debating chamber or other space where members could discuss their work; each published its own 'Transactions'; each maintained a museum or had collections on display; and each sought to establish a library.[78] Allen's pioneering work, largely on English field clubs, has partly been added to by others chiefly in respect of the sites, including the pub, in which botany was promoted amongst artisans.[79]

[74] [Mrs] Gordon, *The home life of Sir David Brewster* (Edinburgh, 1870), 454–5.

[75] 'On recent geographical discovery and research', *North British Review,* XXXIX (1863), 357–78.

[76] Although the article appeared anonymously as was usual, Thomas Chalmers was the author of 'Report on the Poor Laws of Scotland', *North British Review,* II (1845), 471–514, in which he took issues with W. P. Alison. [77] *North British Review,* XXVIII (1858), 70–1.

[78] D. Allen, *The naturalist in Britain: a social history* (second edition, Princeton, 1994), 142–3.

[79] A. Secord, 'Scientists in the pub: artisan botanists in early nineteenth-century Lancashire', *History of Science,* 32 (1994), 269–315; 'Artisan botany', in Jardine, Secord and Spary (eds.), *Cultures of natural history,* 378–93.

Almost no attention has been paid to such institutions within Scotland. This is unfortunate since, I shall argue, understanding *where* such bodies were located and *how* they operated is to be afforded insight into the values contemporaries placed upon science, and offers an example of the historical geography of science understood as a located social concern with natural knowledge. Considering scientific bodies in these terms is also instructive in comparison with what is known of literary and historical societies in this period in promoting national identity in Scotland. It has been argued that historical societies in Scotland have an origin in the foundation in 1780 of the Society of Antiquaries, closely followed by the Royal Society of Edinburgh (1783) and the Literary and Antiquarian Society of Perth (1784). More particularly, the proliferation of historical clubs from the 1820s – such as the Bannatyne Club in 1823, the Maitland Club (1829), the Abbotsford Club (1833) and the Spalding Club (1839) – has been seen as signalling a renewed interest in the history of Scotland broadly understood and marking 'a new confidence in Scottish national identity'. The demise of many of these bodies from the later nineteenth century marked what has been termed the 'strange death of Scottish history'.[80] This is not strictly true, since institutionalised interest in Scottish history and, thus, national identity received renewed impetus from the 1880s.[81] But there are, as I shall hope to show, parallels to be drawn between civic institutions of a historical nature and those bodies whose purpose was likewise national knowledge but whose focus was the natural world.

There were about 40 such bodies in Scotland between the late eighteenth century and the end of the nineteenth century (Table 5.3).[82] Some, such as the Glasgow and Clydesdale Statistical Society, had a limited existence. That body aimed 'to procure, arrange, and publish Facts in that department of Science [statistics], which may be defined [as] "The knowledge of the present state of a country, with a view to its future improvement".'[83] Its President was James Cleland, its Councillors included four Glasgow professors (W. J. Hooker, Botany; James Thomson, Mathematics; William Meikleham, Natural Philosophy; and William Couper, Natural History), and John Leadbetter, President of Glasgow's Mechanics' Institute. Members included Robert Owen of New Lanark, Thomas Chalmers, David Brewster, Robert Jameson and Captain Maconochie of the Royal Geographical Society in London. Admittedly, its concern with *social* statistics may not have been the same as

[80] M. Ash, *The strange death of Scottish history* (Edinburgh, 1980); C. Kidd, '*The strange death of Scottish history* revisited: constructions of the past in Scotland, c.1790–1914', *Scottish Historical review*, 126 (1997), 86–102.

[81] G. Morton, 'Historical clubs', in M. Lynch (ed.), *Oxford companion to Scottish history* (Oxford, forthcoming).

[82] This figure must be regarded as a working estimate: it does not include University student societies, and some bodies such as that Orkney body begun in 1844 – see Table 5.3, p. 185 – which may never have been formally established.

[83] *Transactions of the Glasgow and Clydesdale Statistical Society*, 1 (1837), 3.

Table 5.3 *Field clubs and natural history societies in Scotland, 1771–1887*

| Title of institution | Date of foundation/by whom founded (where known) | Membership in 1873 | Remarks |
|---|---|---|---|
| Aberdeen Philosophical Society | 1840: Rev. John Fleming, Aberdeen | 106 (in 1863) | P; Pu |
| Aberdeen Natural History Society[a] | 1863: Six Aberdeen professors and '35 other gentlemen' | 80 | E; P; Pu |
| Alloa Society of Natural Science and Archaeology | 1862: ? | 97 | A; E; P; Pu |
| Ayrshire Naturalists' Club | 1850: Rev. David Landsborough | (ceased in 1854) | |
| Ayrshire Archaeological and Natural History Society | 1877: ? | ? | ? |
| Berwickshire Naturalists' Club | 1831: George Johnston | ? | A; E; P; Pu |
| Banffshire Field Club | 1880: ? | ? | P; Pu |
| Botanical Society of Edinburgh | 1836: J. H. Balfour, Edward Forbes, William Brand and 9 others | ? | A; E; P; Pu |
| Buchan Field Club | 1887: ? | ? | P; Pu |
| Dumfriesshire and Galloway Natural History and Antiquarian Society[b] | 1862: Drs. Gilchrist, Dickson, Grierson and Gibson (president in 1862 – Sir W. Jardine) | *c.* 100 | E; P; Pu |
| Dundee Naturalists' Field Club[c] | 1863: 'some young men in Dundee' | *c.* 30 | A; E; P |
| Edinburgh Natural History Society | 1782: James Edward Smith, Robert Brown | ? | L; M; P; Pu |
| Elgin and Morayshire Scientific Association (also known as the Moray Society) | 1836: ? | 85 | M |
| Glasgow Philosophical Society | 1802: ? | 540 | Ex; P; Pu |
| Glasgow, Natural History Society of | 1851: William Gourlie, Dr W. B. Lorraine, Robert Gray and others | 205 | A; P; Pu |

| | | | |
|---|---|---|---|
| Glasgow Geological Society | 1856: James Armstrong and 'a number of young men' | ? | P; Pu |
| Glasgow and Clydesdale Statistical Society[d] | 1836: ? (president in 1837 was James Cleland) | ? | Pu |
| Hawick Archaeological Society | 1856: W. Norman Kennedy, Alexander Michie and Francis McKenzie | 129 | P; Pu |
| Inverness Field Club | 1875: ? | ? | E; L; M; P; Pu |
| Inverness Literary Institute | 1870: ? | 80 | P |
| Kilmarnock Glenfield Ramblers' Society | 1884; ? | ? | ? |
| Kirkcaldy Naturalists' Society | 1882: ? | ? | ? |
| Largo Field Naturalists' Society | 1863: Mr Howie | 45 | A; Ex; P |
| Montrose Natural History and Antiquarian Society | 1836: 'members of the Chess Club' | 190 | E; M; P; Pu |
| Nairn Naturalists' Club[e] | *c.*1872: 'some young men of Nairn' | ? | E (intended) |
| Northern Institution [in Inverness] | 1825: George Anderson (in decline by 1834) | | |
| Orkney Natural History Society[f] | 1837: Rev. William Stobbs, Rev. William Clouston | 17 | E; M; P |
| Perth Literary and Antiquarian Society | 1784: Rev. J. Scott | – | M |
| Perthshire Society of Natural Science | 1867: F. Buchanan White | 127 | A; E; M; P; Pu |
| Plinian Society | 1823: ? | – | E; L; P |
| Renfrewshire Natural History Society | 1847 (begun as Renfrew Naturalists' Society) | ? | A; E; L; P |
| Royal Physical Society of Edinburgh | 1771: ? | – | – |
| St Andrews Literary and Philosophical Society | 1838: ? (president in 1873 was the Duke of Argyll) | – | – |
| Stirling Field and Archaeological Society | 1879: ? | ? | E; P; Pu |
| Stonehaven Natural History Society | ?: ? | ? | ? |

Table 5.3 (*cont.*)

| Title of institution | Date of foundation/by whom founded (where known) | Membership in 1873 | Remarks |
|---|---|---|---|
| Strathbogie Field Club | 1883: ? | ? | ? |
| Trifontial Scientific Society [Leith] | 1821: Mr David Thorburn, Mr John Coldstream, Mr James Morton | ? | P |
| Tweedside Physical and Antiquarian Society [Kelso] | 1834: Sir Thomas Macdougall Brisbane | 60 | A; E; L; M; Pu |
| Wernerian Natural History Society [Edinburgh] | 1814: Prof. Robert Jameson | – | – |

*Note:*
A = annual meetings held; E = excursions held; Ex = exhibitions or other displays held; L = library maintained; M = museum maintained; P = programme of meetings and/or lectures; Pu = publications (*Proceedings* or *Transactions*: not including notification of a society's events in the local press); ? = no information found.

[a] It is possible that this was based on an earlier 'Aberdeen Natural History and Antiquarian Society', founded in 1845 by Professor William MacGillivray.

[b] There was also an 'Omnium Gatherum Society', based in Dumfries at the Crichton Royal Institution for the Insane, which was founded in 1867 by Dr Crichton and which concentrated upon natural history.

[c] This was a branch of the Dundee Young Mens' Christian Association.

[d] No further record of this Society beyond 1837 has been traced.

[e] The Northern Association of Literary and Scientific Societies was established in 1881 in Elgin, and membership was open to every scientific society and field club in the counties of Aberdeen, Banff, Elgin (Moray), Nairn, Inverness, Ross and Cromarty, Sutherland, Orkney and Shetland; this Association published its own *Transactions*.

[f] An Orkney Antiquarian and Natural History was begun in 1844 but no later reference or records have been traced for it.

*Sources:* W. Elliot, 'Opening Address', *Transactions of the Botanical Society of Edinburgh* 11 (1873), 19, 242–50 supplemented with information from J. A. Gibson's 'Scottish Report' to the Council Meeting of the Society for the History of Natural History (October 1995) ('Directory of Scottish Natural History Societies').

those societies more interested in natural knowledge, but, in most cases, I suggest, these institutions were motivated by common concerns to do with Scotland's national identity.

Like their historical-literary counterparts, many such bodies had publications. Most had regular programmes of meetings or, in the summer, excursions. A few were offshoots of other bodies – as was the case for the Dundee Naturalists' Field Club, for example – or the reflection as an educative enterprise of one individual's enthusiasms, for example Dr Gilchrist's Omnium Gatherum Society in Dumfries. Early foundations were associated with Edinburgh and Glasgow but, by the 1830s and more noticeably from the later 1850s, smaller towns – Alloa, Kelso, Largo, and Montrose, for example – had their own natural history society. Relatively few maintained a museum or a library, and it may be that some had more regular programmes of lectures and of excursions than is clear from Table 5.3.

Not all bodies were interested in natural science. The Philosophical Societies of Aberdeen and Glasgow and the Perth Literary and Antiquarian Society were more concerned with literary debate and civic improvement through polite and gentlemanly discussions, in the meeting rooms of the urban public sphere, than with studying Scotland's natural products. But the failure of this latter body was in part the result of its not considering the natural world, a fact which was pointed out in the revival of a civic body to such purposes in Perth in 1867 (see pp. 190–1). Yet in looking at 'antiquities', including archaeology, some bodies combined interests in Scotland's human and natural history, and thus shared an intellectual genealogy with earlier chorographical traditions concerned with national identity throughout Britain and Europe as well as in Scotland (see pp. 38–56) whilst blurring the distinctions between 'literary' and 'scientific' bodies. The Kilmarnock and Glenfield Ramblers' Society reflects, for a particular town, that engagement with healthy exercise and 'moral recreation' in the countryside that others have shown to characterise nineteenth-century Britain more generally, and which was evident in Scotland from 1843 in the foundation of the Scottish Rights of Way Society.[84]

Yet one feature in particular connects most of these field clubs and natural history societies: the idea of the local. I mean by this that these institutions should be understood as local *sites* for making natural and national knowledge, that the members valued *local knowledge* for its own sake *and* as part of scientific understanding more generally, and that their members' fieldwork and publications should be seen as the means by which given *localities* were examined. Fieldwork was about the observation, collection and classification of Scotland's nature. In those terms, this widespread popular engagement with

[84] C. Smout, 'The Highlands and the roots of green consciousness, 1750–1990', *Proceedings of the British Academy*, 76 (1991), 237–63.

being 'in the field' did not differ from the professional geological or botanical concerns of university professors. For such professional scientists, seeing, measuring and reporting things in the field played a crucial part in the establishment of rigorous scientific methodology and theoretical debate, whether the field was Scotland or anywhere else. Thus understood, then, fieldwork played a key part in the emergence and discursive constitution of science both as contemporaries recognised it and as scholars like Golinski and Allen have shown in relation to the situated geographies of science.[85]

These bodies likewise used field trips and texts to promote science as a form of collective civic natural enquiry and to instil a sense of local pride. The Berwickshire Naturalists' Club, for example, which Allen considers 'the original ancestor of local natural history societies' in Britain, held all its meetings in the field and in different parts of the Scottish Borders in order to convenience its members. The Society was itself geographically mobile and 'made' in the field. It had, as it were, no 'fixed' debating space other than the outdoors, no 'cultural capital' in the form of either library or museum. It was certainly a social scientific club: meetings began with breakfast and ended, after a day's collecting, with a dinner and discussion of scientific matters in the evening. For Allen, the means employed by the Berwickshire club, even its origin, was the *result* of local geography: 'That it was the Border Country in which this new institution first arose may have been no coincidence; for here, more than in most parts of Britain, the hierarchical divisions in society have for long been untypically ill-defined and loose. What would have been hard for others, and particularly for Englishmen, the men of Berwickshire doubtless found comparatively simple'.[86]

Allen's point highlights the importance of the local and of the social nature of institutionalised natural knowledge that I am addressing here. Yet understanding the historical geography, origins and operating practices of field clubs in Victorian Scotland as rooted in local environmental circumstances *alone* runs the risk of obscuring the features they shared in common and of underplaying the connections between such bodies. There is a danger, too, of advancing a form of geographical determinism as explanation, rather than the facts of social context and individual agency. The Berwickshire Naturalists' Club may indeed have been more socially egalitarian than many of its counterparts. But it originated with John Baird, who, with his two brothers, had been a member of the Plinian Society at the University of Edinburgh at exactly the moment that field classes in botany were introduced into teaching there by Robert Graham, Professor of Botany (as they earlier had been by Robert

[85] J. Golinski, *Making natural knowledge: constructivism and the history of science* (Cambridge, 1998), 91–102; D. Allen, 'Walking the swards: medical education and the rise and spread of the botanical field class', *Archives of Natural History,* 27 (2000), 335–67; see also J. Camerini, 'Remains of the day: early Victorians in the field', in B. Lightman (ed.), *Victorian science in context* (Chicago, 1997), 354–77; J. Clifford, *Routes: travel and translation in the late twentieth century* (Cambridge, Mass., 1997), 52–91; H. Kuklick and R. Kohler (eds.), *Science in the field, Osiris* 11 (1996). [86] Allen, *Naturalist in Britain,* 145.

Jameson in geology and were consequently by Hooker in Glasgow and Henslow in Cambridge[87]). The club took the form it did because of the abiding interest of the Bairds and of George Johnston, a Berwickshire doctor, in their local area.[88] Johnston's *The botany of the eastern Borders,* published in 1853, had a sub-title – '*with the popular names and uses of the plants, and of the customs and beliefs which have been associated with them*' – that indicates several of the ways in which local knowledge was understood and valued.

Such local texts were common from the 1840s.[89] They reflected local endeavours and a growing interest within civil society in science both as a social pursuit and as a means to improve one's self and the nation. They also indicate an emergent subject specialisation within scientific knowledge, as the case of botany illustrates. As the author of one local flora put it in 1838, 'A large and increasing share of attention has of late been paid, by Botanists, to that branch of the science which has been denominated *Local botany*. . . . The publication of these works, by diffusing more widely a knowledge of the local distribution of plants, has tended to the still farther promotion of that branch of Botany called Botanical Geography'.[90] Fieldwork was crucial to such enquiry, as Allen has shown.[91] Of course, the development of biogeography at this time owed greatly to the work of Alexander von Humboldt and Decandolle and, in Britain, to men such as Edward Forbes, whose attention from the 1830s to the local dynamic connections between geology and plants was hailed by contemporaries such as Charles Lyell and Roderick Murchison as marking the birth of a new science of geographical distribution based on the idea of motion.[92] In Scotland, a leading figure was the Yorkshireman, Hewett Cottrell Watson, who had been a student of Robert Graham on his pioneering field classes and who, in three books published between 1835 and 1873, firmly established botanical geography or, to use the title of his 1873 book, *Topographical botany*, as a legitimate subject of enquiry. For those who could not have access to Watson's four-volume *Cybele Britannica, or British plants and their geographical relations,* begun in 1847, Fleming's review in the *North British Review* stressed the importance of such knowledge to science and to knowing Scotland.[93]

Watson's work had its counterparts. The first volume of the *Transactions of the Botanical Society of Edinburgh* in 1844, for example, carried an article on

[87] J. Browne, *The secular ark: studies in the history of biogeography* (New York and London, 1983), 119. Evidence for Jameson using field trips within his teaching, certainly by 1817 if not earlier, comes in a letter from James Cairns to Jameson in which Cairns comments how he had received a letter from his nephew noting 'the pleasure and improvement he derives from your lectures; and above all from your excursions; certainly by far the best way of demonstrating natures [sic] works upon a grand scale'. EUL, MS Gen. 1999, 7 July 1817. [88] Allen, *Naturalist in Britain,* 104–5, 130, 145–6.

[89] W. Hooker, *Flora Scotica, or a description of Scottish plants* (Edinburgh, 1821); R. Greville, *Flora Edinensis: or a description of plants growing near Edinburgh* (Edinburgh, 1824); A. Murray, *The northern flora: or a description of the wild plants belonging to the north and east of Scotland* (Edinburgh, 1836).

[90] G. Dickie, *Flora Abredonensis, comprehending a list of the flowering plants and ferns found in the neighbourhood of Aberdeen* (Aberdeen, 1838), Preface. [91] Allen, 'Walking the swards', 350–62.

[92] Browne, *The secular ark,* 121.

[93] [J. Fleming], 'Geographical and physical distribution of plants', *North British Review,* 20 (1854), 501–22.

the botanical excursions to Skye and the Outer Hebrides in August 1841 of John Hutton Balfour, Regius Professor of Botany at Glasgow, and Charles Babington.[94] Balfour made numerous botanical excursions throughout Scotland between 1846 and 1878, often accompanied by as many as eighty students and fellow scientists. The whole tone of the reports upon these excursions and of Balfour's lectures on the geographical distribution of plants is one of collecting from and examining a nation's countryside – particularly in the Highlands – which is hardly known to scientific enquiry.[95] Understanding the geographical distibution of plants was central to the society's objectives at its foundation: it was for his work in this respect that Humboldt was elected a member in 1837.[96]

The emergence of biogeography as a form of natural science should not be seen, then, in strictly academic spaces and in disciplinary terms. Much of what Balfour, Babington, Watson and others were doing was being done no less rigorously in local civic contexts by members of Scotland's field clubs and natural history societies. The fact that natural knowledge *and* civic pride were together recognised as of importance and were, even, mutually constitutive, is clear from the inaugural address in 1867 of F. Buchanan White, president of the Perthshire Society of Natural Science.

For some time past it has been the desire of the naturalists of Perth that their city, the capital of a county perhaps richer in its fauna and flora than any other in Britain, should have an association, the members of which, by working together and mutually assisting each other, might advance the cause of those sciences which they love; and that the "Fair City" should not, in this age of reform, be totally unrepresented in the scientific body of Great Britain.[97]

What characterised Perth from the late 1860s, Berwickshire from 1831, and the botanically-minded in Edinburgh from 1836, was apparent in Inverness from 1825 with the work of the Northern Institution and its attempts to establish a museum and maintain a library.[98] That body failed. In launching the Inverness Scientific Society and Field Club in January 1876, William Jolly, a local man and HM Inspector of Schools, lamented earlier shortcomings as he sought to revive civic interest in science in the town:

[94] J. Balfour and C. Babbington, 'Account of a botanical excursion to Skye and the Outer Hebrides during the month of August 1841', *Transactions of the Botanical Society of Edinburgh,* 1 (1844), 133–44.

[95] Anon., 'Botanical excursions made by Professor John Hutton Balfour in the years from 1846 to 1878 inclusively', in *Notes from the Royal Botanic Garden, Edinburgh,* VII (1902), 21–49; see also NCL, MS Box 3.1.2, Botany No. 1 [-11], Dictates by J. H. Balfour [n.d., but probably 1847–48] on 'Geographical distribution of plants'; Morrell and Thackray (eds.), *Gentlemen of science,* 494. Such evidence contradicts claims that the 'Golden Age' of Scottish vegetational mapping was in the late nineteenth and early twentieth centuries: see A. Mather, 'Geddes, geography and ecology: the golden age of vegetation mapping in Scotland', *Scottish Geographical Journal,* 115 (1999), 35–52. Further support for my bringing earlier this civic engagement in botanical enquiry comes from F. Horsman, 'Plant distribution patterns: the first British map', *Archives of Natural History,* 26 (1999), 279–86.

[96] *First annual report of the Botanical Society of Edinburgh* (Edinburgh, 1837–40), 41.

[97] *Perthshire Society of Natural Science. Inaugural address by the President* (Perth, 1867), 3.

[98] *Transactions of the Northern Associations of Literary and Scientific Societies,* 1 (1887).

Yet how little have we done for science in Inverness, although we occupy the centre of a wide and interesting field of scientific study! Other towns of far less population and importance have long surpassed us in scientific spirit and scientific appliances. They have had their museums, science schools, scientific lectures, scientific societies, scientific field clubs, and like evidences of scientific activity.[99]

Of course, not every one was a member of such bodies. They were largely male dominated, and subscription levels excluded the less well-off. Yet, just as the transformation of intellectual life in Victorian England was importantly directed by local scientific institutions,[100] so too in Scotland national identity was being made as a civic enterprise through local societies and in the practices they employed.

## Conclusion

In discussing mid-nineteenth-century Scottish nationalism, Morton has argued that 'mid-century civil society meant local government and the associational activity of the urban bourgeoisie. It was the propensity of the urban middle class to organise clubs, societies and associations in the "spaces" in civil society left untouched by the central and local state which, it is hypothesised, was the essential mediating structure between state and nation'.[101] The concepts of 'civil society' and of 'civic nationalism' are, of course, more complicated than is suggested by the 'associational activity' of the urban middle class, as Morton has himself shown, and I do not suppose this chapter to have fully located science in general and geography in particular in the 'spaces' of mid-nineteenth-century Scottish culture and society, or even to have documented all the senses and sites in which geography was differently understood and used. Even so, there are several related senses in which it is possible to 'place' geography as a situated and institutionalised social and scientific activity and, in those terms, as a civic enterprise in the half century from 1830.

The first sense has to do with education and the different nature of geography's teaching sites. The evidence discussed here on private classes in geography, the activities of itinerant self-taught artisans, on geography teaching within Mechanics' Institutions, universities and throughout Scotland's schools points both to a complexity in the historical geography of geography not before documented, and to social and geographical differences in exposure to geography as a form of knowledge. The second sense, closely related to the first, concerns the development of a geographical publishing market, apparent in textbooks from about 1820 and, from the 1830s, in the place of

[99] W. Jolly, 'Opening address: the scientific materials of the north of Scotland', *Transactions of the Inverness Scientific Society and Field Club*, 1 (1875–1880), 5.

[100] Allen, *Naturalist in Britain*; Heyck, *Transformation of intellectual life in Victorian England.*

[101] G. Morton, 'Scottish rights and "centralisation" in the mid-nineteenth century', *Nations and Nationalism*, 2 (1996), 262.

geographical periodical literature in the urban public sphere. The third encompasses the idea of geography as a form of 'civic consciousness', differently apparent, I argue, in the concerns of men like Thomas Chalmers and James Cleland, for example, who thought and acted geographically in order to improve the lot of Scotland's urban poor; in Keith Johnston's publishing and pricing policies; in recognition by scientists and, in the case of James Pillans, by humanists of geography as a matter of academic and popular scientific importance; and, more commonly, in the excursions and lectures of natural history societies through which Scots encountered and promoted their own geographical locality and developed national and scientific knowledge in so doing.

Yet, for all this activity, for all the sense that geography was a means to social emancipation through learning and self-improvement and, in parochial schools, a basis to moral instruction, for all that it was produced and consumed in the 'spaces' of the urban public sphere, the story of geography in mid-nineteenth century Scotland is, in several respects, a story of failure. The one journal that sought to promote geography and other natural sciences in socially emancipatory ways lasted less than two years. There was no professor of geography in this period, even though professors of mathematics, of humanity and of civil and natural history taught the subject, and this despite the fact that, in 1827 anyway, geography was taught to more pupils in Scotland's parochial schools (5,370 pupils) than was the case for Greek and mathematics combined.[102] Geography figured more prominently in Scotland's 'democratic intellect' than has been supposed. Yet the geography books used in Scotland's schools were not the product of original research but works of synthesis, either drawn, like Stewart's long-successful *Compendium of modern geography*, from the work of others or, like Murray's 1834 *Encyclopedia*, compiled by natural scientists from other disciplines for whom a particular element of geography was part of their own more specialised concerns. Specialist scientific societies and popular bodies were numerous. What many did – in their fieldwork, in programmes of local studies – was profoundly geographical. But not one called itself a 'geographical society'. In sum, in the very period in which science in Scottish intellectual culture was, so to speak, democratised, institutionalised and specialised, geography, for all its place in the urban public sphere and inclusion within schools, failed to register in the very ways that could have secured it civic 'space': as a recognised subject with specialist staff in a given institutional setting.

Why this should be so is not clear. I do not mean to dismiss the fact of geography's 'failure': as others have noted, 'failure often exhibits more clearly than success the constraints and resources available for the diffusion of science'.[103] Three reasons suggest themselves. First, geography as a subject was in this

[102] G. Davie, *The democratic intellect: Scotland and her universities in the nineteenth century* (Edinburgh, 1999 edition), 12. [103] Shapin, 'Nibbling at the teats of science', 151–2.

period almost wholly descriptive and concerned more with the *facts* of the spatial location of phenomena and hardly at all with their causal *explanation*. Even when certain empirical and epistemological practices – fieldwork, classification, realism in description, comparative explanation, emphases upon the region – were from the 1830s much more widely employed, these were matters shared by all the natural sciences. A similar disciplinary 'failure' has been noted of geography in France at this time.[104] Geography understood as a matter of location, of distribution and of earth description was, of course, of interest and 'useful' in the several sites explored here, but whilst valued in those terms by specialists in other sciences, it did not have its own subject 'core'. Second, there was no single figure through whom the public profile of geography might be raised, either in the universities or across that network of institutions through which geographical knowledge as a civic enterprise was undertaken. Third, given both these facts and the nature of the curriculum within schools and universities, there was no agreed institutional recognition for geography despite widespread public engagement with it.

It is, I suggest, possible to 'explain' geography's position in this period in terms of either a 'weak' or a 'strong' model. In the first sense, geography's diminishing formal significance (despite such widespread civic engagement with the subject and with local geographical knowledges in the field) may be explained with reference to the discipline's lack of an essential epistemological focus at a time of disciplinary formation within the sciences in general. In the second sense, geography did not so crystallise as a subject because it did not have to. For botanists and geologists, as much as for ameliorative social scientists and churchmen, thinking geographically – about location and about causes in and relationships over space – was, simply, central to their scientific disciplines and to their social and historical concerns.

However we are to understand it, geography's place as a matter of concern to civic society in nineteenth-century Scotland has been recovered here at various scales. In closing his discussion of the Edinburgh Philosophical Association in the 1830s, Shapin argued that 'what one sees in Edinburgh is conflict between groups which had a *national* (or international) cultural horizon and those whose cultural interests were almost purely *local*'.[105] To some degree, geography's place within the 'provincial metropolis' that was Edinburgh would support this view. In other respects, the international *and* the local concerns of the *Edinburgh Journal of Natural and Geographical Science* would deny it. Scientific enquiry as an institutionalised form of civic nationalism had a geography of local sites and institutions. Certain towns and not others had scientific societies, each with particular practices and concerns with certain parts of Scotland as their 'study site'. Exposure to geography in school was not everywhere the same across Scotland. Scientific ideas had different

[104] A. Godlewska, *Geography unbound: French geographic science from Cassini to Humboldt* (Chicago, 1999).
[105] Shapin, 'Nibbling at the teats of science', 170.

geographies of reception even within one city as political and ideological affiliations determined different responses.[106] And as Hugh Mill's experiences in Thurso in about 1865 suggest, we should not lose sight of domestic space as a site for geographical learning:

> A round mahogany box, about four inches across, with a domed lid stood on the parlour mantelpiece in the home of my infancy, . . . It must have been familiar to me since I was four years old, and it had a strange and mysterious significance, for it contained the world. It actually held a dainty little terrestrial globe about the size of an orange, . . . On this my mother had received from her father the first lesson in that rare accomplishment then so common in the prospectuses of Seminaries for Young Ladies – The Use of the Globes; and with its aid she in turn gave all her children their first ideas of Geography. I cannot remember a geography lesson at school, though such there must have been, but the lesson of that little globe was so deeply impressed that I never seem to have wanted more.[107]

The following chapter takes further these connections between 'home' knowledge, geographical enquiry at local and national scales, and the idea of empire.

[106] J. Scowen, 'A study in the historical geography of an idea: Darwinism in Edinburgh 1859–1875', *Scottish Geographical Magazine,* 113 (1998), 148–56; more generally, see D. Livingstone, D. Hart and M. Noll (eds.), *Evangelicals and science in historical perspective* (New York and Oxford, 1999).

[107] H. Mill, *Life interests of a geographer, 1861–1944: an experiment in autobiography* (East Grinstead, 1944), 64.

# 6

# Geography and national identity in an age of high empire, 1884–1930

Attention to the relationships between geography and empire is both relatively recent and one of the main features of the critical histories of geographical knowledge. Since Hudson (1977), who explored the connections between geography, militarism and commerce, European 'national schools' of geography, the development of university geography and the coordinating role of civic geographical societies,[1] numerous studies have examined geography's place in Hobsbawm's 'Age of Empire' [1875–1914].[2] Attention has been paid to the geographical nature of imperialism,[3] to the role of geographical societies in imperial contexts,[4] to imperialist rhetorics of race and the human condition,[5] and to representations of the imperial subject in photography,[6] the music hall,[7] travel accounts,[8] popular literature[9] and geographical textbooks.[10]

[1] B. Hudson, 'The new geography and the new imperialism: 1870–1918', *Antipode*, 9 (1977), 12–19.

[2] E. Hobsbawm, *The age of empire* (London, 1987).

[3] F. Driver, 'Geography's empire: histories of geographical knowledge', *Environment and Planning D: Society and Space*, 10 (1992), 23–40; A. Godlewska and N. Smith (eds.), *Geography and empire* (Oxford, 1994); M. Bell, R. Butlin and M. Heffernan (eds.), *Geography and imperialism, 1820–1940* (Manchester, 1995).

[4] M. Heffernan, 'The science of empire: the French geographical movement and the forms of French imperialism, 1870–1920', in Godlewska and Smith (eds.), *Geography and empire*, 92–114; 'The spoils of war: the Société de Géographie de Paris and the French empire, 1914–1919', in Bell, Butlin and Heffernan (eds.), *Geography and imperialism*, 221–64; G. Sander and M. Rossler, 'Geography and Empire in Germany, 1871–1945', in Godlewska and Smith (eds.), *Geography and empire*, 115–29; P. van der Velde, 'The Royal Dutch Geographical Society and the Dutch East Indies, 1873–1914: from colonial lobby to colonial hobby', in Bell, Butlin and Heffernan (eds.), *Geography and imperialism*, 80–92; J. Mackenzie, 'The provincial geographical societies in Britain, 1884–1914', in Bell, Butlin and Heffernan (eds.), *Geography and imperialism*, 93–124.

[5] D. Livingstone, 'Climate's moral economy: science, race and place in post-Darwinian British and American geography', in Godlewska and Smith (eds.), *Geography and empire*, 132–54.

[6] J. Ryan, 'Imperial landscapes: photography, geography and British overseas exploration, 1858–1872', in Bell, Butlin and Heffernan (eds.), *Geography and imperialism*, 53–80; *Picturing empire: photography and the visualization of the British empire* (London, 1997).

[7] A. Crowhurst, 'Empire theatres and the empire: the popular geographical imagination in the age of empire', *Environment and Planning D: Society and Space*, 15 (1997), 155–74.

[8] E. Said, *Orientalism* (London, 1978); *Culture and imperialism* (London, 1991).

[9] R. Phillips, *Mapping men and empire: a geography of adventure* (London, 1996).

[10] A. Maddrell, 'Empire, emigration and school geography: changing discourses of imperial citizenship, 1880–1925', *Journal of Historical Geography*, 22 (1996), 373–87; 'Discourses of race and gender and the comparative method in geography school texts 1830–1918', *Environment and Planning D: Society and Space*, 16 (1998), 81–103.

Given that the 'craft practices of the emerging geographical professional' – exploration, topographic and social survey, cartographic representation and regional inventory – were 'entirely suited to the colonial project' and that the Church's missionary movements underpinned European imperialism, it is hard to disagree with the claim 'that geography was the science of imperialism *par excellence*'.[11] Despite such attention, there is yet more to know concerning the place of geographical knowledge in the 'Age of Empire'. For Godlewska and Smith, the 'connection of histories *of* geography with historical *geographies* is what needs to be explored'.[12]

This chapter considers these connections in Scotland in the forty years or so following the foundation in 1884 of the (Royal) Scottish Geographical Society.[13] Several scholars have shown how national identity in later nineteenth-century Scotland revolved around empire and imperialism: Scotland's economy was 'the Workshop of Empire'; the union with England was understood as 'the imperial partnership'; and the Scots were 'a race of Empire builders'. Traditions of Scottish militarism secured an enduring symbolic liaison between martial prowess and national identity. The place of Scots in missionary movements, in which context David Livingstone was venerated as the embodiment of Scottish character, promoted Scotland's religious identity. Scots were seen as imperial administrators and considered to have commercial acumen at home and abroad. In politics, the idea that Scotland's union with England was an 'imperial *partnership*' was widely shared. For Finlay, 'the imperial mission commanded a near universal political consensus in the nineteenth and early twentieth centuries'.[14]

The sense in which Scots led rather than simply shared in the idea and the making of the British empire, and that such imperial activity was central to Scotland's own national identity, is clear in the foundation and work of the Royal Scottish Geographical Society (RSGS). For Finlay, the society 'proved to be an ideal forum for the propagation of imperial values and ideas and, under the guise of exploration, geographers and geologists were often at the cutting edge of imperial expansion'. Geography in imperialist terms underpinned Scotland's own identity:

Geography was essential to this process as the exploration of the world was a valuable mirror to reflect Scottish national prowess to an eager domestic audience. The construction and reinforcing of certain 'imaginative geographies' of foreign and distant

[11] D. Livingstone, *The geographical tradition: episodes in the history of a contested enterprise* (Oxford, 1992), 170.

[12] N. Smith and A. Godlewska, 'Introduction: critical histories of geography', in Godlewska and Smith (eds.), *Geographies of empire,* 7.

[13] The Society did not have the epithet 'Royal' attached to it until 1887. The Minute Book of the RSGS Council on 4 July 1887 confirms notification of 'the Queen's gracious permission to the Society to bear in future the title "The Royal Scottish Geographical Society"': RSGS Papers, Minute Books of Council, Vol. 2, 35.

[14] R. Finlay, 'The rise and fall of popular imperialism in Scotland, 1850–1950', *Scottish Geographical Magazine,* 113 (1997), 13–21: quotes from p. 17.

lands enabled the Scots to compare and contrast their own supposed superior national characteristics against those of 'inferior' peoples. . . . Paradoxically, the best way to demonstrate the qualities of the national character was to display them outside of Scotland.[15]

With these ideas of empire, geography and national identity in mind, this chapter is, in part, a work of synthesis. Yet in drawing upon the work of others, I am also concerned to suggest that much current work offers too simple and, even, too celebratory a reading of the connections between national identity and geographical knowledge in 'imperial Scotland'. In part, then, this chapter is about revisionism and a return to complexity. It is also about the nature of that scientific work employed in this period to establish geographical knowledge of the nation, work that I here call 'the sciences of national survey'.

What follows is in three main parts. The first discusses the formalisation of geography in the late nineteenth and early twentieth centuries in relation to the foundation of the RSGS and the establishment of university departments of geography. What were seen by many as almost 'triumphal' moments for geography is contrasted with contemporary evidence concerning the perceived deficiencies of education in geography. The second section examines the ways in which particular sciences directed their work less at the promotion of the essential 'imperial Scot' and more at national self-knowledge. It is true that in oceanography and in the polar explorations of William Speirs Bruce, Scottish science was bound up with the geographical and scientific making of the British Empire. In contrast, survey work in botanical and geological mapping was, I suggest, not about Scotland's place in the British Empire, however much the techniques of field measurement and spatial correlation were shared by different disciplines and apparent as geography's 'craft practices' in the work of Scots and others overseas. Further, whilst the work of those in this period who mapped the racial and linguistic characteristics of Scots had parallels in that racialised knowledge then informing geography's 'imperial mission' and the development of ethnography and of anthropology as particular disciplines, I want to show how such matters of survey and of epistemological practices were questions of *national* as well as imperial identity. The scientific constitution of national knowledge 'in the field' in this period was also connected with the exploration of Scotland by its citizens, as mountaineering clubs and rambling made many aware of their country's geography by walking and climbing in it. The third section turns away from such open spaces to examine the means employed by Patrick Geddes to promote his vision for geographical knowledge in the Outlook Tower in Edinburgh. This site, as I hope to show, was not just an institutional setting for local geographical understanding, including work in the field, but for the promotion through geography of ideas of global citizenship.

[15] Finlay, 'The rise and fall of popular imperialism', 18.

My concern in addressing these issues is not to deny the ways in which 'imaginative geographies' of empire sustained Scotland's national identity from the late nineteenth century. It is to examine the hitherto rather neglected ways in which Scotland was itself geographically encountered in that age of high empire, and to consider the scientific practices and the sites in which geographical knowledge in and of Scotland was undertaken at a time when Scotland was self-consciously part of Britain's 'imperial mission'.

## Institutionalising the subject

For many European nations, the emergence of 'modern' geography is a late nineteenth century phenomenon. Scholars who have studied the formalisation and institutionalisation of geography in this period – the foundation of geographical societies and of university departments – have identified European imperialism, political recognition of the commercial advantages to be had from geographical understanding, and a concern for popular geographical education as amongst the shared leading causes.[16] The first civic societies to appear were the *Société Geographique de Paris* and the *Gesellschaft für Erdkunde zu Berlin*, both founded in 1821. The Royal Geographical Society was established in 1830. The RSGS was established amongst the last of over sixty-two such bodies begun between 1870 and 1890,[17] and, in Scotland, University positions in the subject were not established until the early 1900s. What has been seen as a founding 'moment' for Scottish geography was relatively late in world terms.

### *The foundation and role of the Royal Scottish Geographical Society*

An anonymous commentator noted:

> It has been said that as Scotland has done so much for exploration, colonisation, and the science of geography without a Geographical Society, a Scottish Geographical Society is wholly unnecessary.[18]

This view of this anonymous commentator clearly did not find support. Although contemporaries recognised that there was a long-running genealogy for geography in Scotland, they also considered that the foundation of such a formal body would be a new departure in terms of popular support for geographical work:

> The truth is that Scottish explorers and Scottish geographers have hitherto been working under tremendous disadvantages. They have received no properly organised

[16] T. Freeman, *A hundred years of geography* (Manchester, 1971), 49–68; D. Stoddart, *On geography and its history* (Oxford, 1986), 41–141; H. Capel, 'Institutionalization of geography and strategies of change', in D. Stoddart (ed.), *Geography, ideology and social concern* (Oxford, 1986), 37–79; T.Unwin, *The place of geography* (London, 1992), 79–89. [17] Capel, 'Institutionalization of geography', 57.

[18] Anon., 'Scotland and geographical work', *Scottish Geographical Magazine*, 1 (1885), 18.

assistance from their own country, and they have had to contend with the greatest discouragements, the chief of which have sprung from the ignorance and consequent indifference of their fellow- countrymen.[19]

This contemporary view is important. Scholarship on the society has tended to stress its foundation in consequence of rather than despite popular social support, and has identified support from Scotland's, and notably from Edinburgh's, scientific community as fundamental reasons to have a geographical society.

Lochhead has noted three principal factors in explanation of the foundation of the society and the 'prowess' of Scots in late-nineteenth-century British geography. The first, Scotland's 'intellectual heritage', was evident in the nation's 'democratic intellect'. The second, the 'geographical heritage', consisted of 'a tradition of practical geographical work in the form of cartography and map-making, the compilation of geographical descriptions for encyclopaedias, and of exploration'. The third was the 'scientific background', in which geology, oceanography, meteorology and natural history are identified as 'the principal fields of scientific research'. Thus, 'there were several strands of scientific study and research which must have predisposed their practitioners to at least a sympathetic attitude towards the geographical perspective. It was to the tradition of work in the natural and field sciences in particular that geography in Scotland owed its distinctive strength and character'.[20]

This argument, developed by Lochhead in a later paper[21] and, broadly unchallenged by others,[22] is, I suggest, correct in only some respects. In others, it is open to criticism on several grounds. The first is that of timing. Whilst the contribution to geographical thinking of men like Archibald and James Geikie, Professors of Geology at Edinburgh, the findings of the 1872–6 *Challenger* expedition and the work of Charles Wyville Thomson and of John Murray in oceanography, the building of meteorological observatories and, in natural history, the work of Robert Brown and Marcel Hardy together with the role of Marion Newbigin and Patrick Geddes is not to be disputed, most of their important work *post-dates* the foundation of the society. The RSGS, I want to argue, was part of the promotion of sciences of national survey but not consequent upon them.

The second difficulty concerns the questions of personal influence and of epistemological connection and the relationships between them. Lochhead's argument centres more upon the presumed influence of personal connections than it does in demonstrating the shared nature of scientific enquiry.

[19] Anon., 'Scotland and geographical work', 18.

[20] E. Lochhead, 'Scotland as the cradle of modern academic geography in Britain', *Scottish Geographical Magazine,* 97 (1981), 100.

[21] E. Lochhead, 'The Royal Scottish Geographical Society: the setting and sources of its success', *Scottish Geographical Magazine,* 100 (1984), 69–80.

[22] I. Adams, A. Crosbie and G. Gordon, *The making of Scottish Geography: 100 years of the R.S.G.S.* (Edinburgh, 1984); M. Bell, 'Edinburgh and empire: geographical science and citizenship for a 'New' age, ca.1900', *Scottish Geographical Magazine,* 111 (1995), 139–49.

Archibald Geikie, for example, who 'considered himself a physical geographer – presumably influenced in this by his friendship with Murchison' – is cited as an important figure, and his 1887 *The Teaching of geography: suggestions regarding principles and methods for the use of schools* is noted without consideration, for example, as to its effect in relation to what other evidence suggests was a widespread concern for the state of geography in schools.[23] Of the Geikies' work with the Geological Survey, we are told, without substantiation, that 'It may be that these men brought to the developing society [the RSGS] an added interest in practical research projects, derived from their work in connection with the survey, which was significant in the formation of the character of geographical work in that country'.[24] The idea and practices of survey derived from field measurement were as important in geological enquiry as they were in oceanographical and botanical work, but the social and epistemological context to such work is a complex matter. It may well be true that, to paraphrase Lochhead, 'there were . . . strands of scientific study . . . which *must have* [my emphasis] predisposed their practitioners to at least a sympathetic attitude towards the geographical perspective', but presumption in such terms begs too many questions. Simply, more remains to be known. It is for this reason that I pay attention to the sciences of national survey.

The final difficulty concerns interpreting the circumstances of foundation. I am not disputing the civic origins in the work of John George Bartholomew of the map-making firm or the role played by Mrs A. L. Bruce, David Livingstone's daughter. Neither do I wish to underplay the fact that Bartholomew drew upon his first-hand knowledge of geography in Germany in promoting popular knowledge of geography and the state's imperial interests. I do want to question, however, the established cause and effect explanations of the society's foundation. Given the timing and the nature of the contextual influences upon the foundation of the RSGS, and the evidence for the place of geographical knowledge in Scotland in the fifty years before 1884, it is possible to suggest that the RSGS was established *despite* rather more than simply *because* of any contemporary recognition and support of the new scientific geography.

These claims find support from consideration of what the society aimed to do. Although the RSGS was concerned with both national and imperial geographical issues, it emphasised education and research above exploration. As was noted in the prospectus advertising the society's foundation, 'it is therefore one of the first objects of the Scottish Geographical Society to advance the study of geography in Scotland; to impress the Public with the necessity and inestimable value of a thorough knowledge of Geography in a commercial,

[23] Lochhead, 'Scotland as the cradle of modern academic geography', 101; On Geikie's interests in geography and education and his other geography texts – *Physical geography* (1877), *Elementary lessons in physical geography* (1877), *An elementary geography of the British Isles* (1888) – see W. Marsden, 'Sir Archibald Geikie (1835–1924) as geographical educationist', in W. Marsden (ed.), *Historical perspectives on geographical education* (London, 1980), 54–65, and D. Oldroyd, *The Highlands controversy: constructing geological knowledge through fieldwork in nineteenth-century Britain* (Chicago, 1990), 337–8.

[24] Lochhead, 'Scotland as the cradle of modern academic geography', 101.

scientific or political education'.[25] Among the twelve aims stated at its foundation (four had to do with how the society was to operate), four had to do with the usefulness of geography, the need for exploration, and the desirability of publishing explorers' data with geographical information from and of use to overseas colonies. The first aim, for example, advertised the intention of making people aware of geography and of geography as useful to its audiences: 'To interest the people in the practical uses of Geography, and impress them with its great value in connection alike with science, civilisation, and international commerce'. The other four concentrated upon scientific matters, including promoting investigations into 'the Geography and Statistics of Ethnology, Climatology, &c'. A plea was made for 'special attention' to be paid to Scottish topography 'by encouraging local research'. The society's final declared aim was 'to press for the recognition of Geography as a branch of higher education, and to encourage its study in the schools and universities of Scotland, by offering prizes, or other means'.[26]

Bridges notes that these things 'show that the founders were [as] concerned with the intellectual development of the subject as [with] promoting British imperialism'.[27] It is also the case that such concerns were directed at knowing Scotland, at promoting geographical knowledge in and of the nation, given, in the words of the contemporary above, 'the ignorance . . . and indifference' of Scots to such matters.

The Royal Scottish Geographical Society must further be understood not in terms of a single organisation with a single objective, but rather as a question of local geographies of scientific and civic enterprise. To an extent at its foundation and certainly in its later promotion of the subject, the society in Edinburgh owed its success to the conjunction of scientific interests and disciplines centred upon that city.[28] In the Aberdeen branch, inaugurated on 3 February 1885, imperial and commercial interests were more to the fore as was also the case in Glasgow and, to a lesser extent, the Dundee branch. In Aberdeen especially, interest in having a branch of the society was particularly prompted by Free Church of Scotland ministers and others with missionary interests in Africa. Notable in this respect was George Adam Smith, then minister of Queen's Cross Free Church and later to be Professor of Hebrew at the Free Church College in Glasgow and author, in 1894, of *The historical geography of the Holy Land.*[29] Professor Donaldson, Professor of Humanity in the university, argued that Aberdonians had the appropriate talents and 'hard

[25] *Prospectus of the Scottish Geographical Society* (Edinburgh, 1884), 1. This Prospectus was produced as a four-page leaflet. The one consulted here is inserted in the Minute Book of the RSGS Council, Vol. 1.

[26] *Prospectus of the Scottish Geographical Society,* 2.

[27] R. Bridges, 'The foundation and early years of the Aberdeen Centre of the Royal Scottish Geographical Society', *Scottish Geographical Magazine,* 101 (1985), 77.

[28] Bell, 'Edinburgh and empire'.

[29] On the foundation of the Aberdeen branch, see Bridges, 'The foundation and early years'; on the Dundee branch, see A. Smith, 'The Dundee centre of the Royal Scottish Geographical Society 1884–1985', *Scottish Geographical Magazine,* 101 (1985), 184–6; on George Adam Smith, see R. Butlin, 'George Adam Smith and the historical geography of the Holy Land: contents, contexts and connections', *Journal of Historical Geography,* 14 (1988), 381–404.

heads' suited to the work of such a body overseas. Bridges has hinted that the Aberdeen Philosophical Society, begun in 1840 (see Table 5.3), whose meetings included talks on geography and many of whose members were later to join the geographical body, may have provided a precedent for the civic interest shown in having a branch of the society.[30] In these terms – as situated local circumstances involving other civic associations and connections between social and religious attitudes in different Scottish towns – there is, as Mackenzie notes, 'A great deal more work . . . to be done on the distinctive characteristics of the four branches of the Scottish society',[31] not least given the evidence for the 'general disinclination' expressed in Inverness in 1896 about a branch there,[32] and the short-lived London branch, which began in April 1892 and ended after only two meetings.[33]

The imperial, and, particularly, Africanist emphases of the society were certainly stressed by Henry Morton Stanley (whose own position has been the subject of revision[34]) in his inaugural speeches to the opening of the branches in Edinburgh, Dundee and Glasgow (Stanley did not visit Aberdeen until 1890). Stanley's address in Glasgow, which was co-hosted with the Geographical Section of the Glasgow Philosophical Society, focused upon 'viewing geographical science as it affects Africa'. He enumerated the achievements of Scots in Africa in order to instil a sense of 'imperial enterprise', speaking of the Scots as an imperial race 'whose energy and enterprise has sent Scotsmen to all countries of the world as pioneers of discovery, as founders of thriving colonies, as successful merchants and traders, and as useful missionaries and philanthropists'.[35]

[30] Bridges, 'The foundation and early years', 79.

[31] Mackenzie, 'British provincial geographical societies', 101.

[32] RSGS Papers, Uncatalogued MS, letter from a [Mr] E. Macmillan of the Caledonian Banking Company, in Inverness, on 7 October 1896, to Colonel Bailey of the RSGS. This notes 'I have . . . made enquiry in several quarters as to the feasibility of establishing a Branch of your Society in Inverness but I am sorry to say I have not got any encouragement. The fact is, we have as many Societies of one kind and another that I find a general disinclination to add to their number. It has been suggested to me, however, that possibly arrangements might be made in connection with our Scientific Society and Field Club to have one or two Lectures in the course of the Winter, and if you think that can be done I shall be glad to put you in communication with the Secretary of that Society.' The Inverness Field Club [see Table 5.3), p. 185] held lectures on geographical topics and organised excursions, but it is unclear that these were held in conjunction with the RSGS. An Inverness Branch of the RSGS was not begun until 1999.

[33] RSGS Papers, Council Minutes, Vol. 2, 107. (14 April 1892). Professor James Geikie, Chair, noted that the London Branch membership had risen from 59 to 105. The *Scottish Geographical Magazine,* 8 (1892), 680 notes that 'in view of certain privileges which the Royal Geographical Society of London has accorded to members of this Society, the Council has resolved to withdraw this Branch'. It is unclear what these privileges were.

[34] F. Driver, 'Henry Morton Stanley and his critics: geography, exploration and empire', *Past and Present,* 133 (1992), 134–66.

[35] H. Stanley, 'Inaugural address delivered before the Scottish Geographical Society at Edinburgh, 3rd December 1885', *Scottish Geographical Magazine,* 1 (1895), 1–17. The Geography and Ethnology Section of the Glasgow Philosophical Society was formed on 11 April 1883: see 'Report of the Committee appointed to prepare a draft constitution, &c., for the proposed geographical and ethnological section', in *Proceedings of the Philosophical Society of Glasgow,* 14 (1882–83), 539–43. The Section lasted until at least 1914 and was in regular communication with the offices of the RSGS over lecture programmes: see, for example, *Scottish Geographical Magazine,* 30 (1914), p. 44, at which it was indicated that H. J. Mackinder 'will address the Phil. Soc. on a geographical subject'.

Given such appeals to this imperialist *mentalité* at just the moment of geography's civic recognition, it is perhaps hardly surprising that the imperial focus of the RSGS has been stressed to the relative neglect of its other aims and activities. Yet, as Mackenzie has shown, Scotland's 'provincial geographical societies' between 1884 and 1914 'seemed to succeed rather than precede the developing scramble for Africa', and they certainly shared their interest in 'municipal imperialism' with bodies in Liverpool, Manchester, Hull, Tyneside and Southampton.[36] It is possible, even, that imperialism was no longer as central to the mission of the RSGS by the early twentieth century as it had been in 1884–5. Bell has argued that, in relation to what she calls the 'high stage of empire' (1870–1914), debates on the nature and benefits of imperialism within the *Scottish Geographical Magazine* provided an important link between 'the aggressive imperial expansion of the late nineteenth century and the more introverted ideology of the early twentieth century which, in the aftermath of the South African War (1899–1902), marked a "new" age devoted to national renewal at home and the creation of a Greater Britain overseas'.[37] Certainly, when James Geikie presented the society's case for Treasury support to the Chancellor of the Exchequer in January 1908, his argument made no mention of the society's 'imperial mission' but centred upon its work in 'the advancement of geographical education in Scotland', which included its publications, its lecture programmes, its library, its support for the Botanical and the Hydrographical Surveys, and what he called 'One of their latest moves, which fortunately turned out successfully, . . . the attempt to get a lectureship started in the University of Edinburgh'.[38]

Bell's claims about the advancement of geography and the role of the *Scottish Geographical Magazine* are particularly supported by what is known of the journal under the editorship between 1902 and 1934 of Marion Newbigin.[39] Maddrell has shown that Newbigin 'did much to shape the nature of geography disseminated to readers of the journal, [and] demanded high literary and scientific standards, making the *SGM* a leading British geographical journal'. Newbigin did so because of her views on geography as a form of civic and natural science, and of geographical education through scientific geography as a means to train the worldly citizen. The purpose of the *SGM*, she argued in an early editorial, was to 'make clear the methods and results of modern science' and to 'act as an intermediary between scientific specialists and the general public, and thus to assist human progress by allowing the conclusions of science to be promptly applied in human affairs'.[40] These ideas she

[36] Mackenzie, 'British provincial geographical societies'. [37] Bell, 'Edinburgh and empire', 140.

[38] *The Glasgow Herald,* 14 January 1908, 3.

[39] Adams, Crosbie and Gordon, *The making of Scottish geography*; M. Newbigin, 'Geography in Scotland since 1889', *Scottish Geographical Magazine,* 29 (1913), 471–79; 'The Royal Scottish Geographical Society: the first fifty years', *Scottish Geographical Magazine,* 50 (1934), 257–69.

[40] A. Maddrell, 'Scientific discourse and the geographical work of Marion Newbigin', *Scottish Geographical Magazine,* 113 (1997), 33–41; M. Newbigin, 'Editorial: the value of geography', *Scottish Geographical Magazine,* 21 (1905), 4.

expanded upon practically in her involvement with Geddes' teaching in the Outlook Tower, and in a 1907 paper (which formed the basis to her 1924 Presidential Address to Section E (Geography) of the BAAS) where she argued 'We want in the first instance to train the power of observation and develop the intelligence, with the view not only of making better citizens, but also of increasing the happiness of life; and in the second place we want to give an insight into the methods of science'.[41] It was just such reasoning that motivated the move to establish geography formally within the universities.

### *The foundation of teaching posts in geography in Scottish universities, 1908–1919*

The history of geography as a taught subject within British universities and other institutions of higher education is both relatively recent and remarkably incomplete. Few have followed the work of Scargill on geography at Oxford or of Stoddart on geography's formal beginnings at Cambridge.[42] Outside Oxford (in 1887) and Cambridge (1888), the first teaching position in geography was in Manchester in 1892. In England and Wales, similar appointments followed from the early twentieth century. Yet by 1920, there were only six chairs of geography. Honours classes leading to formal qualifications came to Liverpool in 1917, to the London School of Economics and to Aberystwyth in 1918, to University College London and to Cambridge in 1919, Manchester in 1923, Sheffield in 1924 and, in 1933, to Oxford. For Stoddart, 'The recency of many of these developments . . . is salutary, and the protracted history of developments at most of these institutions gives some indication of the difficulties which the subject often faced'.[43]

Considered in this context, the establishment of geography teaching positions at Edinburgh (1908), in Glasgow (1909), in Aberdeen (1919) and at St Andrews (1935) should be understood as both recent and late (see Figure 6.1). The first is true in terms of the university's identifying a formal educational place for the subject. The second is true given that much longer tradition of geography teaching in Scotland's universities documented here. The formalisation of geography teaching in Scottish universities follows the pattern identified by Stoddart for England and Wales: an initial appointment, of a single person, teaching the subject at diploma level or as part of the general MA or BSc degree, before the establishment of a honours school. In Aberdeen, the university principal's promise in 1909 that the university would have a teaching

[41] M. Newbigin, 'The study of weather as a branch of nature knowledge', *Scottish Geographical Magazine*, 23 (1907), 627–62: quote from pp. 627–8.

[42] D. Scargill, 'The R. G. S. and the foundation of geography at Oxford', *Geographical Journal*, 142 (1976), 438–61; D. Stoddart, *On geography and its history*, 41–127.

[43] J. Keltie, 'Geographical education – Report to the Council of the Royal Geographical Society', *Supplementary Paper of the Royal Geographical Society*, 1 (1886), 439–594; Stoddart, *On geography and its history*, 46.

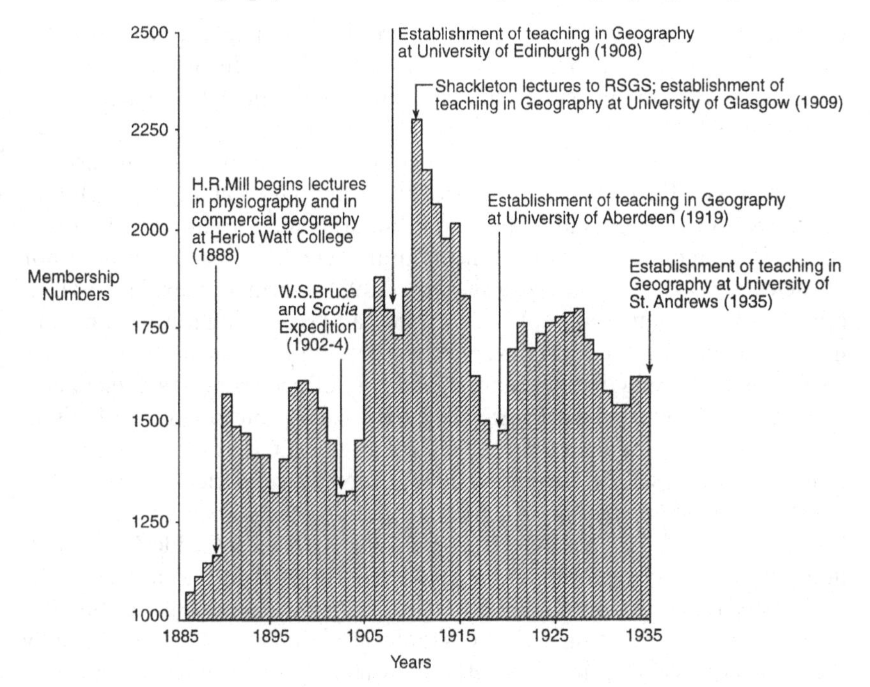


6.1 Civic interest in geography, 1884–1935. The shaded columns indicate the membership levels of the Royal Scottish Geographical Society (RSGS). The dates of foundation of formal geography teaching posts in Scotland's universities and of significant events in the public promotion of geography are also indicated.

post in geography was not forthcoming until 1919, even though a university lecturer in education, John Clarke, had been giving geography classes and publishing on the subject since 1898.[44]

In 1889, the council of the RSGS had prepared a memorial to the commissioners appointed by parliament then investigating the state of Scottish universities in order 'to press on them very urgently the claims of Geography to be recognised as a department of the Higher Education'.[45] Individual universities and other places of learning throughout Britain were contacted late in 1901 by Sir Clements Markham, president of the Royal Geographical Society, acting on behalf of delegates of British geographical societies, in which he insisted 'upon the desirability of recognising geography in their Curricula'. John Scott Keltie reported in his letter of April 1902 how 'in the case of the University of Edinburgh, and of Yorkshire College, Leeds, it is stated that steps are being taken to establish Chairs of Geography; but in all cases except

[44] Bridges, 'The foundation and early years', 82.

[45] *Memorial of the Council of the Royal Scottish Geographical Society to the Commissioners appointed by Parliament in the Universities (Scotland) Act, 1889,* 1.

Edinburgh University, regret is expressed that nothing can be done for Geography without Financial assistance from outside'.[46] In fact, at Edinburgh no teaching position in geography was established until 1908 (a chair had to wait until 1930).

As he recognised in opening the first course of geography at Edinburgh in October 1908, 'the first lecturer in geography appointed at any Scottish university' was George Chisholm, born in Edinburgh and a graduate of that university in both philosophy and in the natural sciences.[47] Chisholm was author of the *Handbook of commercial geography* (1898), a formative text in the development of economic geography. Chisholm had been giving lectures on commercial geography for the University of London's extension courses since 1896. It is interesting that in tracing the history of geography teaching in universities, Chisholm looked to Barbie du Bocage's professorship in Paris in 1809 as such a beginning, rather than to the circumstances of his own country or university. His self-description as 'first lecturer' is true, strictly speaking, but Chisholm made no mention of earlier geography teaching by university professors. Neither did he mention Hugh Mill, who, at Patrick Geddes' invitation, had lectured in physiography and other subjects in Perth in 1887 as part of the University Extension movement and who, from 1888 until 1892, lectured in physiography and in commercial geography at Heriot Watt College (formally the Edinburgh Mechanics' Institute), then located in the adjacent street to Edinburgh's Old College and Chisholm's office. Both Mill and Chisholm emphasised commercial geography, Mill teaching from an *Elementary commercial geography* published in 1888. Mill then left Edinburgh to become librarian at the RGS. Teaching in 'Industrial and Commercial Geography' was continued by his assistant William Couper, by A. J. Herbertson in 1896–7, by William Speirs Bruce in 1899–1900 and by Caleb Cash in 1902–3.[48]

The sense that, at its foundation and in the following decades, geography teaching in Scottish universities and with Mill at Heriot Watt was more concerned with knowing the world as a commercial environment and with understanding the nature of geography as a 'new' discipline than it was with inculcating imperialist attitudes is supported by the formal beginnings of geography teaching in the University of Glasgow. Both Professor J. W. Gregory, professor of Geology at Glasgow from 1903 and Glasgow's principal, Sir Donald MacAlister, had been involved in the founding of geography at Cambridge, and the relationship between the two subjects in Glasgow

[46] RSGS Papers, Uncatalogued MS, J. S. Keltie, letter to the Secretary of the Scottish Geographical Society, 16 April 1902. (On the letter, the Secretary of the RSGS has added, in ink, the letter 'R' to the institutional address as given by Keltie).

[47] Chisholm's view of his purpose and position is clear from his inaugural address: G. Chisholm, 'The meaning and scope of geography', *Scottish Geographical Magazine*, 24 (1908), 561–75. On Chisholm's influence upon economic geography, see M. Wise, 'A university teacher of geography', *Transactions of the Institute of British Geographers*, 66 (1975), 1–16.

[48] This is clear from the information and course outlines in the *Heriot-Watt College Calendar* for the years 1887–8 to 1902–3.

parallels the Cambridge experience. In 1909, H. G. Lyons, former director general of the Egyptian Survey, was appointed to teach geography in Glasgow. By 1913, 'the Glasgow School of Geography had already proved both its quality and its public usefulness, and efforts were in progress to place it on a footing commensurate with its importance as a branch of liberal education, by procuring for it a suitable endowment'. That year was the centenary of the birth of David Livingstone. As MacAlister reported in 1921, 'The proposal was made and cordially adopted, that among the local memorials to be established in his honour, as missionary, explorer, physician, and pioneer of empire, a Livingstone Chair of Geography in the University should be included'. This never happened: 'Perhaps because the the effort was dispersed among a multiplicity of objects . . . The War and the Peace have diverted men's minds to other needs and other memorials'.[49]

At the same time, beyond the universities' walls, it was clear to some at least that geography was not strongly placed.

### *'A parlous plight'? Geography in Scottish education 1885–c.1928*

Keltie's 1885 report on the state of geographical education has principally been considered in relation to England. Its impact was no less in Scotland, partly given its concidence with an 'Exhibition of educational appliances in Geography', held in Manchester and London and displayed in Edinburgh in the city's Industrial Museum in 1886. There was widespread concern not only that geography should be taught as a means to public education, but also with *how* it was taught: model-making and observational skills were being talked about, not just the rote learning of capes and bays.[50]

In Aberdeen in September 1885, Keltie argued in a paper on 'Geographical education' that 'so far as our schools are concerned, the two great weaknesses are want of knowledge in the teachers and want of organisation in the programmes and methods'.[51] In response to Keltie's lecture and to more general concerns, the RSGS advertised its 'Scheme for the Encouragement of Geography in Scottish Schools' in June 1886, in which prizes were to be given for, amongst other things, essays on 'The best method of teaching geography in elementary schools'.[52] At the same time, Professor S. S. Laurie, Professor of Education at Edinburgh, had been developing what, in a paper to the RSGS in July 1886, he called 'the preliminaries of method' regarding geography as a

[49] D. MacAlister, 'Geography at the University of Glasgow', *Scottish Geographical Magazine,* 37 (1921), 53–7: quote from p. 56; see also Glasgow University Library, MS Minutes of Senate, MS GUA 26720, Senate Minutes, 5 March 1908 for official notification 'that the Senate are in favour of increased attention being paid to the study of Geography in the Universities'.

[50] T. Ploszajska, 'Constructing the subject: geographical models in English schools, 1870–1944', *Journal of Historical Geography,* 22 (1996), 388–98.

[51] J. Keltie, 'Geographical education', *Scottish Geographical Magazine,* 1 (1885), 504.

[52] Scottish Geographical Society, *Scheme for the encouragement of geography in Scottish schools* (Edinburgh, 1886), 3.

school subject.[53] Stressing geography's morally educative role, Laurie's scheme involved moving from an understanding of one's locality to an 'extended knowledge of the native country' which should be concurrent, he emphasised, with 'an extended knowledge of the world as a whole, and especially of the British Empire'. Laurie's proposals were laid out fully in his 1888 *Occasional addresses on education subjects*, and they provided the basis to the Scotch [sic] Education Department's 1891 Code of Regulations of Geography, with its attention to six 'standards' – in effect, different scales – from standard I, which involved a plan of the school and playground, to standard VI, 'the geography of the world generally, and specially of the British colonies and dependencies'.[54] At the same time, the RSGS held courses for teachers on, amongst other things, map-making and historical geography.

Recognising the importance of Keltie, the RGS and the RSGS on the *purpose* of geography teaching and the work of Laurie and others on the *methods* should not obscure the fact that geography teaching in Scottish schools was not compulsory. It was this that prompted A. J. Herbertson's alarmist review in 1898 of 'the parlous plight of geography in Scottish education' in which he argues, whilst claiming that Scotland was probably better provided for than England, that the real problem lay in the fact that a child's exposure to geography reflected both schools' attention to written papers and stultifying classroom teaching, when what was wanted was the study of processes in the field. Good school teaching depended upon having good teachers but training was variable. As Herbertson remarked: 'At present the teacher looks in vain for a Normal School or a University in Scotland where he can get some scientific training in geography'. It was such concerns, of course, that motivated the plans for lectureships in geography in the universities. Herbertson additionally stressed the importance of commercial geography: 'It is quite impossible to exaggerate the importance of a sound training in economic geography', and concluded his anxious review '[Thus] even the practical value of geography is lost sight of owing to the general apathy to the subject common throughout our educational system'.[55]

Herbertson's attempts to remedy this perceived deficiency included experimental geographical excursions aimed at local and regional survey, which were continued by Robert Smith, the botanist, and which later influenced the curriculum on 'Nature Knowledge' in the Education Department.[56] But such provision, innovative and welcomed as it was, remained entirely a local matter. As one commentator hinted in 1902, where one studied geography determined what sorts of geography one was taught: 'Country and town pupils begin their

[53] S. Laurie, *The training of teachers and methods of instruction* (Edinburgh, 1901), 200.

[54] S. Laurie, *Occasional addresses on education subjects* (Cambridge, 1888), 200–14.

[55] A. Herbertson, 'The parlous plight of geography in Scottish education', *Scottish Geographical Magazine,* 14 (1899), 81–8.

[56] H. Beveridge, 'School excursions in Scotland', *The Geographical Teacher,* 1 (1911–12), 79–82.

study of geography with different eyes and minds'.[57] This sense of local variation in the teaching of geography is apparent in other material and yet was accompanied by growing recognition in Scotland of geography's role in promoting citizenship.

The 1912 *Memorandum on the teaching of geography in Scottish primary schools*, prompted by debates set in train by Herbertson's 1898 paper and related concerns about school geography teaching, continued not to prescribe what should be taught: 'it has for years past been the policy of the Department . . . to refrain from prescribing in detail the courses of instruction in the various school subjects'. The *Memorandum* stressed the importance of geography in the primary school: 'The Primary School is the training-ground for future citizens. Geography has a place in its curriculum, inasmuch as the study of it contributes to good and intelligent citizenship . . . the characteristic methods of geography are of constant service in ordinary life.' The teaching of Elementary Geography (primary school geography) was to be approached through observational and descriptive methods in three stages which, effectively, meant attention to the local, to observed facts put into world context and then attention to more detailed themes 'with an eye to its practical bearings on life and citizenship'.[58]

As a document seeking to regulate what was taught and, to some degree, how – there are sections on model-making, map-reading and on the use of texts and maps, for example – the 1912 *Memorandum* merits further attention in relation to contemporaries' claims about geography as a means to understand Scotland and the world. Two related issues must be noted here. The first is the sense amongst many contemporaries that, given the lack of prescription, there was, in effect, geographical variation within Scotland as to what was and, indeed, what *could* be taught. As one commentator acidly noted in 1912, 'The teacher whose school is in a Glasgow slum will probably feel that . . . the injunction to apply the scholar's knowledge "to explain the relations of mountains, rivers and plains to each other, and to the distribution of various forms of life, particularly of human life, as governed by the evolutionary process," is a mere trifle'.[59] The second has to do with the fact that geography was a compulsory part of the leaving certificate in Scottish schools only in association with English: too few pupils received a geographical education beyond the intermediate stage.

Geography's position slowly improved from 1912 – in terms of numbers of teachers with formal qualifications in geography, numbers of teacher trainers and regularising the curriculum – but such changes were distinctly uneven in

[57] Anon., 'The syllabus question', *The Geographical Teacher*, 1 (1911–12), 83.

[58] These quotations are from the Scotch Education Department, *Memorandum on the Teaching of Geography in Scottish Primary Schools* (London, 1912).

[59] Anon., 'The Scotch Educational Department and the teaching of geography', *Scottish Geographical Magazine*, 28 (1912), 321.

their effects across Scotland and the subject was not compulsory. Even by 1928, concerns continued to be raised about the state of geography teaching, its variability across Scotland and the lack of shared standards between the institutions involved.[60] National practices of geographical education in this period, recently examined for English schools,[61] must not neglect the ways in which different schools differently worked to promote different emphases within their geographical curriculum. Neither should such local geographies ignore the ways in which national knowledge was a matter of more precise scientific investigation.

### Science and the nature of national survey

In his 1885 paper, as we have seen, H. A. Webster not only noted 'there are whole departments of geographical investigation at which we have only begun to work in a serious and fruitful manner', but also remarked that 'nothing perhaps is more striking to one who has not hitherto thought much about the geography of his native country than to learn at how recent a date we have got to know with scientific accuracy the physical features of this little fragment of an island'.[62] Webster's claims might be tempered given that longer history of geographical enquiry noted here, of course, but he was correct in one important sense. Only from the later nineteenth century can we refer to scientific specialisation and institutionalisation, in which context the discipline of geography with its 'departments of geographical investigation' – in both the institutional and the discursive senses – was set.

My purpose in this section is to turn away from contemporary concerns with geography *per se* – that is, with debates about what was understood as the discipline itself – and to consider the ways through which particular branches of science sought to know Scotland through geographical enquiry in the field. I want to suggest that science helped forge a sense of what, geographically, Scotland was in the later nineteenth century and that it did so both because of particular endeavours – specialised enquiry in given subject areas and particular institutional sites – *and* because many of the practices used in knowing the nation – such as reliance upon instrumental accuracy and field measurement with a view to the spatial correlation of geographical features – were shared. In support of these contentions, what follows discusses three sets of examples of the 'sciences of national survey' at work: the first deals with land, the second with water, and the third with that question central to questions of geography and of national identity in the age of empire – race.

[60] Anon., 'The teaching of geography', *Glasgow Herald,* 11 September 1928, 12.

[61] T. Ploszajska, *Geographical education, empire and citizenship: geographical teaching and learning in English schools, 1870–1944* (Cambridge, 1999).

[62] H. Webster, 'What has been done for the geography of Scotland, and what remains to be done', *Scottish Geographical Magazine,* 1 (1885), 487.

*Terrestrial knowledges: geological and botanical surveys, 1883–1913*

David Oldroyd has emphasised the importance of the northwest Highlands of Scotland in the construction of geological knowledge through fieldwork and stressed the role of Archibald Geikie and of the Geological Survey and the professional surveyors, Benjamin Peach and John Horne, between 1883 and their 1907 publication *The geological structure of the north-west highlands of Scotland.*[63] For Oldroyd, the story of geological work is a complex one of methodology, epistemological warrant and social relationships, a story of empirical measurement, of the importance of observation, of field-based inferences from surface features, of negotiation between field evidence and theory-led explanation, and, not least, of the physicality of scientists' direct encounter with the outdoors. Such issues were realised in more precise forms of scientific representation in which, for example, interests in understanding the surface topography of Scotland – in ways which closely followed established traditions of portraying landscape – were accompanied by a concern to understand (if only by inference) what Scotland looked like beneath the ground (see Figures 6.2 and 6.3). For Peach in particular, encountering Scotland's geological history as a matter of national knowledge was an aesthetic concern with the problem of accurate representation.[64] And just as the camera had been earlier employed in the moral surveillance of Scotland's urban geographies, so it was used in the later nineteenth century to represent Scotland's natural environment, both in terms of painterly landscape traditions, as in the works of George Washington Wilson, and in the field as a trusted instrument of accurate scientific record (Figure 6.4).[65]

The conjunctions of scientific interest and aesthetic sensibility that underlie the idea and representation of terrestrial survey in this period have, of course, earlier parallels in the mapping work of the Military Survey (see pp. 150–2). In geological survey, the works of J. R. McCulloch on the geological classification of the Highlands (in which he combined an interest in the natural scenery with descriptions of local people) and his involvement with Jameson and Murchison in state-sponsored geological mapping, or the place of Hugh Millar as 'antiquary of the natural world', were precedents to the later nineteenth- and early twentieth-century work of Peach and Horne and others.[66] Before 1883, however, Scotland had not been subject to accurate scientific scrutiny and representation in the above terms. In advancing understanding of

[63] Oldroyd, *The Highlands controversy,* 266–99.

[64] A. Anderson, *Ben Peach's Scotland: landscape sketches by a Victorian geologist* (Inverness, 1980).

[65] C. Withers, 'Picturing highland landscapes: George Washington Wilson and the photography of the Scottish Highlands', *Landscape Research,* 19 (1994), 68–79.

[66] J. McCulloch, *The highlands and western isles of Scotland* (London, 1824); *Memoirs to His Majesty's Treasury respecting the geological map of Scotland* (London, 1836); J. G. Paradis, 'The natural historian as antiquary of the world: Hugh Miller and the rise of literary natural history', in M. Shortland (ed.), *Hugh Miller and the controversies of Victorian science* (Oxford, 1996), 122–50.

6.2 Landscape sketch, 'Moraines at head of Strath dionard' (Sutherland), by the geologist H. M. Cadell, 1885. Cadell was a prominent and active member of the RSGS in its early years.

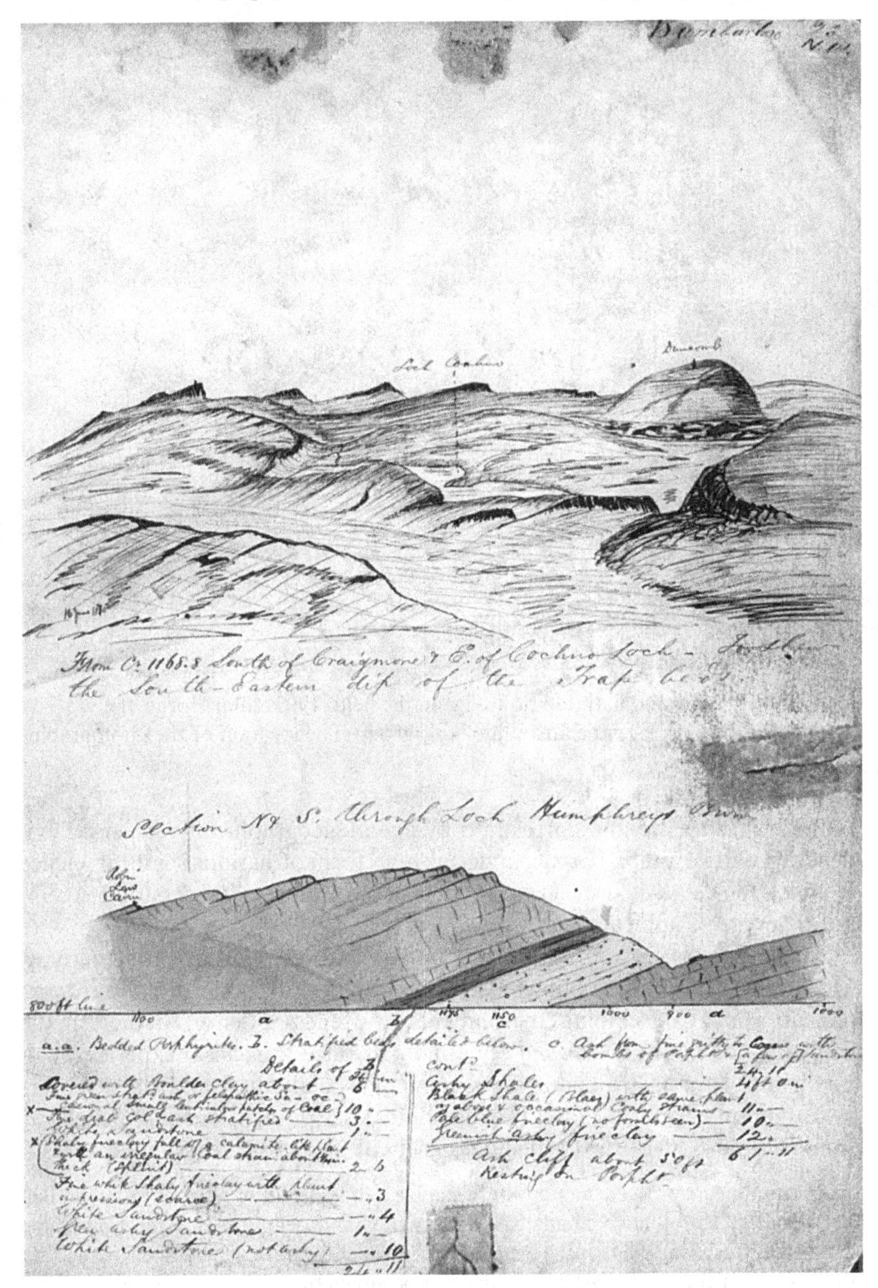

6.3 Geological section of part of Dunbartonshire, *c.*1870, from the manuscript field notebooks of the geologist R. L. Jack. The conjunction of topographic portrayal, inferential interpretation of sub-surface features and accurate textual description and measurement illustrates the ways in which Scotland was the subject of more precise scientific accuracy by the later nineteenth century.

6.4 Photography and natural knowledge in the field, 1903. John Horne, the geological surveyor, is to the far right of this group. The location of the photograph is not known.

geological sequence and correlation by age of geographically-dispersed evidence, geologists were, I argue, undertaking a form of national natural science in which the geological structure and age of Scotland was understood in temporal and geographical terms.

Such work was also a matter of popular and localised scientific activity. Mountaineering and rambling reflected interest in the moral value of outdoors recreation and a developing environmental awareness.[67] Being 'out there' in the field was, for many, social *and* scientific – at once a physical experience, an opportunity for instrumental measurement, social ritual and the scientific investigation of one's locality. The following description of one summer excursion (in July 1885) of the Perthshire Society of Natural Science hints at this:

> the mountaineers proceed, and at length stand on the summit of *Stuc a Chroin* [on Ben Vorlich, Perthshire]. Here, amidst the driving mist and pelting rain, a meeting of the Mountain Club was held, at the exact hour for which summonses had been issued. In the absence of the Cairnmaster, the cairn was occupied by the Scribe and Annalist, and

[67] C. Smout, 'The Highlands and the roots of green consciousness, 1750–1990', *Proceedings of the British Academy*, 76 (1991), 237–63; C. Withers, 'Contested visions: nature, culture and the morality of landscape in the Scottish Highlands', in A. Buttimer and L. Wallin (eds.), *Nature, culture and human identity in cross-cultural perspective* (Dordrecht, 1999), 271–86.

the Geometer having declared the altitude of the hill to be 3189 feet above sea-level, a number of new members were initiated, the Quaighbearer being in attendance with the Quaigh. After the toasts peculiar to the Club had been duly celebrated, and other business transacted, the members proceeded to investigate the botany of the peak so far as the mists permitted.[68]

In contemporary and more strictly professionalised botanical survey, attention also centred on national knowledge through description of the geographical distribution of Scotland's vegetation. In considering what he called 'plant associations' in discussing the Botanical Survey of Scotland, Robert Smith drew upon conceptions of vegetation study that had their roots in Humboldt and Decandolle and R. K. Greville. He acknowledged, too, the local floristic mapping of men like H. C. Watson in the 1830s. Patrick Geddes, who worked with Smith, considered Smith's work a vital part of contemporary geographical knowledge: 'Here, then, is a vital beginning in Scottish, nay in general, geography; for these sheets [Smith's botanical maps, drawn by the Bartholomew firm] represent what fairly promised to be the fullest and most thoroughly detailed botanical survey as yet of any region of the earth'.[69] Partial further work was undertaken by Smith's successor, Marcel Hardy, between 1900 and 1905. The fact that such work was never complete – shortage of funds, lack of a secure organisational basis, the failure of Geddes to realise his own vision for geographical knowledge in Scotland and Hardy's quitting Scotland in 1905 were all contributing factors – should not diminish the fact that vegetational mapping was, like geology, conceived of as a national survey and a key 'department of geographical investigation'.[70]

## *Empiricism and imperialism: oceanography, bathymetry and Scottish polar exploration, 1872–1921*

The *H. M. S. Challenger* expedition of 1872–6, and the publication of its scientific findings under the leadership of Charles Wyville Thomson and, from 1881, of John Murray, is often credited as beginning the modern science of oceanography, although it had predecessors in the marine researches of William Scoresby and of Edward Forbes.[71] Contemporaries regarded the work as the

[68] *Proceedings of the Perthshire Society of Natural Science Session 1881–82, 1885–86* (Perth, 1886), July 18, 1885, p. 271.

[69] R. Smith, 'Botanical Survey of Scotland I. Edinburgh District', *Scottish Geographical Magazine,* 16 (1900), 385–416'; 'Botanical Survey of Scotland II. North Perthshire District', *Scottish Geographical Magazine,* 16 (1900), 441–67; P.G. [Patrick Geddes], 'Robert Smith', *Scottish Geographical Magazine,* 16 (1900), 599.

[70] M. Hardy, 'Botanical geography and the biological utilisation of the soil', *Scottish Geographical Magazine,* 18 (1902), 225–37; 'A note upon the methods of botanical geography', *Scottish Geographical Magazine,* 18 (1902), 406–13; A. Mather, 'Geddes, geography and ecology: the golden age of vegetation mapping in Scotland', *Scottish Geographical Journal,* 115 (1999), 35–52; P. Stott, 'History of biogeography', in J. Taylor (ed.), *Themes in biogeography* (London, 1984), 1–24.

[71] W. Herdman, 'Oceanography', in J. Ashworth *et al.*, *Edinburgh's place in scientific progress* (Edinburgh, 1921), 126–44; J. Tait, 'Oceanography in Scotland during the XIXth and early XXth centuries', *Bulletin de Institute du Oceanographie de Monaco,* 1 (1968), 281–92; J. 'Robert Jameson and the explorers: the search for the north-west passage part I', *Annals of Science,* 31 (1974), 21–47.

greatest contribution to geographical knowledge since the early modern voyages of discovery. The *Challenger* expedition was British in conception, international in terms of its coverage and findings – seventy-six authors contributed to the fifty-volume *Report* – and yet strongly Scottish in personnel and production: three of the five scientific team were Scots – Thomson, Professor of Natural Philosophy at Edinburgh, James Young Buchanan, chemist and physicist and John Murray, naturalist. Siting the Challenger Office in Edinburgh allowed that city to become an international centre for marine science.[72]

As Deacon has shown, who should handle the scientific evidence and produce the findings – and, thus, *where* marine science should develop – was the source of considerable tension between the London-based and the Scottish-based scientific communities.[73] In that sense, the development of marine science as a matter of national competitiveness mirrors tensions over polar exploration between the British interests of Sir Clements Markham and London scientists and the Scottish interests of William Speirs Bruce. The Challenger Office in Edinburgh acted as a centre of calculation in the emergence of shared systematic procedures in marine science, for the coordination of others' national surveys – the Royal Indian Marine Survey begun in 1884, for example – and for promoting recognition of the commercial benefits that would follow scientific understanding of the oceans. This concentration and institutionalisation of international oceanographic research also allowed scientists to turn to Scotland's own interests and waters as subjects of scientific attention. In regard to Scotland's 'imperial mission' as part of her national identity, this was true of Murray's support for the 1902–4 Scottish National Antarctic Expedition under Bruce. In regard to national identity understood as the systematic empirical investigation of Scotland's geography, it was apparent in Murray's survey of the ocean floor and waters between the Shetlands and the Faroes in 1880 and 1882, in the foundation of the Granton Marine Station in 1884, in the bathymetrical surveys of Scotland's lochs by Murray and others, and in the establishment by Bruce of a Scottish Oceanographical Laboratory in Edinburgh in 1907.

The Scottish Marine Station for Scientific Research, established at Granton near Edinburgh in 1884, was the centre for investigation of the chemistry and biology of the Firth of Forth and the adjacent North Sea. It was brought together as a training and teaching site by, amongst others, Patrick Geddes, William Speirs Bruce and Hugh Mill. Although it was relatively short lived, closing in 1903, the Granton Station is important for several reasons. Apart from being the first permanent laboratory in Great Britain for the study of marine biology, it acted as a focal centre for the systematic study of Scotland's coastal waters. It was regarded by contemporaries as a site of national endeavour: as one popular journal noted, 'the heartiness with which the appeals to

[72] M. Deacon, *Scientists and the sea 1650–1900: a study of marine science* (London, 1971), 370.
[73] Deacon, *Scientists and the sea*, 366–406.

the public have been responded to by donations of money, apparatus, and material, shows how thoroughly the people of Scotland realise the importance of the work which is being done'.[74] Further, and at a conceptual level, it provided a key space for making national science in the same way as, for example, Huxley's rooms in South Kensington or, as I show later in this chapter, Geddes' Outlook Tower in Edinburgh.[75]

Granton was also an important site for training in bathymetry. Studies of Scotland's lochs began in the later 1700s, and some lochs had been surveyed in the 1860s by the Admiralty's Hydrographic Department. However, coordinated work with a view to a national understanding dates only from (unsuccessful) appeals to the government in 1883–4 by the Royal Society of Edinburgh and the Royal Society in London, and, effectively, from the work of Murray and Pullar and others between 1897 and 1909. The 'Bathymetrical Survey of the Fresh-water Lochs of Scotland', published from 1903 onwards, was a major advance in the systematic understanding by survey of Scotland's nature.[76] More detailed understanding of *how* the Bathymetrical Survey worked reveals, however, that the national coverage was far from complete, and, like their geological and botanical counterparts, the officers of the Survey had to negotiate social access to knowledge of national space: 'As a matter of fact, all lochs were surveyed on which boats could be found at the time the work was being carried out. . . . To transport a boat to many of the remote lochs in the Highlands would have entailed much labour and difficulty, not to speak of the objections of proprietors, keepers, and others, who do not wish to have grouse moors and deer forests disturbed at a time when the lochs are most accessible'.[77] It is clear, too, that such survey work afforded opportunities for intellectual exchange on natural national knowledge. W. A. Herdman, a student of Archibald Geikie, recounted in 1923 his experiences whilst on a field excursion in the Cairngorms which had Peach, John Murray, and John Thomson, the explorer and photographer, amongst the party: 'On the long tramps, there were hot discussions . . . On the tops of these ancient mountains of Scotland we could and did consider the changes of continents and the supposed permanence of ocean basins'.[78]

[74] *Chambers's Journal of popular, literature, science and art,* No. 30, Vol. I, 468 (26 July 1884); on Granton Marine Station, see W. Boog Watson, 'The Scottish Marine Station for Scientific Research, Granton, 1884–1903', *Book of the Old Edinburgh Club,* 23 (1969), 5–8; R. Currie, 'Marine science', *Proceedings of the Royal Society of Edinburgh,* Series B, 84 (1983), 231–50.

[75] S. Forgan and G. Gooday, 'Constructing South Kensington: the buildings and politics of T. H. Huxley's working environments', *British Journal for the History of Science,* 29 (1996), 435–68; S. Forgan, '"But indifferently lodged . . .": perception and place in building for science in Victorian London', in C. Smith and J. Agar (eds.), *Making space for science: territorial themes in the shaping of knowledge* (London, 1998), 195–215.

[76] P. Maitland, 'Freshwater science', *Proceedings of the Royal Society of Edinburgh,* Series B, 84 (1996), 171–210.

[77] J. Murray and J. Pullar, *Bathymetrical survey of the freshwater lochs of Scotland* (Edinburgh, 1910), volume I, xvii.

[78] M. Deacon, 'The *Challenger* expedition and geology', *Proceedings of the Royal Society of Edinburgh, B,* 72 (1972), 145–53: quote from page 151.

The fact that this conjoint survey work and the social network through which it operated is not better known may be because contemporary public and scientific attention was focused upon the activities of William Speirs Bruce and the Scottish National Antarctic Expedition (SNAE) of 1902–4. Given widespread European interest in polar exploration at that time and the fact that, in Scotland, interest and pride in the voyage of the *Scotia* was intense, it is noteworthy that Bruce should be so little known today.[79] Understood in its contemporary context, however, and in relation to the ways in which scientific activities sought national identity, Bruce's expedition is, I contend, of importance for several reasons. It was part of a longer tradition of polar exploration, one in which Scots played a major part. There is no doubt that contemporary reaction towards the SNAE considered it a heroic accomplishment consistent, for example, with Livingstone's in Africa. The fact that the expedition took place at all was the consequence of Bruce's own social and intellectual local context: John Murray's support; friendship with Geddes and Mill; training in Edinburgh, Granton and in the meteorological station on Ben Nevis; and Bruce's earlier membership of the Dundee whaling expedition of 1892–3. The expedition was motivated, too, by exactly those concerns with exact and systematic science – in which geographical enquiry was centrally implicated – then sustaining the advance of scientific knowledge and of imperialism. It was both an imperial *and* a national endeavour. Bruce intended the SNAE as a *Scottish* expedition, not a British one, at a time when many European imperial nations were undertaking similar expeditions and when, politically, many considered Scotland part of the British Empire. As the printed record *Voyage of the Scotia* put it: 'It remained for Scotland to show that as a nation her old spirit was still alive, and that she could stand beside the other nations and worthily take her place in this campaign of peace. It was in this spirit that the Scottish National Antarctic Expedition was planned, organised and carried through by its indefatigable leader, Mr. William S. Bruce'.[80]

Prominent amongst Scotland's earlier polar explorers was John Rae, whose several Arctic expeditions between 1846 and 1854 were notable in being the first to confirm the fate of Franklin's expedition on the *Erebus* and the *Terror* and because, unlike others, Rae adopted native survival and travel techniques. Rae's explorations and methods added significantly to Arctic understanding, but because he suggested in his confidential report to the Admiralty that the survivors of the Franklin expedition had resorted to cannibalism before dying, and because he was more 'native' than 'hero', his contemporary reputation suffered greatly.[81] I am not claiming that these facts significantly coloured the

[79] P. Speak (ed.), *The log of the* Scotia *expedition*, 1902–4 (Edinburgh, 1992); 'William Speirs Bruce 1867–1921', *Geographers' Biobibliographical Studies*, 17 (1997), 17–25.

[80] J. Conroy, 'The Scottish National Antarctic Expedition', *The Scottish Naturalist*, 111 (1999), 159–82: quote from p. 182.

[81] V. Stefansson, 'Arctic controversy: the letters of John Rae', *Geographical Journal*, 120 (1954), 486–93; R. Richards, *Dr John Rae* (Whitby, 1985); I. Bunyan *et al.*, *No ordinary journey: John Rae, Arctic explorer, 1813–1893* (Edinburgh, 1993).

later views of London scientists towards Bruce – there is no evidence that they did – but this example makes the point that scientific understanding as a matter of national identity, perhaps especially exploration in the field, depended greatly on *how* it was done and by *whom*, not simply *that* it was done.[82] Credibility in these terms was no less an issue for Speirs Bruce (or Rae) than it was for Martin and Sibbald and their dealings with London two hundred years earlier.

Credibility as a matter of national identity was also a question of which nation would lead the way in polar research. Between 1893 and 1905, the Antarctic was the focus for national expeditions from Norway, Belgium, Germany, Sweden and France, as well as the British Antarctic Expedition of 1898–1900 under Borchgrevink and the British National [sic] Antarctic Expedition of 1901–4 under Scott. The polar activities of Europe's imperial powers had been prompted by the 1895 International Geographical Congress, held in London, at which a resolution was passed (initially framed by Hugh Mill) 'that the exploration of the Antarctic regions is the greatest piece of geographical exploration still to be undertaken'.[83] By 1899, a Joint Antarctic Committee had been established in London, under the chairmanship of Sir Clements Markham and without any representation from the Scottish scientific community. It was Markham who announced proposals for a National Antarctic Expedition at the 1899 International Geographical Congress in Berlin. By that time, however, Bruce had secured funding from Scottish industrialists, notably from the Coats textile firm of Paisley, for his own expedition.

As Bruce wrote of the Scottish National Antarctic Expedition and the voyage of the *Scotia,* 'While "Science" was the talisman of the Expedition, "Scotland" was emblazoned on its flag'[84] (see Figure 6.5). Markham's letter to Bruce of 23 March 1900 in which he remarked 'I am very sorry to hear that an attempt is to be made at Edinburgh to divert funds from the Antarctic Expedition, in order to get up a rival enterprise' – what Markham in the same letter termed 'this mischievous rivalry' – provides a clear indication that Markham and others saw polar exploration as British, to be coordinated in London and in competition with others' imperial interests.[85] In promoting his expedition as Scottish, in getting industrial not government funding – 170 firms and individuals are listed as supporters in the *Scottish Geographical Magazine* for November 1903 – and in proposing collaboration with other national expeditions, Bruce was perceived by Markham and others as diminishing the credibility of *British* polar science and of the scientific authorities promoting such imperialist endeavour.

[82] This is a point Rae made in correspondence with Edward Ellice, MP, in a letter about the polar expeditionary qualities of civilians *vis-à-vis* naval officers: National Library of Scotland, MS 15048, J. Rae to Edward Ellice, 4 December 1854.

[83] H. Mill, *Life interests of a geographer* (East Grinstead, 1944), 113.

[84] Three of the Staff, *Voyage of the Scotia* (Edinburgh and London, 1906), viii.

[85] Clements Markham to W. S. Bruce, 23 March 1900, quoted in Speak, *The log of the* Scotia *expedition, 1902–4*, 31.

6.5 The 'Scottishness' of the Scottish National Antarctic Expedition.

### *Recording people: the 1907 Pigmentation Survey of Scotland, racial anthropometry and the environmental basis to identity, 1900–1927*

During the nineteenth century, the idea of what 'race' was held to signify shifted from loose association with ethnic lineage towards a greater scientific 'precision' as a marker of physical development and, notably, of intellectual and moral capacity. One of the principal concerns of the discipline of geography in the later nineteenth century was the association between geographical environment and racial type, in which context native peoples subject to the imperial rule of European nations were classified as inferior to the white man.[86]

From the later nineteenth century, the measurement of racial characteristics – of central interest to the 'science' of physical anthropometry – was increasingly used to justify imperialism. Measurement of physical indicators like 'cephalic index', 'cranial capacity', hair type and skin colour were used to sustain beliefs that definable external traits indicated an internal racial essence *and* a fixed moral capacity. Ranking of racial type in such ways produced 'moral statistics' which were used to legitimise the racialised hierarchy of white supremacy. In geography, such environmental determinism received its most extreme expression in debates on the moral effects of climate and in the work of Ellen Churchill Semple, Griffith Taylor and Ellsworth Huntington.[87] In British geography, the work of H. J. Fleure in the early 1900s examined racial differences, notably within Wales, in relation to cephalic index and pigmentation, although Fleure's interest in racial types centred more upon the effects of locality than upon a prior belief in racial type and cultural superiority. In Britain generally, these concerns, particularly between 1892 and 1899, were part of ethnological and linguistic debates about a Celtic racial stereotype and part of institutionalised concerns within British science about national identity as a matter of racial ancestry, local population variation and environmental influence.[88]

Within Scotland, such ideas were apparent in the work of John Gray, James Tocher and John Beddoe between 1900 and 1907. Beddoe had been much involved in the Ethnographic Survey of Britain in 1892–9. In 1900, Gray and Tocher had published a survey of the 'pigmentation statistics' of adults and school children in east Aberdeenshire, and further work followed on pigmentation, surnames and hair colour. There is an important connection to be made here with the activities of field clubs and other civic institutions of natural knowledge (see pp. 182–91) since Gray and Tocher's work was initially based on surveys undertaken in association with the Buchan Field Club: physical measurements were obtained from Buchan folk attending the Mintlaw Gathering in 1895, and at similar events in Aberdeenshire in 1900.

What interested Gray and Tocher was geographical difference within

[86] P. Jackson, *Race and racism: essays in social geography* (London, 1987); Livingstone, 'Climate's moral economy'; 'The moral discourse of climate: historical considerations of race, place and virtue', *Journal of Historical Geography,* 17 (1991), 413–34; '"Never shall ye make the crab walk straight": an inquiry into the scientific sources of racial geography', in F. Driver and G. Rose (eds.), *Nature and science: essays in the history of geographical knowledge* (Cheltenham, 1992), 31–43; N. Stephan, *The idea of race in science: Great Britain 1800–1960* (London, 1982).

[87] Livingstone, '"Never shall ye make the crab walk straight"'; *Geographical tradition,* 230–6.

[88] J. Urry, 'Englishmen, Celts and Iberians: the ethnographic survey of the United Kingdom, 1892–1899', in G. Stocking (ed.), *Functionalism historicised: essays on British social anthropology* (Wisconsin, 1984), 83–105; P. Gruffudd, 'Back to the land: historiography, rurality and the nation in interwar Wales', *Transactions of the Institute of British Geographers,* 19 (1994), 61–77; '"A crusade against consumption": environment, health and social reform in Wales, 1900–1950', *Journal of Historical Geography,* 21 (1995), 39–54.

Scotland's population in terms of physical characteristics – 'divergences' (as they termed it) – which hinted at longer-term racial distinctions within the Scottish people. These issues prompted interest in a Pigmentation Survey of Scotland. Between 1901 and 1907, Gray oversaw the measurement of the hair and eye colour of nearly half a million Scottish school children, a survey discussed in Gray's 1907 'Memoir on the Pigmentation Survey of Scotland'. Maps presented there were intended to reveal the racial make-up of Scotland: which areas were made up of 'native Scots', represented by darker types, and which by Scots of Norse stock, with lighter pigmentation. Gray's pigmentation survey work was the basis for Beddoe's 1908 work in which a 'Compound Index of Nigrescence' was produced for Scotland[89] (see Figure 6.6).

Such racial interests – insofar as they were concerned with differences between Celts and others in Scotland – were foreshadowed by the linguistic surveys of Scotland by Ernst Ravenstein and J. A. H. Murray in the 1870s and 1880s.[90] That the human geography of Scotland was in these ways a matter of national survey lends weight to my argument that specialist practices were being brought to bear in order, to use Webster's words, 'to get to know with scientific accuracy this little fragment of an island'. Specialist scientists encountered Scotland's human past in the field (see Figure 6.7). For Newbigin, the issue of race and national identity in Scotland was a consequence of environmental difference. Differences between Highlander and Lowlander, for example, were 'less the result of the racial composition of the population in the two areas, than the different natural surroundings'. She later considered that natural conditions in the Highlands hindered human occupance such that the inhabitants there 'always tend to lag behind the rest of Scotland in social evolution'.[91] For Maddrell, Newbigin's attention to racial difference and hierarchy in such terms derived from her biological training, from interests in the scientific basis to social evolution and, importantly, from her belief in geography as a both a constituent of and a means to social order.[92]

It is also true that Newbigin and others were defining Scottish national identity as a historical quality consequent upon the racial and environmental nature of her geography. Geography was, simply, the determinant of national character. This claim finds support in the disciplinary attention given in the early twentieth century to the historical geography of Scotland, notably in the work of W. R. Kermack. Ignored by modern historians of the subject,

[89] J. Gray and J. Tocher, 'The physical characteristics of adults and school children in East Aberdeenshire', *Journal of the Royal Anthropological Institute,* 30 (1900), 104–24; J. Gray, 'Memoir on the pigmentation survey of Scotland', *Journal of the Royal Anthropological Institute,* 37 (1907), 375–401; J. Tocher, 'The necessity for a national eugenic survey', *Eugenics Review,* (July 1910), 124–41; J. Beddoe, 'A last contribution to Scottish ethnology', *Journal of the Anthropological Institute of Great Britain,* 38 (1908), 212–20.

[90] J. Murray, *The dialect of the southern counties of Scotland* (London, 1873); E. Ravenstein, 'On the Celtic languages in the British Isles: a statistical survey', *Journal of the Royal Statistical Society,* (1879), 579–643.

[91] M. Newbigin, *A new regional geography of the world* (London, 1929), xv.

[92] Maddrell, 'Scientific discourse and the geographical work of Marion Newbigin', 38–40.

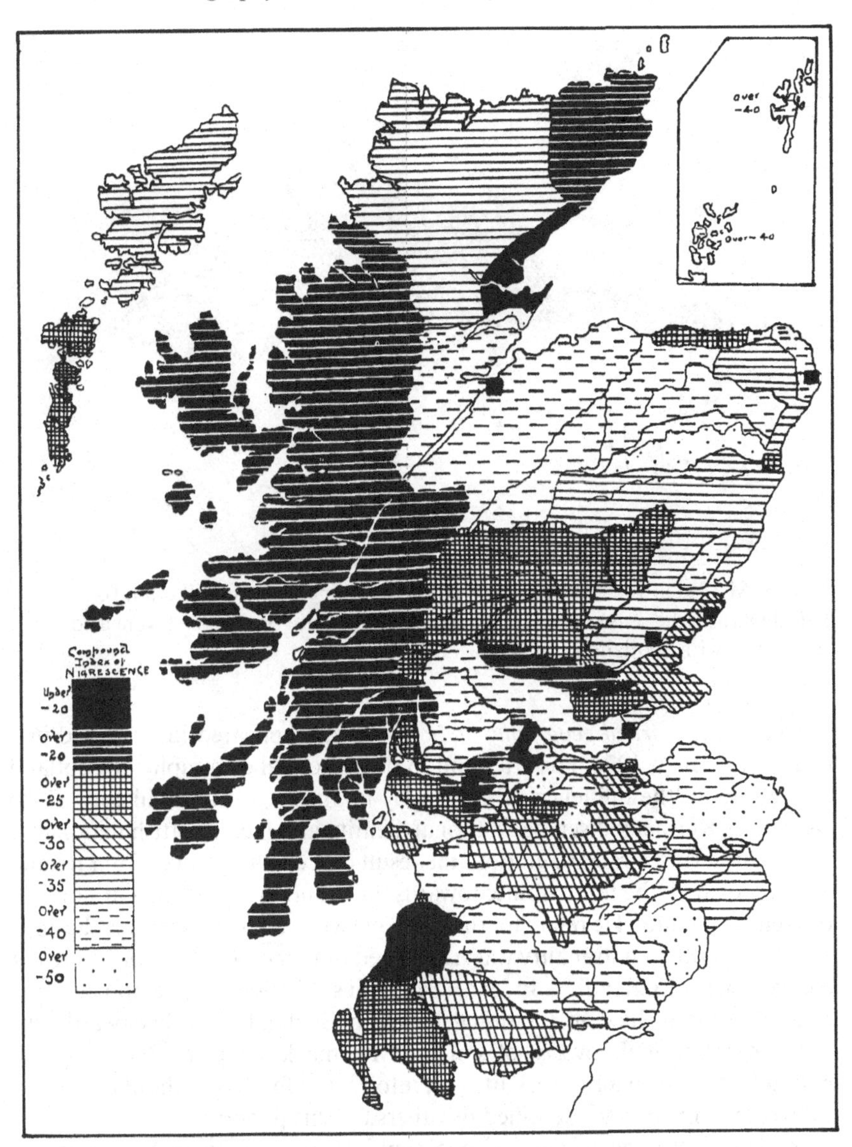


6.6 Scotland's 'racial geography': the 'compound index of nigrescence' from Gray (1907).

6.7 H. M. Cadell's sketch of a 'Picts House at Stoir' (Stoer in Sutherland), 1886. This illustrates the sense in which field studies of natural phenomena were also encounters with Scotland's human past.

Kermack's *Historical geography of Scotland* first appeared in 1913 and in a revised edition in 1926. His attention to the historical geography of Scotland centred upon 'the influence of geography upon the course of history'. Like Newbigin, Kermack discussed national identity and race. Scottish nationality was, in his terms, a racial question, the result of Celtic influences from Ireland which provided 'the first impulse towards unity and the development of a cultured national life'. He refers to 'that struggle with the Norsemen' as 'the anvil on which Scottish nationality was hammered out, and prepared for its greater conflict with England'.[93] Indeed, Scotland's historical development was throughout described in relational terms to England – as 'backward', the result of a 'national poverty' of soil (and, Kermack hints, of intellect) – and, in that sense, Kermack's work fits well into that 'inferiorist' school of historical writing in Scotland identified by others for this period.[94]

In relation to the history of geographical ideas, Kermack's reading of Scottish national identity is consistent with that emphasis on environmental explanation then marking the discipline more generally. In his 1927 *Human environment and progress*, subtitled *The outline of world historical geography*, Kermack wrote that 'the aim of Historical Geography is to explain all aspects

[93] W. R. Kermack, *Historical geography of Scotland* (second edition, London, 1926), 21.

[94] C. Beveridge and R. Turnbull, *The eclipse of Scottish culture: inferiorism and the intellectuals* (Edinburgh, 1989), 16–30.

of the life of man in the past, political, economic, social, intellectual, as they were affected by, or as they affected, his geographical environment at different periods'.[95] His book throughout stresses the environmental conditioning of human existence and the essentially reductionist role of geography in explaining history. At the same time, there were those whose vision was neither of separable 'departments of geographical investigation' nor, strictly speaking, of an environmentalist reading of Scotland's identity, but of geography as a holistic means to Scottish and global citizenship.

## Patrick Geddes, geographical visions and 'the Temple of Geography'

Patrick Geddes' wide-ranging influence within the human sciences and his vision for regional survey in the geographical imagination in Britain between 1918 and 1939 has been discussed by others.[96] Central to Geddes' work in these contexts was the Outlook Tower in Edinburgh's Castlehill, founded by him in 1892 as 'a type museum of Geography, History, and Sociology, and as a centre of civic and regional study'.[97] Although others have recognised how the Outlook Tower was central to Geddes' work, little attention has been paid to the geography of the tower itself, to the different spaces it contained, and to the social and intellectual functions they performed in promoting geography. Made up of several rooms, the Outlook Tower was, simultaneously, observatory, exhibition hall, classroom, library, site for instrumental and cartographic display, a model for an International Geographical Syndicate and what one contemporary termed a 'temple of Geography'.

It is important to recognise, however, that in having these functions the Outlook Tower extended earlier sites of popular geographical enquiry. One such was the diorama. Karen Wonders has argued that dioramas and panoramas declined in popularity after the mid-nineteenth century and only returned to popular favour after the Paris Exposition of 1900.[98] Certainly, the Paris Exposition was a major influence upon Geddes' use of the Tower. But dioramas were in use before then in Scotland to promote geographical knowledge and national understanding. Mr Alexander Lamb's 'Royal Diorama of Scotland' of 1878 was, indeed, a site of wonders: 'More interesting pictures and splendid scenery we never saw produced'. In the first of several pictures, the crowd was treated to 'a fine view of Glasgow', with the scene changing 'to afford various aspects of the city'. After scenes from the Firth of Clyde, Mr Lamb 'proceeded to conduct his audience to Loch Leven, the fair city of

[95] W. R. Kermack, *Human environment and progress* (London and Edinburgh, 1927), 1.

[96] D. Matless, 'Regional surveys and local knowledges: the geographical imagination in Britain, 1918–1939', *Transactions of the Institute of British Geographers*, 17 (1992), 464–80; H. Meller, *Patrick Geddes: social evolutionist and city planner* (London, 1993).

[97] This form of words appears on a series of pamphlets, dating from about 1892 and appearing regularly until Geddes' death in 1932, entitled *The outlook tower*.

[98] K. Wonders, *Habitat dioramas: illusions of wilderness in museums of natural history* (Uppsala, 1993), 12–22.

Perth, Taymouth Castle . . . Dundee, Aberdeen, Inverness, and the romantic scenery of Cape Wrath at sunset'. From there, the audience was treated to views of Argyll, Balmoral, the Borders and Scott's residence at Abbotsford before ending with 'Edinburgh from various points of view'. In an echo of the triumphal processions of 1604, 1633 and 1661, Scotland's identity was performed as well as geographically visualised: 'To give life and reality to the painted scenery, Scottish characters were from time to time introduced by a talented company of performers, wearing highland costume'.[99] The diorama understood as a visual display of one's country has clear connections with the traditions of landscape painting, of topographic survey and with the embodied display of geographical identity. Dioramas also have a parallel in the ways national exhibitions presented displays of nations in the later nineteenth and early twentieth centuries. The 1886 International Exhibition of Industry, Science and Art held in Edinburgh was used as a display site by many Scottish geographical companies – Bartholomew the map-makers, the Morison firm of globe makers from Glasgow, as well as the RSGS – and the 1911 Scottish Exhibition of National History, Art and Industry emphasised the work of the Scottish National Antarctic Expedition. Between November 1908 and June 1910, the RSGS repeatedly lobbied the liquidators of the 1908 Scottish National Exhibition for surplus funds to be used to promote 'national training in Geographical Science' given that 'the Society has now been recognised by the State as a Scientific Society doing national work'.[100] Geddes' plans for the Outlook Tower built upon these earlier expressions of situated institutional support for the subject.

The Outlook Tower had five storeys. On top was the terrace roof or 'Prospect', affording views of Edinburgh and surrounding areas and housing a camera obscura. The camera was seen by Geddes as a key instrument in promoting the powers of observation within geography and for extending his vision for geography and regional survey. For Stoddart, Geddes' camera 'came to symbolise a form of observation and a definition of subject matter which has governed field studies since that time'.[101] Such emphasis upon the moral and instrumental basis to observation was not new. Geddes was using an existing camera obscura which had been open on the site since 1867; this site was one of several such sites using a camera obscura for promoting popular understanding of science,[102] and his attention to the power of local vision as a basis

[99] *Galloway Advertiser and Wigtownshire Free Press,* 8 August 1878, 4c.

[100] P. Kinchin and J. Kinchin, *Glasgow's great exhibitions 1888, 1901, 1911, 1938, 1988* (Bicester, 1990), 19, 95–6, 111; *International Exhibition of Industry, Science and Art: Official Catalogue* (Edinburgh, 1886), 23; RSGS Papers, Uncatalogued MS letter, signed by James Geikie [President], J. G. Bartholomew [Hon. Sec.], R. Richardson [Hon. Sec.], and G. Chisholm [Sec.], to Mr. W Paterson, 8 March 1909.

[101] Stoddart, *On geography and its history,* 144.

[102] A photograph of Edinburgh's Castlehill *c.*1867 by William Donaldson Clark shows the precursor Observatory to Geddes' Outlook Tower.

to geographical knowledge mirrors Colin Maclaurin's unsuccessful petition to Edinburgh's town council in 1741 (see p. 122).

Below the camera obscura and the Prospect was the Edinburgh Room, which had a large-scale model of the city and its region and displays about the city's historical development. The middle storey was occupied by the Scotland Room, 'in which an analagous collection, illustrative of the history and geography of Scotland, of the present state of the country, and of its possible advances, is in progress'. A map of Scotland was painted on the floor, correctly orientated to the points of the compass. The storey below was devoted to the British Empire and to the English-speaking countries. Below that was the Europe Room and, on the ground floor, the World Room.

The Tower was organised, then, as a set of different but connected geographical spaces, each containing a variety of means – instrumental observation, scale models, wall paintings, floor and wall maps, exhibitions of national artefacts – by which Scotland's public could be geographically educated. The building was an instrument for local and national knowledge and global understanding: 'The exhibition of material was designed in a sequence to lead the visitor from the local to the regional, to the national and the global'.[103] Geddes used the Outlook Tower as the base for his Edinburgh Summer Schools (begun in 1886 and first held at the Granton Marine Station). These were taught by local teachers, professional geographers and scientists such as P. N. Rudmose-Brown (a member of the *Scotia* expedition), Marcel Hardy, A. J. Herbertson, and by visiting academics. Geddes gave a lecture course on 'The Higher Uses of Geography', as that subject was central to his vision of harmonious unity between people and nature. He gave other courses, including one on 'Nature Study and Geography in Education' which depended upon using the Outlook Tower as a site for and an instrument of learning. The first class, designed to promote the 'Art of Seeing', was entitled 'The Artist's Outlook' and used the camera obscura in picturing the locality and as a device of geometric measurement. Classes followed on the 'Astronomer's Outlook', the 'Geographer's Outlook', the 'Zoologists' Outlook', and the 'Historian's Outlook', before ending with an excursion designed to promote the 'Personal Outlook'.[104]

His advocacy of the subject was also an advocacy of the spaces necessary for it to be produced and consumed. His centre of geographical calculation was a site for the coordination of popular geographical understanding. In one introduction to his lectures, Geddes argued: 'Life needs environment and instruments; the man of affairs needs his clearing house, his bureau of action. So the student needs a no less comprehensive study, a clearing-house and a

[103] Meller, *Patrick Geddes*, 102.

[104] Strathclyde University Archives [hereafter SUL], MS T-GED 13/1/17, 3; Patrick Geddes, *Course of nature study and geography in education* (Edinburgh, c.1904).

summary of the world of thoughts, a visualised encyclopedia.' He noted two such institutions 'actually incipient:- I. The Outlook Tower, Edinburgh' [and II] Elisee Reclus' 'Great Globe', designed for the 1900 Paris Exposition, 'but at present suspended in execution'.[105] A later undated manuscript in Geddes' hand, possibly from 1914, reveals the several uses to which the spaces of the Tower were put. The Edinburgh and Europe rooms were in use for talks by C. C. Fagg and Geddes on regional survey, and the Scotland Room was used for practical demonstrations and as the coordinating site for local and regional fieldtrips, whether in Edinburgh's old town, on Blackford Hill and the nearby Pentland Hills, or to local market towns and the coast at Port Seton. Geographical apparatus was displayed with a view, as Geddes noted in introducing one such display in 1906, to promoting 'various methods of teaching Home-Geography'.[106] Local schools were encouraged to display their models. Links with schools were further strengthened by programmes of 'Excursions for the Study of Local Geography', supported by courses in regional geography for teachers, some of which were jointly run with the Royal Scottish Geographical Society.

The role of the Outlook Tower as a teaching space was reinforced by Geddes' use of the building as a museum. Geddes' occupation of the building in 1892 was coincident with the establishment that year of a National Museums Association and widespread public debate over the role of civic museums as educative spaces.[107] Geddes was also influenced by the Musée Social in Paris and by Paul Otlet's work in the International Bibliographical Institute in Brussels. Geddes' intentions for a single site that would be more than a museum in any traditional sense depended for him upon the unification of the tower with the globe projects of the French geographer, Elisée Reclus. At the 1889 Paris exposition, a globe of 120 foot in circumference had been displayed and, by 1898, Reclus had plans for one twice that size for the projected World Exposition in Paris in 1900. Both men saw the globe as a symbol of global geographical knowledge and of their shared conceptions of world citizenship through understanding that 'unity to the world' that underlay national diversity.[108]

The 'Tower and the Globe' notion, as it was termed by 1904, was thus central to Geddes' conception of the Outlook Tower as an instrument and site for geographical enquiry, or as he put it, '[something] more than an ordinary museum, it is not only an encyclopedia but an *Encyclopedia Graphica*'.[109] Geddes' ideas for a single building along these lines were outlined in plans for

[105] SUL, MS T-GED 13/1/17, 3.

[106] SUL, MS T-GED 7/8/61, [Pamphlet entitled '*Exhibition of apparatus useful in the teaching of regional geography*', 6–14 July 1906].

[107] S. Forgan, 'The architecture of display: museums, universities and objects in nineteenth-century Britain', *History of Science,* 32 (1994), 139–62.

[108] Meller, *Patrick Geddes,* 103–17.

[109] SUL, MS T-GED/4/18, 2.

a 'Nature Museum' as part of the intended Carnegie reconstruction of Dunfermline in 1904 in his *City development: a study of parks, gardens and culture-institutes.* There he recognised contemporary interest in geography – 'everyone must be more or less aware how great an improvement is going on in geographical education' – and argued that the development of Dunfermline provided the chance to have in Scotland buildings along the lines of those in Paris and Brussels: 'practically a new class of exhibition and museum in one, in which the veteran explorer and geographer, the trained artist and the simplest visitor, rustic or child, could and did find alike the most keen and active pleasure, with genuine and enduring instruction to boot'. As Geddes put it, 'Here, then, is a concrete opportunity in Dunfermline: the knowledge, the skill, and experience which produced these panoramas is largely still available . . . Let us import some of it here: . . . pending the creation of a first-class geographical institute, . . . there would be no place for seeing the world like Dunfermline'.[110]

Geddes' plans involved the creation of just such an institute. He was prompted not just by ideas of civic renewal but by the plans of others, notably of Halford Mackinder who, with the support of the RGS, had been planning a new geographical institute at Oxford from 1892.[111] Manuscript sketches by Geddes show him conjoining tower and globe in several ways, as, for example, a tower of regional survey in combination with celestial and terrestrial globes (Figure 6.8). These sketches were the bases to Geddes' plans for a building which he considered 'the cosmic presentment of Universal Geography', for which the Outlook Tower was the model in both form and function:

> [P]roceeding from the visible and immediate prospect, and by its study by help of all the sciences of observation . . . we descend, storey by storey, through City and Province and region or State to Nation and Empire, . . . to the larger Occidental Civilisation . . . to the Oriental and Primitive sources – the facts of geography and history, the problems and possibilities of useful activity being also represented, as far as may be, upon each level. Various elements of this arrangement have been already in intermittent progress in the Outlook Tower in Edinburgh.[112]

In his unpublished essay 'Plans for an International Geographical Syndicate', Geddes makes clear that the proposed national institute, modelled on this local Edinburgh site, should be seen as the basis to all future geographical buildings. His proposals found support from John Bartholomew, Sir Clements Markham and Sir Archibald Geikie, amongst others. The plan for the building was drawn up from Geddes' sketches by Galeron, the French globe-maker who had exhibited celestial globes in Paris in 1889 and 1900 (Figure 6.9). John Bartholomew, in a memorable statement of support, drew together the views

110 P. Geddes, *City development: a study of parks, gardens, and culture-institutes* (Birmingham, 1904), 117.
111 Meller, *Patrick Geddes*, 131–4.
112 P. Geddes, 'Note on draft plan for institute of geography', *Scottish Geographical Magazine*, 18 (1902), 143.

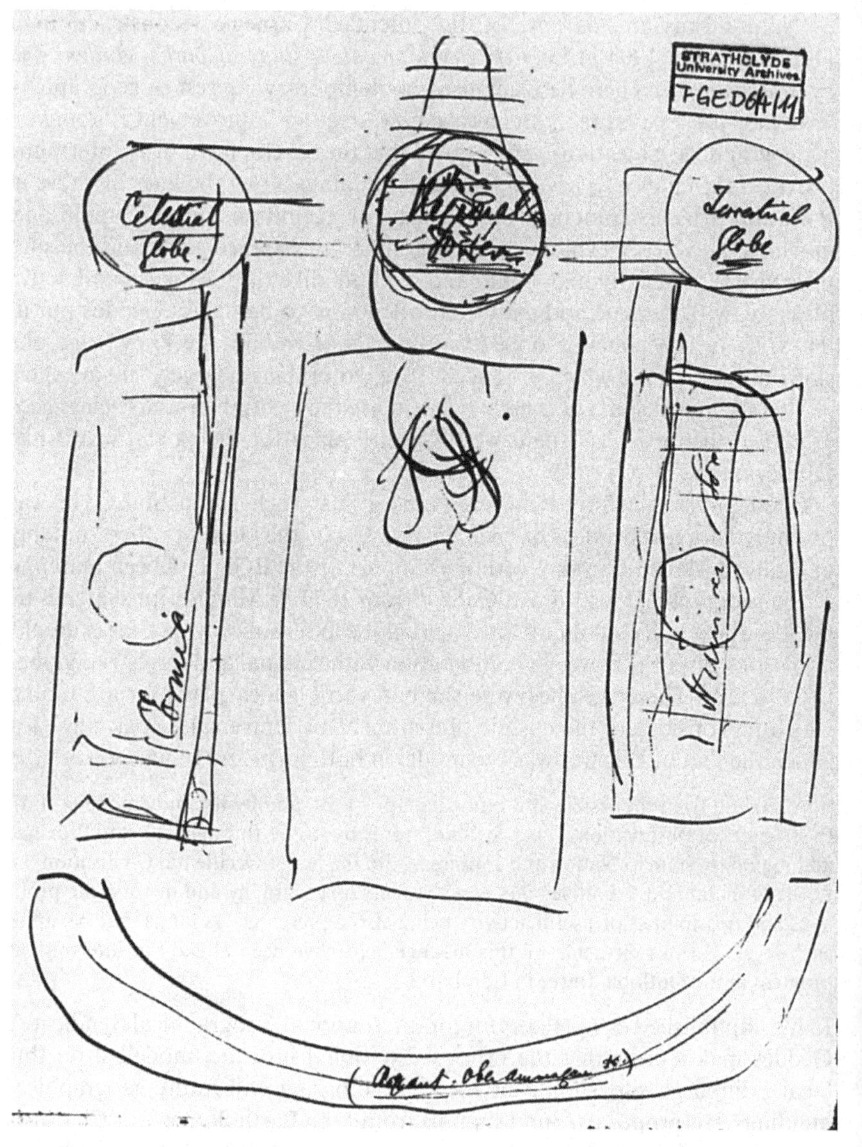


6.8 Patrick Geddes' sketches for a geographical institution, based on the conjunction of 'the Tower and the Globe' and the human and cosmic worlds.

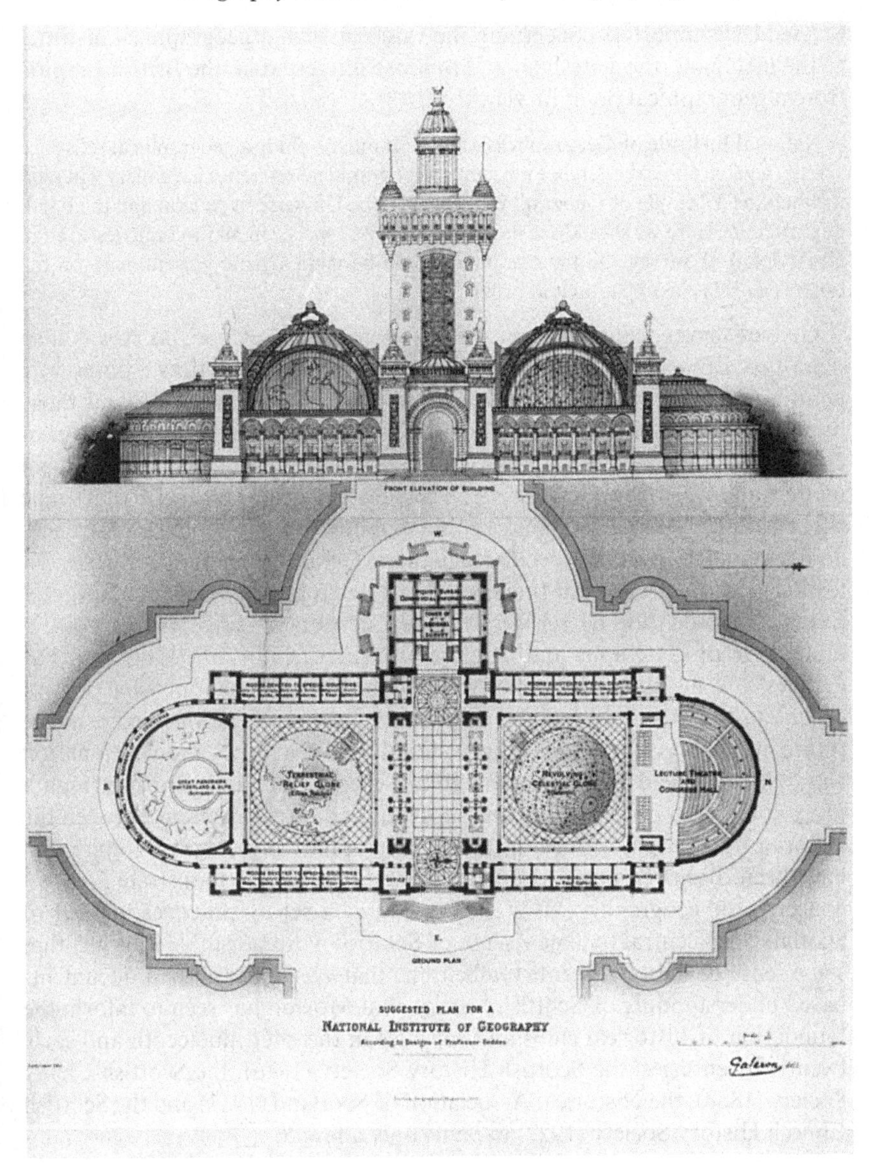


6.9 The proposed design for the National Institute of Geography as it appeared in the *Scottish Geographical Magazine* (1902).

of Geddes' supporters concerning the value of such a geographical institute to the merchant, the statesman and to those interested in 'the British Empire, from a geographical point of view':

> A National Institute of Geography is wanted. It must include a geographical reference library, a geographical reference museum, a geographical reference art gallery – in fact, it should be a Temple of Geography, devoted to the Universe in general and the Earth in particular. Here we should see the world in all its aspects, in all its countries, even in all its detail of survey. On the one hand would be vivid artistic presentment, on the other scientific classification and order.[113]

Geddes' never-realised vision for this 'temple of science', as it was also described, drew, then, not just upon given intellectual practices – observing, painting, displaying, teaching and classifying – but upon survey and those more specialised sciences informing national identity and the development of geography at the time. Local knowledge was a means to understanding global matters and *vice versa.* Geddes' vision was sustained by the belief that a single site, an 'Encyclopedia Graphica', could embody those practices and sciences and contain the spaces needed to teach them. Geddes' intentions for what, following Foucault, we might think of as this heterotopic site of global significance,[114] should not be seen only as a question of who was to control knowledge of geography and of empire in the early twentieth century. For Geddes and his supporters, the National Institute was about what sort of spaces and sites should be used to promote geographical knowledge in the future, what sort of geography there should be in the future and which nation was to take the lead in such matters. In those terms, Geddes' visions exemplify those complex social, personal and institutionalised connections between the history of geography and its historical geographies in the age of empire that have been discussed here. As Meller shows, they also demonstrate Geddes' concerns for geography – as a subject and as a set of practices applied to Scotland – as central to 'a new sense of Scottish nationalism'.[115] It is just that same sense of national identity, albeit one that was apparent in a document-based understanding of Scottish history, that Morton has seen to inform the foundation of historical clubs and societies in the later nineteenth and early twentieth centuries: the Scottish History Society (1886), the Scottish Clergy Society (1888), the Historical Association of Scotland (1911) and the Scottish Church History Society (1922) to name only a few.[116]

[113] J. Bartholomew, 'A plea for a national institute of geography', *Scottish Geographical Magazine,* 18 (1902), 146.

[114] M. Foucault, *The order of things* (London, 1970), xvii–xviii; see also A. Ophir and S. Shapin, 'The place of knowledge: a methodological survey', *Science in Context,* 4 (1991), 13–14.

[115] Meller, *Patrick Geddes,* 98.

[116] G. Morton, 'Historical clubs', in M. Lynch (ed.), *Oxford companion to Scottish history* (Oxford, forthcoming).

## Conclusion

In his account of public science in Britain between 1880 and 1919, Turner claimed that, after about 1875, public science changed: 'Instead of being promoted as an instrument for improving the student morally and bringing greater physical security or personal profit to humankind, science came to be portrayed as a means to create and educate better citizens for state service and stable politics, and to ensure the military security and economic efficiency of the nation'. Whilst contemporary views toward science emphasised scientific education as a means to citizenship, other concerns included the role of scientific thinking in solving social problems and the relative state of British national science in comparison with France and Germany. In short, science was, by the late nineteenth century, a matter of national political importance: as professional and institutionalised disciplines demanding funding; as sets of specialised procedures which, increasingly, divorced practitioners from public understanding; and, finally, as a question of national identity in relation to ideas of citizenship and of empire, given the practices being used to explore the world (including one's own national space), and widespread specialist and popular concern over the comparative state of national scientific education.[117]

In relation to those claims, the position of geography in Scotland between 1884 and 1930 – as a university discipline, within school education and as a matter of institutionalised civic concern – and of those practices of geographical knowledge sustaining both the discipline and Scotland's sense of itself, may be understood in political and instrumentalist terms. Geography was an instrument of national and of imperial identity. Geography's 'instruments' varied, from the camera in the field, Geddes' Outlook Tower and, earlier, the Granton Station to formal institutions with their publications, popular clubs with their outings, and observational survey methods designed to describe, explain and correlate the varied spatial distribution of natural and human phenomena.

In considering the nature of the discipline at the end of the nineteenth century, Lacoste distinguished between two apparently different forms of geography: 'la géographie des états-majors', as that kind of strategic and imperial geography carried out as a means of reinforcing state power; and 'la géographie des professeurs', or the published record of the discipline as undertaken by academics.[118] As Lacoste argued more generally, and as this chapter has shown for Scotland, it is not appropriate to distinguish so clearly between these two versions of the history of geography. We should not think of firm distinctions between the professional geographical community and the

[117] F. Turner, 'Public science in Britain, 1880–1919', *Isis,* 71 (1980), 592.
[118] Y. Lacoste, *La géographie, ça sert, d'abord, à faire la guerre* (Paris, 1976), 19–21.

geographical 'laity'. In Scotland, public audiences for geography and for geographical knowledge existed before the institutionalisation of the subject in universities and in dedicated civic societies. The work of such bodies was not everywhere the same or immediately successful and when it was, it depended upon professionals certainly, but not upon geographers alone. Geographical knowledge differently served different public and specialist audiences.

In discussing the connections between science and the public, Shapin has shown how the the idea of a 'canonical account' of the relations between the two – with, in a modern context, an understanding of science as a specialised pursuit more or less free from public understanding – is not appropriate. What is needed is attention to particular circumstances and to the fact that 'the scientific community' and 'the public' are, as Shapin notes, 'ceaselessly at work in defining themselves, . . . In the course of so doing they also define what is to count as knowledge and the proper means of doing it'.[119] In relation to questions of national identity and imperial authority, what the place of geography and of geographical knowledge in Scotland in the forty or so years from 1884 highlights is indeed the question of *self-definition*: self-definition as a geographical community, in terms of what it meant to be a geographer, in given discursive practices and in relation to Scotland's geographical publics and in the ways in which these issues came together to focus upon what Scotland geographically was. Recognising that geography, like science and the idea of the nation itself, was undergoing 'self-definition', concerns in Scotland about geography as a subject centred upon its place in education and upon the view that geography was a means to political citizenship and to environmental awareness. The development of an academic geography that would sustain such concerns demanded engagement with particular scientific knowledges, what I have here called the 'sciences of national survey'. It depended, too, upon promoting a knowledge of the British empire that took, for example, one discursive form in commercial geography and another, practical, form in a national expedition.

Such practices have been too often associated with particular people: Chisholm with the commercial geography of empire; Bruce with the Scottish National Antarctic Expedition; Geddes with citizenship, education and regional survey; Smith with botanical survey. What this chapter has attempted to demonstrate is both the complexity of the connections between the histories of the different sorts of geography at work in the late nineteenth and early twentieth centuries and the importance of institutional spaces and of geographical difference within Scotland in understanding that complexity. Chisholm's attention in one institutional setting to commercial geography was shared by Bruce and Herbertson and others elsewhere a decade before. The

[119] S. Shapin, 'Science and the public', in R. Olby, G. Cantor, J. Christie and M. Hodge (eds.), *Companion to the history of modern science* (London, 1990), 998–9.

efforts of Geddes and others at using local sites in and around Edinburgh to promote geographical understanding reflected for the capital's geographical audiences what, for example, the Perthshire Society of Natural Science was doing in its civic promotion of local natural history at much the same time. Doing local geography meant different things to different schoolchildren within Scotland and may not even have been possible for some. The Outlook Tower was a crucial site for the promotion of geography within Edinburgh, but even within that city other institutions worked to the same purpose in different ways. The Edinburgh Geographical Institute, as the Bartholomew's cartographic firm described its premises, was regarded by John Bartholomew as a leading site for the promotion of national geographical knowledge.[120] The preface to their 1895 *Atlas of Scotland* argued that Scotland had been symbolically brought to order as a result of scientific survey and civic enterprise: 'The completion of the publication of the Ordnance Survey Maps, the advanced progress of the Admiralty and Geological Surveys, and the large collections of date, produced by the recent activity of Scottish scientific societies, . . . make the present time most favourable for the production of an Atlas in which that valuable work is summarized and incorporated'.[121] Similar concerns informed their 1912 *Survey atlas of Scotland.*

Understanding the connections, then, between the history of geography and national identity as a matter of geography and of science in the age of empire not only demands that we take seriously their situated historical geographies, but that we show what those historical geographies were and how they worked, for whom they had given purpose and meaning. The historical geography of geographical knowledge in Scotland in the period 1884–1930 has been shown here to be dependent upon the inter-related use of the emergent sciences of national survey to promote both knowledge of the nation and knowledge of the 'sciences' themselves, including geography, geology, botany and Scottish history, as national products.

[120] The Edinburgh Geographical Institute was so named only from *c.*1888–9 (the building in Edinburgh's Duncan Street still bears that title, despite being closed in 1987 and converted into flats in 1997: a commemorative plaque was opened at the site in July 1997). It is clear that Bartholomew thought of his mapping firm as a crucial site for the making of geographical knowledge. In one of its previous locations, the EGI was situated in 'Gibbet Loan': Bartholomew considered this an inappropriate address and he unilaterally changed the name to Park Road on the next Edinburgh town plan: see L. Gardner, *Bartholomew: 150 years* (Edinburgh, 1976), 37.

[121] J. Bartholomew, *The Royal Scottish Geographical Society's Atlas of Scotland* (Edinburgh, 1895), Preface.

# 7

# Conclusion: a historical geography of geographical knowledge

In 1997, the Scottish Tourist Board ran a publicity campaign to encourage Scots to get to know their nation. One of the illustrations used showed two people rambling in the Scottish countryside (in Glencoe), their eyes drawn upwards towards uncertain horizons (Figure 7.1). In its original published form, this illustration was overlain with the words 'Rise up *now.* And see your nation again'. Such a call to know one's country is, in one sense, moralising and populist.[1] In another, this appeal to national self-knowledge by active engagement with the landscape is entirely consistent both with the activities of professional scientists 'in the field', for whom scientific knowledge demanded practical encounters (Figure 7.2), and with those other longer-term means to geographical knowledge and national identity that have been the subject of this book.

At different times and for different people in different places, geographical knowledge *in* and *of* Scotland has been a matter of, variously, being out there for yourself, of being told by credible others what was out there, of painting, surveying and mapping the 'shape of the nation' and of imparting such knowledge to others. Such geographical knowledge has been part of the emergence of 'modern' science. Geographical knowledge has been both a means to and the subject of polite civic discourse within Scotland's public sphere since at least the late seventeenth century, as it was in England and in other European nations. Geography in one form or another was taught in the universities since the later sixteenth century and in many of Scotland's schools from the eighteenth century. In these and in other sites, such knowledge was part of the means by which the nation came to 'know itself'. Geographical knowledge has been shown to have had a historical geography – of particular sites, given intellectual practices and of different audiences.

My attention to the historical geography of geographical knowledge has

[1] It echoes, perhaps deliberately, the words of 'Flower of Scotland', a modern folk-song which has achieved popular status as Scotland's 'national anthem'.

7.1 Encountering Scotland's geography in the 1990s – popular practices.

7.2 Encountering Scotland's geography in the 1900s – professional practices. A party of surveyors from the Geological Survey cross the River Dee near Braemar, 1903. John Horne is second from the left, Benjamin Peach second from the right.

suggested that Scotland's national identity was constituted in and through such knowledge. Rather than considering national identity to reside in terms of ethnic characteristics or in language or in 'traditional' culture, some or all of which attributes somehow symbolise Scotland, I have tried to show how the very notion of Scotland itself was geographically made. National identity was, then, never the simple and direct consequence of a securely-known geographical space. It was, like that space itself, constituted by the actions of certain people at certain moments working in certain ways in certain sites to 'give shape' to a given territory.

Proposing a historical geography of geographical knowledge in even these general terms is to distinguish this work both by what it is and by what is not. Whilst it shares some affinity with attempts to understand the place of geography in Britain, it has not been my aim to write a history of geography (however understood) as such in Scotland. Yet I hope that on the basis of the material discussed here we may now be able to move with some confidence away from the essentialist and normative claims of, for example, Robertson in 1973 and his attention to Scotland's 'protogeographers', by which undefined term he understood geographical practitioners before the formal establishment of geography teaching in the universities,[2] and away from the claims of Lochhead as to academic geography's origins in later nineteenth-century Scotland.[3] The material discussed here should allow us to distance ourselves both from Chisholm's view of geography's history, in Edinburgh anyway, which documents it as a matter of exploratory endeavour in the work of men like James Bruce and Mungo Park,[4] or, in more modern context, from other 'great men' views of Scottish and British geography.[5]

In offering an assessment of the situated and social nature of geographical activity in Scotland over the *longue durée*, this book has both contributed to a developing interest in the critical history of geographical knowledge and built upon it. It has extended the work of those who have seen geography as a formal academic discipline motivating ideas of national identity largely, if not only, from the later nineteenth century,[6] although it shares with Hooson a belief in the power of history and of geography to promote what he calls 'identity-consciousness'.[7]

In examining geographical knowledge as having both a history and a geography, a *historical* geography as a situated practical activity, I have tried, too, to site geography more firmly within work in the history of science – in which context in Scotland it is notable usually by its absence.[8] I have tried to connect concerns about geographical knowledge in historical context with the work of historians of science who have stressed both the socially constructed nature of knowledge and the importance of certain epistemological criteria in given settings to an understanding of the nature of scientific knowledge. If, then, as a work of historical geography this book reflects the broad notions of social constructivism in the history of science, it has also been an attempt to understand, for one 'national' context, what we might think of as questions of self-definition and of geographical constitution: the situated and mutual making of 'the field' of geographical knowledge and 'the territory' of Scotland.

[2] C. Robertson, 'Scottish geographers: the first hundred years', *Scottish Geographical Magazine* 89 (1973), 5–18.

[3] E. Lochhead, 'Scotland as the cradle of modern academic geography in Britain', *Scottish Geographical Magazine* 97 (1981), 98–109.

[4] G. Chisholm, ''Geography', in J. Ashworth *et al.* (eds.), *Edinburgh's place in scientific progress* (Edinburgh, 1921), 168–75.

[5] R. Rudmose Brown, 'Scotland and some trends in geography: John Murray, Patrick Geddes and Andrew Herbertson', *Geography* 33 (1948), 107–20.

[6] See D. Hooson (ed.), *Geography and national identity* (Oxford, 1994), particularly the following chapters: M-C. Robic, 'National identity in Vidal's *Tableau de la géographie de la France*: from political geography to human geography', 58–70; G. Sandner, 'In search of identity: German nationalism and geography, 1871–1910', 71–91; K. Takeuchi, 'Nationalism and geography in modern Japan, 1880s to 1920s', 104–11; V. Berdoulay, 'Stateless national identity and French-Canadian geographic discourse', 184–96; M. Garcia-Ramon and J. Nogue-Font, 'Nationalism and geography in Catalonia', 197–211; O. Spate, 'Geography and national identity in Australia', 277–82.

[7] D. Hooson, 'Introduction', in Hooson (ed.), *Geography and national identity,* 2.

[8] D. Thompson, 'The history of science in Scotland', in J. Wright and N. Snodgrass (eds.), *Scotland and its people* (Edinburgh and London, 1942), 241–72, in which there is no mention of geography or of geographical questions.

The following section summarises the main features of this historical geography of geographical knowledge by attention to its chronology, its principal discursive practices, the sites of its making and its audiences. The final section considers the wider implications arising from so treating the historical geography of geographical knowledge.

## Making the nation, making geography: Scotland's geographical constitution

The idea of a self-sustaining Scottish identity is first apparent by the later thirteenth century, and a self-conception by Scots as separate and different was clearer still from the early 1300s. There is, however, no clear sense at that time of a 'geographical imagination' at work to sustain that nascent national identity other than, perhaps, that sense of territorial affiliation to kingdom and *ethnos* that underlay the use of chronicle tradition and origin myth in early medieval Scotland. Certainly, there was no one, within what was to become understood as Scotland, for whom the constitution of geographical identity was directly a matter of mapping, textual enquiry, direct personal scrutiny or unequivocal trust in others' accounts.

In these terms (and recognising the difficulties attaching to any statement about definitive beginnings), the first expressions of geographical work recognised by contemporaries as a means to understand Scotland as a historical *and* a geographical entity are to be found in the texts of late Renaissance humanists such as Hector Boece, John Major and George Buchanan. The humanist narrative tradition of '*descriptio*', effectively synonymous with descriptive geography and chrorography, was particularly apparent in the introductory sections of Buchanan's 1582 *History.* In thus 'writing the nation' (as Helgerson puts it of England then and as others have seen of the early modern historical imagination[9]), Buchanan was not the only one doing so. Nor was textual description the only means to Scotland's geographical self-fashioning. Buchanan relied upon others' travel accounts, notably upon Dean Monro's 1549 description of Scotland's western isles. It is clear, too, that he and Major and Boece actively debated extant ancient and classical accounts of Scotland's geography. In addition to travel accounts and regional descriptions, geography in one form or another (chiefly as mathematical geography in association with astronomy and geometry, and as chorography) was being taught in the universities of Edinburgh, Glasgow and St Andrews by the mid-1580s (and earlier in Glasgow), and geography texts were held in the libraries of leading Scots churchmen. Maps of the later sixteenth century reflect James V's partial

[9] R. Helgerson, *Forms of nationhood: the Elizabethan writing of England* (Chicago, 1992); R. Langlands, 'Britishness or Englishness? The historical problem of national identity in Britain', *Nations and Nationalism* 5 (1999), 53–69; D. Sacks and D. Kelley (eds.), *The historical imagination in early modern Britain: history, rhetoric, and fiction, 1500–1800* (Cambridge, 1999).

circumnavigation of Scotland, undertaken to 'know his bounds' and to demonstrate royal authority. In Timothy Pont's work, notably in his mapping between *c.*1583 and 1596, and in what textual chorographical descriptions survive (of Cunningham between *c.*1604 and 1608), we have the first known attempt to delineate Scotland as a whole.[10] James VI's manual of kingship, his 1599 *Basilicon doron*, is underscored by what I would want to think of as a 'geographical consciousness' in relation to Scotland's civil administration, chiefly in terms of what was then (and would more dramatically become) the 'Highland problem'.

This range of geographical activity is consistent with the use of geographical knowledge in constituting national identity throughout Europe at this time and with what has been shown of geography *per se* in England between 1580 and 1620. Geographical knowledge was not just an intellectual and textual exercise. Symbolic representations of Scotland in the early 1600s were produced in consequence of that particular geographical and political creation, 'Great Britain', and courtly masques and triumphal processions presented geographical knowledge – indeed, literally performed and enacted it – as a form of royal authority in association with ideas of *imperium* which drew upon older-established myths and, more substantially, upon direct encounters with the newly-emergent British empire overseas.

As James VI and I well knew, geographical knowledge about one's nation is a means to effective governance of it. Such claims are clear from the fact that, from the 1630s, the Scottish Church, together with several leading members of the gentry such as Sir James Balfour, was concerned enough about the state of the nation to propose schemes for Scotland's geographical description. Sir John Scot of Scotstarvet was important in this respect. He was a crucial link, together with Robert and James Gordon, in rescuing Timothy Pont's manuscript maps to become the basis to the first 'modern' maps of Scotland, in volume V of Blaeu's 1654 *Atlas Novus.* In his 'Greetings to the Reader' with which he prefaces his *Atlas,* Blaeu informs us:

> In the Scottish Synod he [Scotstarvet] brought it about that individual ministers of Churches were instructed by public edict to direct their attention to the description of Scotland, to inquire more diligently into the lands in which each was settled, and to observe more attentively what was relevant to Geographers.

These intentions were, as we have seen, never realised. As Blaeu put it:

> But since neither rewards nor penalties created careful assuidity, compliance languished, with only a few fulfilling their duty.[11]

[10] C. Withers, 'Pont in context: chorography, map-making and national identity in the late sixteenth century', in I. Cunningham (ed.), *The nation survey'd* (East Linton, 2001).

[11] This is taken, in translation, from 'Johannes Blaeu's Greetings to the Reader' which forms part of the prefatory material to Vol. V of his 1654 *Atlas Novus.*

Unrealised maybe, such recognition by contemporaries in the 1630s of national geographical knowledge as a basis to civil adminstration should not be ignored.

A 'map' of geographical knowledge of men in different places differently at work in fashioning Scotland is first discernible in the last two decades of the seventeenth century. In 1682, Sir Robert Sibbald was appointed Geographer Royal and began his enquiries into Scotland's geographies and, in turn, appointed John Adair to undertake new maps of the country. For both men, this mapping impulse was motivated by its intrinsic importance and by their perceptions as to the out-datedness of Pont's and Blaeu's work. Both Sibbald and Adair were funded (albeit never adequately) by the Privy Council and acts were passed in Adair's favour. In 1681, James Paterson's *A geographical description of Scotland* appeared and other related works of natural knowledge were published. At much the same time, Robert Wodrow (like Sibbald more notably and Adair less successfully), was circulating geographical queries to secure reliable information from reliable people about Scotland (and other parts of the world). Martin Martin was, with Adair, reporting upon Scotland's natural diversity by being in 'the field', by taking seriously what he saw with his own eyes and, unlike Sibbald, by relying upon the testimony of the 'common sort'.

The picture thus afforded of geographical activity at the end of the seventeenth century is rather more of a competition for credibility *between* the different routes to geographical understanding than it is of a credibility for geographical enquiry as a whole. Sibbald's warrant as Geographer Royal was, effectively, an intellectual warrant for his subject and a personal one for him. It was less clearly an epistemological licence for any given method. Sibbald's largely Baconian geographical enquiries depended upon his being the recipient of others' information, information that had been requested in a standard form and which he directed at men and at institutions deemed credible. But it also depended upon his using his authoritative position as Geographer Royal to ensure that men practising other forms of geographical enquiry – Adair surveying in the field, Martin collecting and conversing with the natives – could be brought under his overall control. Sibbald's geographies *and* his personal authority to undertake that work depended upon his being a 'centre of calculation' into which geographical information moved (as it did from many correspondents: Appendix, pp. 256–62), and from which should have been generated (but was not) his two-volume geographical synthesis on Scotland's geography.

I would not wish to overplay the different practices of geographical enquiry manifest in the work of men like Adair, Martin, Paterson, Sibbald, Wodrow and, earlier, the Gordons or even Scot of Scotstarvet and Sir Andrew Balfour with his 'Topographical Descriptions'. After all, Adair utilised an essentially chorographical 'List of Queries' as well as being a map-maker, in which context a more strictly mathematical basis to geographical knowledge was employed to provide maps in order to know the 'true shape' of Scotland and to serve as a basis to commerce. Sibbald's attention to exact distances suggests an interest in mathematical precision within his otherwise broadly qualitative empiricist enquiries. The difference is one of emphasis within a shared concern – evident at the highest levels of political and royal authority – to know Scotland geographically as a matter of useful natural and national knowledge.

Neither these men nor their practices were peculiarly Scottish. What they were doing was consistent with geographical activity in France, with the work of chorographically-inclined natural historians in Ireland, Wales and England, and with what geography was being taught in later seventeenth-century Oxford. These men were Europeans. They had connections with the Royal Society in London. For Martin Martin, geographical fieldwork was motivated as much by the curiosity of London-based natural philosophers and *virtuosi* as it was by the needs of Highland patrons and the Edinburgh-based Geographer Royal.

If, then, as I want to contend, geographical knowledge in these ways in the 1680s and 1690s was central to the utilitarian philosophy underlying the move to enlightenment and improvement in eighteenth-century Scotland, it is also true that much of it was a failure. Why this should be so is unclear, but several reasons suggest themselves. Sibbald's works were never published in full. Despite repeated parliamentary legislation, neither he nor Adair ever had the funds they needed. In one way or another, Sibbald, Adair, Martin and Slezer all squabbled over contracts. Perhaps they were, as a whole, too few. The effects of Darien, the move to union and the failure to establish a Royal Society of Scotland probably also contributed in the failure to secure the right environment, despite political support for geographical knowledge.

Widespread civic interest in such knowledge was more clearly apparent in and from the eighteenth century: in geography's books and in education, in schools, in the universities and in the audiences for geographical subjects in Scotland's public sphere. Geography's audiences were first centred in Edinburgh and in Glasgow and were principally apparent in the larger towns, only appearing in smaller centres of population from the later eighteenth century. For some people and at particular moments in given settings, it has been possible to show exactly what sort of geographical knowledge was imparted to Scotland's urban public sphere and why, if less clearly to know to whom. Consider, for example, the connections between geography and navigation taught by James Hall in Portsoy in 1791 and by Alexander Ingram in Leith's Constitution Street between 1815 and 1828; the emphasis on geography as a means to polite learning suitable for women given by Alexander Monro to his daughter at home in 1739; the place of geography in a gentleman's education (in association with civil history and mathematics) that John Mair instilled in his pupils in Ayr and in Perth between 1727 and 1769; and the classes given by Ebenezer MacFait at home in the East Close of Edinburgh's Meal Market in 1754. At the same time, the eighteenth century witnessed two metrological projects at the national scale designed to know Scotland in different ways: the mapping project of the Military Survey (and its concomitant visualisation of the Scottish landscape in the work of Paul Sandby); and that inductive work of political economy and of geographical anatomy, the *Statistical account* (1791–9).

Geography's practitioners could sometimes be unenthusiastic about their subject. John Pinkerton, author in 1802 of *Modern geography* and, in 1815, of the *Modern atlas*, was one such. In 1785 and writing under the alias 'Robert Heron', Pinkerton attempted in an essay 'On the Scale of Fame' to prioritise twenty categories of writing in order of worth, beginning with 'I. Epic Poetry II. Dramatic Poetry III. Moral Philosophy', and ending with 'XX. Geometry. Chronology, Geography'.[12] Even admitting its arbitrary nature, it is testimony of a particular sort to the variation in attitudes towards geography which underlay the different uses of geography in civil society in late eighteenth-century Scotland.

The sense in which geography as a subject, however understood, and of geographical knowledge in broader senses were respected areas of intellectual enquiry and of polite and popular public interest is clear in the foundation (and the title) of the *Edinburgh Journal of Natural and Geographical Science*. The short-lived nature of that journal makes it difficult to establish any direct connections between it and the many natural history and field societies active in Scotland from about the 1840s. It is possible to suggest, however, that we may think of this associational activity and civic engagement with natural knowledge as, in many ways, profoundly geographical. Certain towns and regions had active field clubs and other socio-scientific bodies whose remit was the exploration of their local 'territory'. It is possible that the constitution of some of these bodies actively reflected the social mores and political interests of the towns and areas in which they were established. They certainly reflected the intellectual and practical aspirations of local Scottish publics seeking, variously, to know the geography, botany and antiquities of their local area, or to promote moral education through being in the country, or to see geographical knowledge of other countries as a basis to commercial exchange within the British empire. Some societies did so collectively, as the joint field outings of the Natural History Societies of Perthshire, Dundee, Largo and Kirkcaldy attest.[13] Both the Aberdeen Philosophical Society and the Glasgow Philosophical Society (through its Geography and Ethnology Section) actively promoted knowledge of geography and of empire in order to secure Scotland's place in imperial markets and did so before the foundation in 1884 of the (Royal) Scottish Geographical Society. On a local level, the

[12] R. Heron [John Pinkerton], *Letters of literature* (London, 1785), 504–5.

[13] *Proceedings of the Perthshire Soicety of Natural History Session 1881–82, 1885–86* (Perth, 1886), 270.

Corstorphine Literary and Geographical Society was established in 1880 in what was then a village to the west of Edinburgh. Surviving notebooks from 1912 show its activities included lectures on the geography of Africa, debates about contemporary social issues and outings to local sites. Geography was recognised as a basis to popular understanding of one's nation before the establishment of formal geographical bodies.[14]

The nineteenth century was also characterised by an increasing specialisation of geographical knowledge, and by sectoral and social differences within the public's interest in geography. This is evident in the work of the Ordnance Survey (even allowing for the slow progress of mapping Scotland overall), in the use of particular geography books within Scotland's schools, in the attention given to geographical texts including atlases from the 1830s, and in the place of geography in different institutions of higher learning. From the later nineteenth century and in the early twentieth century, more specialised disciplinary surveys – bathymetrical, geological, ethnographic, linguistic and botanical – reflected both the increased professionalisation and specialisation of scientific knowledge, and yet, despite such different 'fields', they collectively constituted Scotland as a national space.

In those terms, the formalisation and institutionalisation of geography as an academic discipline in Scotland is late – 1884 in terms of a civic geography society, 1908 in terms of university teaching leading to a qualification in geography. Such institutionalisation should be seen not so much as the immediate and successful direct result of shared endeavour within a national 'scientific background', but as a failure to build upon an earlier established geographical heritage. From the early twentieth century, Newbigin in her editorial role and in her own work, together with men such as Kermack and the RSGS through its public lectures, actively used the geography of Scotland as means to extend geographical knowledge more widely. Patrick Geddes and his colleagues in the Outlook Tower from 1895 to 1932 certainly used local field sites as a means to extend the 'Geographer's Outlook', and such attention was also accompanied by Geddes' view that geographical education as a matter of regional survey was a route to citizenship of the world, not just to knowledge of one's own locality or nation.

## National identity and geographical practice

Several recent studies of the making of geographical knowledge in national context have stressed the institutional setting in which geography was promoted and focused their attention almost exclusively on events since the later nineteenth century. Wardenga has discussed the role of the Central Commission on the Regional Geography of Germany from 1882 to 1941, for

[14] I am grateful to the Secretary of the Corstorphine Literary and Geographical Society for making available the Minutebooks relating to the Society's activities since 1912.

example, and has shown how that Committee began its work to secure a national geography, between 1882 and 1916 particularly, by attempting (and failing) to coordinate the activities of earlier regional geographical societies.[15] In Argentina, attempts by the state between 1863 and 1916 to promote national identity failed from a lack of organisation amongst the country's nascent geographical community.[16] Attempts in twentieth-century Poland to replace 'national consciousness' with class consciousness foundered on continued practices of regional identity.[17]

Scotland's geographical making has been no less a matter of situated institutionalised activity. Yet this study has documented in some detail the epistemic and discursive characteristics of Scotland's geographical self-fashioning. In summary, the geographical practices were surveying, mapping, textual description, symbolic visualisation, and what I want to call polite and public instruction. This is not to suppose clear distinctions between such practices. Maps are texts as well as symbols, and surveying in the strictly technical sense has close connections with mapping. Further, these practices share epistemological concerns – about how in various ways reliable knowledge was secured – and have connections with the sites of their making, as well as themselves changing in nature over time.

It is, however, worth making the distinction between the *idea* of territorial survey in which context geographical knowledge, understood as a collective discursive enterprise might encompass the above, and the different *practices* of survey effectively realised. In general intent, Sibbald's chorographical enquiries sought a geographical survey of the whole nation, but his plans for 'an Exact Description' did not use the term 'survey' to describe either the means or the end. The idea of survey understood as an overview of the nation's geography arrived at through organised investigation at a consistent administrative scale – the parish – is suggested in Balfour's and Scotstarvet's concerns in the 1630s, was clearly expressed in 1708 and was repeated at moments throughout the eighteenth century: in Macfarlane's geographical enquiries of the 1720s, in Maitland's in the 1750s, and in Erskine's in the 1780s.

The first effective realisation of the idea of national survey so understood was Sinclair's in the 1790s. This was an inductive enquiry on Baconian principles and, as a work of geographical knowledge and natural history, was reliant, like Sibbald's and the Earl of Buchan's, upon the accumulation of knowledge from reliable 'authorities' themselves resident in the places enumerated.

[15] U. Wardenga, 'Constructing regional knowledge in German geography: the Central Commission on the Regional Geography of Germany, 1882–1941', in A. Buttimer, S. D. Brunn and U. Wardenga (eds.), *Text and image: social construction of regional knowledges* (Beitrage zur Regionalen Geographie, Leipzig 1999), 77–84.

[16] M. Escolar, 'Promotion and diffusion of geographical knowledges: Argentine editorial policies and the nation's geographical body representation (1863–1916)', in Buttimer, Brunn and Wardenga (eds.), *Text and image*, 91–9.

[17] W. Wilczynski, 'Gatekeeping and regional knowledge in Poland', in Buttimer, Brunn and Wardenga (eds.), *Text and image*, 111–21.

The scale of enquiry was the parish, from which scale a national picture was built up. The reliable informants were the ministers working to a more or less standard brief. Similar practices determined the form and content of the *New* or 'Second' *Statistical account of Scotland* between 1831 and 1845. The work of many of Scotland's field clubs throughout the nineteenth century was also a form of local survey. In contrast, the more detailed scientific surveys of the later nineteenth century largely reflected scientific specialisation and were dependent upon accredited professionals in those sciences, or their appointed representatives, themselves working in the field to secure knowledge through personal observation and those warranted practices accepted as markers of disciplinary credibility: instrumental accuracy, temporal and spatial correlation, the search for causal connections and for explanation upon general principles.

There is, then, theoretically at least, an important distinction to note between geographical knowledge about the nature of one's nation derived from reliable authorities, who are themselves resident in the area under investigation, and knowledge derived by being there in person. Both issues are to do with being 'in the field'. In the first context, geographical knowledge was secured by the geographer not being 'out there', but by being a fixed 'centre of calculation' reliant upon the movement in to a central point or person of others' information arrived at through standard means such as a circulated list of queries. That, essentially, was what geographical knowledge in one form of survey was about – in Sibbald's work and in Sinclair's. In the second context, geographical knowledge about the shape of the nation was derived by the geographer being 'out there' and securing credible information through direct encounter – by observing it or hearing it told or, in other senses, measuring it one's self. This sense of survey more adequately describes the later nineteenth-century work of Peach and Horne in geology, of Robert Smith and Marcel Hardy in botany, of Murray and Pullar in bathymetry and the field-based craniometry of Gray and Tocher.

Knowledge was also secured through a combination of these practices. John Adair, for example, issued and circulated queries and, as a map-maker, actively moved in and across the country. Thomas Pennant secured his knowledge of Scotland in 1769 and 1772 by using a standard set of questions entitled 'Queries, addressed to the Gentlemen and Clergy of North-Britain, respecting the Antiquities and Natural History of their respective Parishes, with a view of exciting them to favour the World with a fuller and more satisfactory Account of their Country, than it is the Power of a Stranger and transient Visitant to give'. He also depended upon a Gaelic-speaking native (and Highland parish minister), the Rev. John Stuart of Killin, to provide assistance in the Gaelic parts of Scotland. Pennant thus provides trustworthy textual descriptions not because they are based on his observations alone or, even, because he spent a lot of time 'in the field'. His credibility rests, rather, upon the conjunction of his sources – Scots of a certain social status and linguistic ability – and upon his method, which, in managing geographical difference in standard ways, ensured that native knowledge rather than direct and prolonged experience provided the basis to sound geographical understanding.[18]

Mapping Scotland demanded being 'out there', but not always to the same purpose. So far as we can be certain, Timothy Pont's surviving manuscript maps and textual descriptions suggest that his concern in the 1590s and early 1600s to delineate Scotland was more strictly chorographic than cartographic, more a qualitative appraisal of what was where than an accurate geometric representation. In contrast, Blaeu to some degree and Adair much more notably were concerned to depict the true shape of the nation through 'exact Maps'. Adair's concern was motivated both by the honour and by the utility such accuracy would bring. Yet even in later periods, what Cowley had observed in the 1730s as 'Disagreement among Geographers' was still a matter of concern, although the utility of geography to an understanding of Scotland's national identity was not in doubt.

Consider, for example, the evidence of David MacPherson's *Geographical illustrations of Scottish history*, published in London in 1796. The subtitle to the book declared its intentions to offer '*explanations of the difficult and disputed points in the historical geography of Scotland*'. MacPherson began his *Geographical illustrations* with the remark 'If . . . Geography is one of the Eyes of History, it may be with great truth be affirmed, that the history of Scotland has in all ages been blind of at least one eye'. Accurate mapping should provide better geographical knowledge as a basis to a fuller historical understanding of Scotland's identity:

> Nothing can be more distressing to a reader of history [noted MacPherson] than the difficulty, frequently insuperable, of discovering the situation of places, where the events recorded have happened . . . Many of them are not to be found in modern general maps, or in any maps whatsoever. Hence the reader, who is pretty well acquainted with geography, after a laborious search in such maps as he can procure, finds himself obliged, either to go on without satisfaction, or to abandon his pursuit with disgust.

This, continued MacPherson, 'is only a small part of the geographical embarassments, which attend the study of Scottish history'. He then outlined the circumstances behind the production of the map in his book (Figure 7.3):

> THE MAP
>
> will appear upon examination not to have been copied from any map of Scotland already published, but to be, as much as any general map can possibly be, an original work. The geographical part (or plan work) of it has been constructed from the best actual surveys and charts, corrected by all the celestial observations I could procure,

[18] C. Withers, 'Travel and trust in the eighteenth century', in J. Renwick (ed.), *L'Invitation au voyage: studies in honour of Peter France* (Oxford, 2000), 47–54.

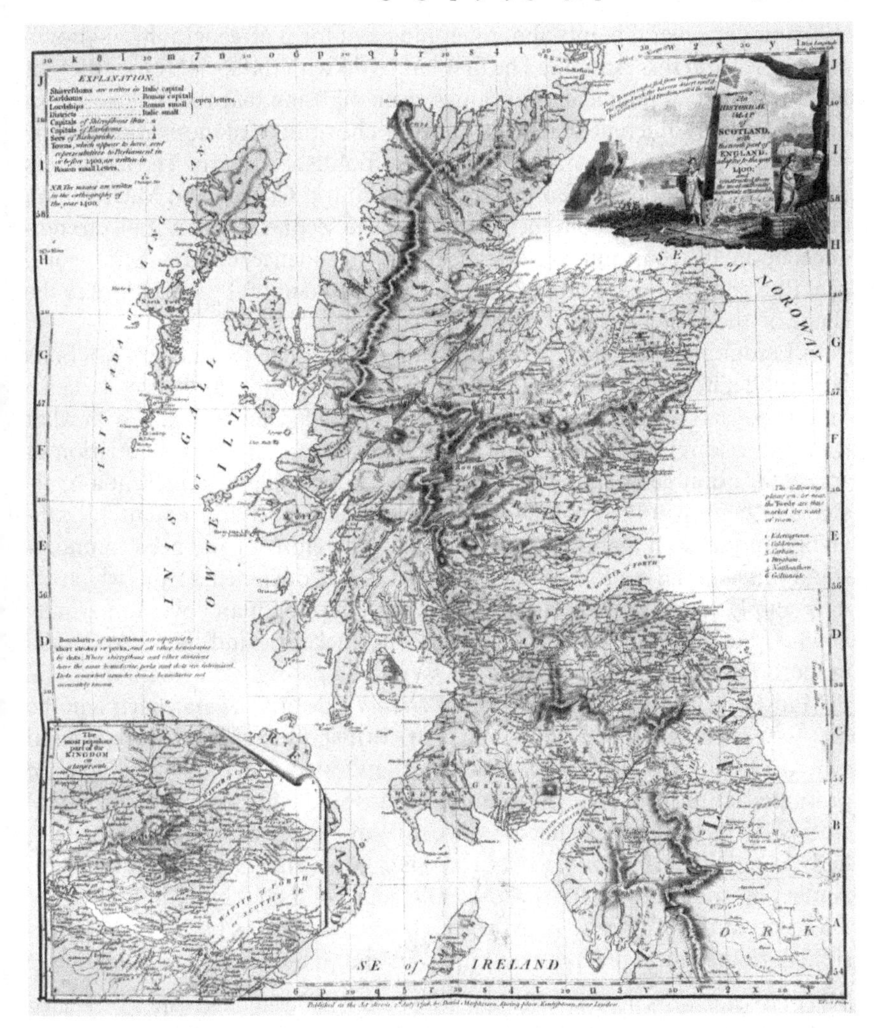


7.3 David MacPherson's 'Historical Map of Scotland . . . adapted to the year 1400', from his *Geographical illustrations of Scottish history* (1796).

and carefully collated with every particular map, antient or modern, which could afford any information. Hence I may venture to say, that the coast-line comes very near to the truth; and even in the small indentings of the headlands and inlets of the sea it will be found very little inferior in minute accuracy to many of the larger maps. But I must acknowledge, that much of the interior country in the north part of the kingdom is laid down upon very slender authority, and I am very sensible, that, with every possible attention to correctness, I have not been able to avoid many errors and uncertainties.[19]

[19] D. MacPherson, *Geographical illustrations of Scottish history* (London, 1796), Preface [iii-iv].

Two more general points about mapping as a form of geographical knowledge arise from this example. The first has to do with the social recognition of maps and of mapping as useful things, even allowing that the philosophy of utility varies in different practical contexts. The utilitarian motivation underlying Adair's work was a matter of national political concern. The usefulness of maps and mapping differently motivated Mark, Cowley and MacPherson and the audiences to whom they appealed. The second relates not just to reliance upon instrumental accuracy but to dependence upon others' work. MacPherson drew upon others' maps but still could not fully be truthful to the shape of the nation.

In Latourian terms, then, the 'truth' of a map, its fidelity, may rest not in its being understood as an accurate 'material trace' for distant audiences, but also in terms of its context and *how* the map corresponded epistemically with what it purports to represent. Asking if a map is of use in the local negotiation of space, for example, is simply different from supposing it to represent a 'true' shape or to have utility in other ways. These issues – of differential utility, reliability and trust in mapped knowledge and, therefore, of the need to understand maps in the context of their making not as documents from which we may simply infer a more accurate representation of Scotland over time – may be illustrated through the case of Murdoch Mackenzie and James Anderson in the 1780s.[20]

Mackenzie was the author in 1750 of *Orcades*, an atlas of sea charts and the first to use triangulation methods to chart British waters. Later work around Lewis was done less sytematically than the Orkney work.[21] This difference prompted an attack upon Mackenzie's maps and, thus, upon his personal credibility, by Dr James Anderson who, in using Mackenzie's maps in the field, had found them 'extremely erroneous'. Mackenzie, in responding to Anderson, sought to establish trust in his maps and credibility in himself by reference to the political context through which his maps had been commissioned, to the *in situ* practices he employed and to the testimony of others:

[H]e begs leave to assure the public, that with accuracy, fidelity, and care, he executed the trust reposed in him, of carrying on a nautical survey, in consequence of instructions received from the Lords of Admiralty. . . . And he submits the merits of his survey, the work of many years, upon the only proper test, the approbation of those *alone* who are most interested, and the best qualified to judge, *seamen*, the masters of vessels of all kinds, who after the continued use of the said surveys from ten to thirty years, have never once found fault, so far at least has come to his knowledge.[22]

Geographical knowledge was here being made through the map not only because of networks of political patronage, the actions of trained surveyors

[20] D. Webster, 'A cartographic controversy: in defence of Murdoch Mackenzie', in F. MacLeod (ed.), *Togail Tir marking time: the mapping of the Western Isles* (Stornoway, 1989), 33–42.

[21] D. Smith, 'The progress of the *Orcades* survey, with biographical notes on Murdoch Mackenzie senior (1712–1797)', *Annals of Science*, 44 (1987), 277–88.

[22] Quoted in Webster (1989), 34.

and the trust placed in them by distant audiences, but also because natives could see for themselves how the map did and, as importantly, *did not* correspond to reality.

Martin Martin secured geographical knowledge about an 'unknown' Hebridean Scotland for himself and for distant others by talking with the natives. He could do so: he was a native Gaelic speaker. Others, like Pennant, had to have such knowledge translated by intermediaries. The idea of translation is linguistic. In mapping, for example, linguistic translation was central to the work of the Ordnance Survey in the Scottish Highlands in the later nineteenth century, as that body 'authorised' written English versions of spoken Gaelic names.[23] Translation is also epistemological. The public audiences for geographical knowledge in Scotland had to have such knowledge represented to them in various ways: in symbolic pageantry; in paintings, texts and atlases; in dioramas and via displays in exhibition halls.

The evidence here reviewed also suggests that Scotland's geographical public was made aware of geographical matters by hearing about them. For some people in some places, hearing about geography was reinforced by instrumental display: using the globes and wallcharts in schoolrooms, or consulting an atlas at home. For others, knowledge was arrived at through linking what one read and was told with what one saw for one's self. Recognising geographical knowledge as a situated spoken *and* embodied activity is to demand closer attention to the sites of its making.

## Sites and audiences

This study has shown that geographical knowledge was constituted in a variety of different sites: school classrooms, lecture theatres, the field, exhibition halls, coffee houses, scientific meetings and, of course, by encountering geography in the home. It has also shown the importance of various analytic spaces – books, atlases, the spoken word, didactic instruction in the countryside, the reading of newspapers – and of particular institutions. Film is largely without my period of concern, but it is clear, too, that 'the cinematograph' was understood as a means to extend geographical knowledge.[24]

One formal institution central to the promotion of geographical knowledge of Scotland as a means to national understanding was the Scottish church, certainly for the two centuries before 1800. Pont's chorographic work was probably the result of the need for knowledge of the kingdom in order to effect ecclesiastical administration. Scotstarvet's and Balfour's work in the 1630s

[23] C. Withers, 'Authorising landscape: 'authority', naming and the Ordnance Survey's mapping of the Scottish Highlands in the nineteenth century, *Journal of Historical Geography*, 26 (2000), 532–54.

[24] Anon., 'Geography and the cinematograph', *Scottish Geographical Magazine*, 30 (1914), 100–1; Anon., 'The use of the cinema in the teaching of geography', *Education Times*, February 1921, 85–6; Anon., 'The use of the cinema in the teaching of geography', *Geographical Teacher*, 10 (1920), 280–2.

was certainly so motivated. Sinclair's survey depended upon the credibility of parochial ministers; Maitland's and Buchan's plans intended likewise. Although the evidence of Thomas Chalmers' thinking geographically and using geography as a subject in order to address urban social problems and of parochial schoolmasters using geography texts in the later 1830s suggests that the Scottish church's representatives drew upon geography, the church's role as a commissioning agent of geographical knowledge had, arguably, ceased by 1799.

The Parliament of Scotland and the Privy Council was active in supporting geographical enquiry as a means to national utility from the 1680s but, as a result of union in 1707, that body and, thus, the Scottish state, was never fully successful in such work. The first single formal public body designed to promote geography was established in 1884, although there were many civic bodies throughout Scotland who incorporated geographical knowledge within their activities. The universities included geography within their teaching curriculum from the 1560s.

It is possible, therefore, to chart over time and over space the place of geography in different institutions and to recognise both that the 'maps' of geographical knowledge in Scotland were more complex than current histories and geographies allow and that further work remains to be done on such situated geographical activity. Different sites acted to constitute geographical knowledge in different ways. The symbolic geographical representation of Scotland during triumphalist processions in the seventeenth century was, of course, different in content from the moving images of nineteenth-century diorama or of documentary film in the 1930s. But in general intention, what connects the use of the street in 1604, 1633 and 1661, the national and international exhibitions of 1886, 1901, 1908, 1911 and 1938, the Royal Diorama of 1878 and the use of instruments by geography's public lecturers, is their place as sites for the *display* of geographical knowledge. Other sites – notably the home, schools, universities, private academies and those other spaces in which geography in a variety of forms was taught – were sites of geographical *instruction*. Yet other sites, such as Sibbald's working spaces where he coordinated the geographical correspondence of others, or the Edinburgh Castle offices of the Military Survey, or the Bartholomew firm's Edinburgh Geographical Institute, may be thought of as sites of geographical *calculation*. This typology should not be thought of as discrete. The heterotopic spaces of Geddes' Outlook Tower, for example, depended upon the display of geographical instruments, active engagement with geography's content through books, lectures and fieldwork and upon Geddes' belief that that site was the forerunner to a wholly new type of international geographical institute. One scheme employed to teach world geography in the mid-1930s involved 'adopting' a tramp steamer and understanding world affairs by following its route: amongst the enthusiastic participants in this scheme was the education officer of the Black Watch stationed in Glasgow's Maryhill Barracks, 'anxious to give his men the benefit of this novel and exciting method of imparting geographical knowledge'.[25]

Understanding the different sites of geography's making is to allow, too, that the audiences have been different. My attention to the question of audience, from the early eighteenth century at least, has been in general terms informed by the idea of the Habermasian public sphere. It is clear, however, both from the evidence presented here and from recent attention by others to the public sphere, that general claims about 'the public' must themselves recognise geographical differences and attend to local specificities. Let me briefly illustrate this point with reference to more-or-less contemporaneous evidence for geographical instruction in the home of three different Scots women in the early nineteenth century. One exercise book (of unknown provenance) shows a Miss Jemima Arrow, in December 1815, learning the geography of Europe by hand-colouring the outlines of European countries.[26] In July 1827, in her home near Galasheils, Miss Eliza Loch was learning geography by writing down as a numbered list 148 statements of geographical fact.[27] In 1804 (in Inverness-shire) and in the summer of 1805 (in her London residence), Elizabeth Grant of Rothiemurchus was instructed at home by 'our Mr Thompson', who came to teach her and her brother geography, astronomy and history.[28] Capturing such interest in things geographical and knowing that different practices were at work in the home as one site of geographical instruction is not to know, however, what use was made by these women of their geographical knowledge, with whom they discussed it, and whether they ever came together with like-minded others to form a 'geographically-aware' public, even if only for a short time and in a given institutional context.

Domestic instruction in geographical knowledge was quite widespread, notably in the larger towns, from the later eighteenth century. It was so amongst those who could afford it and in homes where space normally used for domestic purposes could be temporarily designated 'teaching space'. It is one thing, then, to document the historical geography of geographical knowledge as a matter of individual members of the public in Scotland engaging with such knowledge. It is quite another to claim either that 'the public' as a whole in Scotland was everywhere so engaged or to know how the public sphere for geography related to other audiences exploring, say, botany or geology.

Was Scotland's geographical public constituted only in the fact of individuals with shared interests coming together – in which case we ought to distinguish between public and private engagement – or should we see such

[25] Anon., 'Geography without tears', *The Glasgow Herald,* 6 November 1936, 12g; *The Glasgow Herald,* 31 March, 8g. [26] NLS, MS 14100 [Geography exercise book, 37pp, 22 December 1815].
[27] NLS, MS 14138 [School geography exercise book, 10ff., July 1827].
[28] E. Grant, *Memoirs of a highland lady* (Edinburgh 1988 edition), 56.

individuals *as* the public sphere? For Habermas, the latter obtained: 'The bourgeois public sphere may be conceived above all as the sphere of private people coming together as a public'.[29] It did so, moreover, in different institutions. Yet recent criticisms of Habermas' theory have stressed that this conception of the public sphere is too narrow.[30] Others have shown that the public sphere should be thought of as 'spatialized'. The evidence presented here would suggest that we can go beyond Mah's sense of 'spatialization': as he puts it, 'I refer only to the most general characteristic of conceiving of the public sphere as a space or domain that one enters, occupies or leaves'.[31] This study has documented the place of geography *in* Scotland's public sphere. It has also illuminated the idea of different historical geographies *of* the public sphere and, thus, of the importance of local difference. In these terms, there is much scope for considering further the historical geography of knowledge in terms of the differing public and private geographies of its reception.

## Historical geographies of knowledge

This study of the constitution of 'the field' of geographical knowledge and of 'the territory' of Scotland since 1520 in relation to such knowledge has, I suggest, wider implications for historical geographers, historians of geography and historians of science.

One has to do with recognising that ideas and practices have a practical and situated historical geography, not just an intellectual history. My argument here that the sites and practices of geographical knowledge have been important in constituting national identity is dependent upon an acceptance of the 'relativist turn' in the understanding of science as a social construction and, more particularly, of the 'geographical turn' within the history of science. But even if one were to adopt an essentialist and universalist stance in relation, for example, to science's making, or to the nature of Darwinism, the question of such practices having a geography – a geography of provenance, a geography of reception and so on and, in turn, of their having a *historical* geography – is, I contend, central to explanation of the phenomena in question.

A further implication has to do with recognising that there is, of course, a world of difference between using a geographical rhetoric to understand the situated nature of knowledge making and showing how that setting has a constitutive role in the shaping of the knowledge in question. Showing *how and why* any given knowledge has the (historical) geography it does may depend

[29] J. Habermas, *The structural transformation of the public sphere: an enquiry into a category of bourgeois society* (Cambridge, Mass., 1992 edition), 35.

[30] C. Withers, 'Towards a history of geography in the public sphere', *History of Science,* 37 (1999), 45–78; T. Broman, 'The Habermasian public sphere and "science *in* the enlightenment"', *History of Science,* 36 (1998), 124–49; H. Mah, 'Phantasies of the public sphere: rethinking the Habermas of historians', *Journal of Modern History,* 72 (2000), 153–82.

[31] Mah, 'Phantasies of the public sphere', 160.

upon the scale at which any such enquiries are conducted, upon exactly which sites are the subject of enquiry and upon knowing *for whom* the sites and the practices undertaken there had the meanings they did. Sites of making and of meaning cannot be isolated either from the social networks and epistemic practices that sustained them or from their audiences. That this is so means we ought not to agree too readily with Shapin's claims about being 'local enough', since questions of local meaning are always made in relation to their wider significance.

Attention to the historical geography of (geographical) knowledge ought not to privilege the spatial at the expense of either the social or the temporal. Capturing at one moment the complexity of, say, geography's making should also demand that we recognise the prior context, the potentially longer-term 'processes of becoming'. For that reason, questions of strict origin may be misleading and matters of context are all important. Asking the question 'What is meant by the historical geography of knowledge?' does not, initially, permit of a very precise answer since it will depend upon stating more clearly what exactly the knowledge in question is, at what scale of analysis it is to be understood and on the focus of enquiry – over time, at a 'moment', in given sites, of certain methods, in particular social networks and so on.

This study of the historical geography of geographical knowledge has emphasised the central place of sites, social networks, the making and movement of knowledge, and of particular institutional imperatives in understanding the constitution of Scotland. It has done so in ways which suggest we should not privilege the making of knowledge in the academy. Historically, knowledge has been made in a variety of sites and social networks other than in strictly disciplinary spaces: through encounters in the field, in the classrooms using models, as a matter of survey or of mapping, in popular forms for a variety of publics and as a matter of more strictly definitional enquiry in textual form. It has also suggested that there is merit in understanding as a matter of scholarly attention the facts of failure, of failed intentions to realise the production of knowledge. Understanding the production of knowledge as a spatial and a historical matter demands attention to geographical difference, to knowing how forms of knowledge become legitimised (or not), and to knowing how local circumstances had, if they did, wider geographical and historical meaning.

# Appendix

# Principal respondents to Sir Robert Sibbald's geographical queries

The material here enumerated dates from *c.*1683 to *c.*1700 and, in part, from earlier material, most probably from the 1640s and the attempt of the General Assembly to undertake 'geographical descriptions' following the prompts in 1641–2 of Sir John Scot of Scotstarvet (see p. 50). Sibbald was also drawing upon the topographical descriptions of parts of Scotland by Sir James Balfour (see p. 49). The list is principally based upon Sibbald's 'account of thes I pleased to be done for me' (NLS, Adv. MS 33.3.16, ff.14v–16v), cross-referenced against Adv. MS 34.2.8 in Sibbald's MSS and lists of those upon whom Sibbald drew given in Nicholson's *The Scottish historical library* (London, 1702). The prefatory identifying number is for ease of reference and does not appear in the original MSS. Hew Scott's *Fasti Ecclesiae Scoticanae* (London, 1915) has been used to identify or corroborate parish ministers as authors: references appear as *FES*, volume, page number (NKA = no known author). Many of these manuscript entries were collected together by Walter Macfarlane in the eighteenth century and are printed in Mitchell's *Geographical collections.* Where so printed, reference is given as *GC*, volume, page number.

1 Matthew Markell (also Markail, Markaille) Apothecary, M.D.; 'Orkney Islands'. *GC*, 3, 1–7, 10–11.
2 As above: 'Lead mines in Fife'. Adv. MS 34.2.8, f. 347 'has relation anent Natural Products from my Good Friend M.J.K.M. [Matthew Markell]'. *GC*, 3, 14.
3 Viscount of Tarbat: 'Isles of Hirta [St Kilda] and Rona'. Adv. MS 34.2.8, ff. 26–7 has this by Sir George Mackenzie 'as he had it from intelligent persons dwelling on the place'. *GC*, 3, 17–9, 21–4, 28–9.
4 NKA: 'Memorial of the Burgh of Haddington'. Adv. MS 34.2.8, f. 29 gives the authors as 'ye magistrates yr. of'. *GC*, 3, 65–7.
5 William Dundass of Kinkavill: 'Description of Caithness'. Nicholson (1702) has 'Answers to Queries about Cathness; by Mr William Geddes,

Minister at Wick, and Mr W. Dundas, Advocate'. Adv. MS 34.2.8, ff. 78–83. Geddes, a native of Moray, died in 1694 and prepared for publication a work entitled *Geographical and arithmetical memorials*: *FES,* 7, 141. *GC,* 3, 82–7.

6 Mr Skene (sent by Baillie Walter Robertson): 'Survey of the Town of Aberdeen'. Adv. MS 34.2.8, f. 83 terms Mr Skene 'late Bailie there'. *GC,* 3, 87–91.

7 [Verses on same by Skene]. *GC,* 3, 91–2.

8 Dr Alexander Penycuik of Newhall: 'Tweddal'. Nicholson (1702) notes that a Mr. Forbes was also an author here. This formed the basis to Penycuik's (usually given as Pennecuik) work, *A geographical and historical description of the shire of Tweedale* (Edinburgh, 1715) which was to have had map illustrations by Adair. These were never forthcoming. *GC,* 3, 140–54.

9 Lord Jedburgh: 'Shyre of Roxburgh or Teviotdale'. Nicholson (1702) notes 'as also by Sir William Scot of Harden, and Mr. Andrew Carr of Sinless'. Adv. MS 34.2.8, f. 161 notes 'Ker of Sinlaws' and dates the work at 27 December 1649. It is likely that this (and others of the same date: see nos. 9, 21 here), was prepared for the General Assembly of the Church of Scotland following the request of 1641–2 to undertake geographical descriptions: see p. 50 above. *GC,* 3, 135–8.

10 'Dr Cranstoun gave me an account of ye gentry yr' (Teviotdale).

11 Mr Eliot, minister: 'Sheriffdom of Selkirk or Etrick Forest'. Nicholson (1702) has 'by W. Elliot of Stobbes and Walt. Scot of Orkilton' and a separate entry on 'Ettrick Forest by Mr Elliot'. Adv. MS 34.2.8, f. 163 dates this to 21 December 1649. Eliot/Elliot is probably William Elliot of Yarrow, who died November 1685 (*FES,* 2, 96). *GC,* 3, 138–40.

12 Mr Veech, minister: 'Berwickshyre'. Nicholson (1702) has 'Berwickshire or the Merse, by Mr Elliot, and Mr Veech, ministers there'. Adv. MS 34.2.8, ff. 198–215. This is the same Elliot as above (No. 11). *FES,* 2, 165 has a John Veitch, minister of Westruther, born 1620, outlawed as an Episcopalian, 1680, restored 1690. *GC,* 3, 169–185.

13 Dr Archibald: 'wrote for me ane Account of Dumfries, Galoway, and Anandale'. Nicholson (1702) notes 'Shire of Dumfrese, by Dr. [George] Archibald, with his Account of the Natural Products of Galloway and Dumfreseshire' and makes reference to 'The Stewardry of Anandale, with a Map of the Countrey; by Mr. Johnstone, a Minister there'. *GC,* 3, 185–96.

14 Mr Blake minister at Penpont: 'upper part of Sheriffdom of Nithsdaill'. Nicholson (1702) has 'A brief description of the Upper part of the Sherrifdom of Niddisdale; by Mr Thomas Black. Minister at Penpont'. *FES,* 2, 332 has no Thomas Black for Penpont at this time. *GC,* 3, 196–210.

15 Mr William Dunlop, late (formerly) Principal of the University of Glasgow. 'Description of the Shire of Renfrew'. Sibbald notes of Dunlop

that he was 'well travelled in that countrey and well versed in Charters of the Gentry of yt Bounds'. This material is also reproduced in G. Crawford, *History of the shire of Renfrew* (Paisley, 1782): see also J. Fullarton (ed.), *Descriptions of the sherrifdoms of Lanark and Renfrew compiled about 1710 by William Hamilton of Wishaw* (Edinburgh, 1831). See also Adv. MS 33.2.27. *GC*, 3, 210–18; 2, 201–10.

16 Mr Robert Cleland: 'Burgh and Priory of Pittenweem'. Adv. MS 34.2.8, ff. 261–4. *GC*, 3, 218–221.

17 'John Adair Hydrographer Royal wrotte for me a description of Stormont, Gowrie, and the adjoyning part of Anguse'. *GC*, 3, 221–3 (see also no. 31 below).

18 Rt Hon and Countess of Errol: 'Several descriptions of Buchan'. This is probably *GC*, 3, 233–8 although Mitchell gives no author in his printed version.

19 Garden of Troup: 'A description of the North East'. Adv. MS 34.2.8, f. 350 has 'Account of ye North Side of ye coast of Buchan by Alexander Gordon of Troup'. *GC*, 2, 133–44.

20 Dr Cockburn, late (formerly) minister of Ormiston: 'Description of Buchan'. Sibbald also notes 'He delivered to me a description of peterhead' (also noted in Adv. MS 34.2.8, f. 354). Cockburn was a prolific author of theological texts. He was East Lothian-born (1652), educated at Edinburgh University and at King's College, Aberdeen, a tutor to Lord Keith in 1673, minister of the Episcopal Church in Amsterdam (1698–1709), and died in 1729: *FES*, 1, 340. *GC*, 3, 228–31.

21 Lord Keith: 'Delivered to me ane description of the parish of Dunotyr [Dunottar] written by Mr John Keith Minister yr'. Nicholson (1702) has 'A Note of Remarkables in the Sheriffdom of Mernes, with a more particular Description of the Parish of Dunotter, by Mr John Keith, Minister there'. Adv. MS 34.2.8, ff. 279–84 notes 'John Keyth'. The resident minister in Dunottar parish in 1683–4 was one Gilbert Keith (died 1709). A John Keith was minister there 1605–*c.* 1654 (*FES*, 5, 459). Unless Sibbald is confusing names, it is possible that this predates 1683 and may, like others he used, have been produced in the 1640s (see nos. 9, 11 above). *GC*, 3, 231–6 has the Dunotir text (with no authorship or date); *GC*, 3, 236–40 has 'Keyth on the Mernes'.

22 Mr Geddes, minister: 'Description of Strathspey'. Adv. MS 34.2.8, ff. 284–94 has no author. This is probably the same Geddes as in no. 5 above. *GC*, 3, 240–4.

23 Mr John Auchterlony of Quind; 'wrote for me in answer to my Queries ane information about the Shyre of Forfar, full and accurate'. Nicholson (1702) terms 'Aughterlanney' [sic] 'an ingenious Gentleman of that Countrey'. Adv. MS 34.2.8, ff. 200–9. *GC*,2, 21–51; 3, 244–8. The surname of this correspondent is also spelled as 'Ochterlonie'.

24 Revd Mr [Robert] Edward, minister of Murehouse [Murroes], Angus: 'yn

answer to my Queries wrott a full accurate Description of the Shyre of Angus and of the Town of Dundee'. Edward's *A description of the county of Angus to accompany the map of Angus* was published in 1678. Edward (1616–96) had the map engraved by Gerald Valk and Peter Schenk of Amsterdam at the expense of the Earl of Panmure, whose arms adorn the map and to whom the work is dedicated. Initially, the map was engraved with the imprint of Janssonius Waesbergke, Moses Pitt, and Stephanus Swart. It was altered to Valk and Shenk in a later edition. Why and when Panmure commissioned Edward to undertake this work is not known.

25 Revd Mr Johnston, minister in Anandale: 'Stewartry of Annandale' (see no. 13 above).

26 Rt Revd Murdo Mackenzie, Bishop of Orkney: 'recommended to the Ministers of his diocese to give me descriptions'. A few 'Notes towards Descriptions' are then forthcoming, with no attributions given. Mackenzie is the grandfather of the eighteenth-century hydrographer, Murdoch Mackenzie.

27 'John Mar, Master of a Ship, did deliver to me a General Geographical Description of Zetland [Shetland], written for me by Master Hugh Leigh, minister of the Gospell in Brassay [Bressay] and Buro [Burra]'. Adv. Ms 34.2.8, f. 328 has 'John Marr'. It is probable that this John Marr is the author of *The coasting pilot* (1714). Leigh, a graduate of King's College, Aberdeen, was minister in this parish 1670–1714 (*FES,* 7, 280). *GC,* 3, 248–55 has 'A General Geographical Description of Zetland, by Hugh Leigh, Member of the Gospel in Brassie and Burs through John Marr', with no date given.

28 Robert Stuart: 'information of Argyleshire'.

29 Master Forbes, minister there: 'parish of Traquair'. This is probably George Forbes, educated at Marischal College, Aberdeen, deposed in 1690 for drunkenness and negligence' (*FES,* 1, 292–3).

30 NKA: 'A Short Description of Angus'.

31 'John Adair wrote for me a description of Strathern in 5 pages'.

32 Mr Geddes; 'Description of Caithness "in answer to my queries"'. (See nos. 5 and 22 above.)

33 Mr Wright, minister of Alloa: 'localities of the River of Forth'. This is probably James Wright, MA at Glasgow University 1656, minister there from 1664 until deprived by the Privy Council in 1689 (*FES,* 4, 292).

34 Mr Charles Littlejohn, minister at Largs: 'ane Account of the Shyre of Aire'. Formerly a regent at the University of Glasgow, he was minister in Largs 1680–90, until deprived at the restoration of the Presbyterians (*FES,* 3, 215).

35 'Reverend Mr. Andrew Simpson, sometyme Minister at Kirkinner Galloway and lately at Dunglass wrote for me a large and accurate description of Galoway'. Simpson had been minister there since 1663 (*FES,* 2, 365). (*GC,* 2, 51–133).

36 Mr Duglass, chaplain to the Earl of Sutherland: 'ane account of Sutherland'. Nicholson (1702) has 'Andrew Dunglass'. Adv. MS 34.2.8, ff. 54–6 has 'Account of Sutherland of the Earle of Sutherland'; ff. 124–39 has 'Description of the Province of Sutherland with the Commodities thereof'.

37 Mr Beaton, minister in Mull: This is John Beaton, minister of Kilninian and Kilmore parish from 1679. He was 'outed' in 1700 for 'immoral conduct' (*FES,* 4, 114). Beaton was the last learned member of the famous Beaton family, formerly hereditary physicians of the Lords of the Isles. He was a correspondent with Edward Lhuyd and with Robert Wodrow in answer to their geographical enquiries about the Western Isles: see p. 86.

38 [John] Morison, 'by the procurement of my friend Mr Colin Mackenzie teacher to the Earl of Seaforth wrote for me a description of the Lewis'. Adv. MS 34.2.8, ff. 193–4v has 'John Morison'.

39 Mr Martin MarMartin [sic] 'wrote for me a description of the Hirta or St. Kilda wt ye isles adjacent'. This is Martin Martin, author of works on St Kilda and on the Western Isles: see pp. 88–9. It is peculiar that Sibbald persistently spells Martin's name wrongly given his knowledge of him, that Martin provided an account of Skye for Sibbald and that the latter was recommending him to the Royal Society: see p. 87.

40 NKA: 'A description of Strathern and the Countries adjacent'. It is possible that this is by John Adair (see no. 17).

41 Dr. [George] Archibald: 'a description of the Presbytery of Dumfriess to which is joyned the description of the bounds of Lochmaben parish wt one description of Dumfries in these Shyre' (see no. 13).

42 Revd Mr James Wallace, 'parson at Kirkua' [Kirkwall] 'drew a map of Orkney which he sent to me and wrot for me a description of the Orkney Isles' (see p. 81). James Wallace was a native of Banffshire, born in 1642 and educated at King's College, Aberdeen. His work towards a geographical description of the northern isles was curtailed with his death, through fever, in 1688. His son, also James, published under his own name *An account of the islands of Orkney* (1710), which is in all respects his father's work with minor modifications: no reference is made to his father's work (*FES,* 7, 222).

43 'Item. Severall letters he [James Wallace] wrott me relating to the Description of ye Isles of Orkney and Shetland'.

44 Mr – Wilson, conjunct minister at Kirkwall: 'ane account of the places in the Orkney Isles fittest for Fishing'. Wilson also sent Sibbald some 'naturall curiousities' from Orkney. This is probably John Wilson, minister of Kirkwall second charge in 1683–4, and again from 1687–9. In 1684–7, he was minister at Stronsay and Eday parish in the Shetlands (*FES,* 7, 223).

45–56 Mr William Geddes, son to Andrew Geddes, 'minister first at Wick in Caithness afterwards at Urquhart in the East of Murray, wrott and answer to my queries especially describing Caithness'. There then follows a list of eleven respondents who, via Bishop of Orkney (see no. 26), sent in brief descriptions for the parishes of Zetland: Yetland; Zetland [by 'M. T. V.']; Dunrossness and the Fair Isle by Mr James Key, minister at Dunrossness; Daleting; Yetland; Northmavin; Island of Burray (1684); Bressay (1684); Isles of Ness (1684). There is also a note that Mr Ross, minister at Northmaven, in a letter dated 7 April 1685, wrote to Mr James Wallace with 'an account of several particularities relating to the Description of Zetland' (*FES,* 7, 312).

57 Revd Mr John Fraser, Dean of the Isles: 'wrote of the Isles and a short description of the Isle of I [Iona] or Icolumbkill'. Adv. MS 34.2.8, ff. 195–9. Nicholson (1702) notes 'The Isles of Terry [Tiree], Conna [Canna], Colle [Coll], and Columbkill [Iona]'. *GC,* 2, 216–19.

58 Mr Marmartin: 'Description of the Isle of Sky' ('in which', notes Sibbald, 'he was born'). This is Martin Martin: see no. 39, and pp. 87–96 above.

59 Mr Henry, 'formerly a minister yn Galoway and lately at Corstorphin', 'wrott for me a description of the sheriffdome of Ayre'. This is George Henry, of Stoneykirk parish, who was transferred to Corstorphine in 1672: *FES,* 2, 353.

60 Mr –, minister of Maybole: 'Kyle'. Nicholson (1702) has 'A Description of the Bayliwick of Carrick by Mr Abercromby, Minister at Maybole'. *FES,* 3, 52 has a William Abercromby, educated at Glasgow University, deprived by Privy Council in 1690, and who died in Edinburgh in 1722. *GC,* 2, 1–21.

61 Hamilton of Wigha: 'Sheriffdom of Lanark'. Adv. MS 33.2.37 (see also no. 15).

62 Mr Andrew Sympson, late [formerly] minister at Duglass; 'ane account of the parishes and ye patrons'.

63 James Gordon of Rothiemay: *Description Aberdoniae topographia.* James was the fifth of eleven sons of Sir Robert Gordon of Straloch. A fine antiquarian, he was minister in Rothiemay from 1641 to 1686: *FES,* 6, 331.

64 Robert Gordon of Straloch: 'Adnotata ad descriptionen duarum praefecturarum in Scotia ultra-montana Aberdoniae et Banfiae'.

65 NKA: 'Noats and observations of Dyvers parts of ye hielands and Iles of Scotland'. Adv. MS 34.2.8, f. 361 notes 78 separate items; *GC,* II, 509–613 lists 91 separate items.

66 'J. G.' [James Gordon of Rothiemay]: 'Description of bothe towns of Aberdeene' (Probably to accompany his map of 1661 of the two Aberdeens).

67 NKA: 'Ane Descriptione of Certaine pairts of the Highlands of Scotland'. *GC,* 2, 144–92.

68 Sir William Baillie of Lamentoun 'and his namesake of Carphin': 'Sherrifdom of Lanark'. Adv. MS. 34.2.8, ff. 157–61.

69 Sir Andrew Agnew and David Dunbar of Baldoon: 'Sheriffdom of Wigtoun'. Adv. MS 34.2.8, ff. 154–6.
70 Captain John Smith: 'Account of the Island of Shetland and the fishing thereabout'. This dates from *c.*1633, and part appears in printed form in Smith's *England's improvement revived, in six books* (London, 1670), in which Smith compares Shetland's fishing with that of the Dutch. Smith, employed by the Earl of Pembroke, spent a year in Shetland in 1633–4: *GC*, 3, 60–5.
71 NKA: 'Description of East Lothian'. Adv. MS 34.2.8, ff. 139–42.
72 NKA: 'Description of Mid-Lothian or Edinburghshire'.
73 Mr James Montgomerie of Weitlands: 'Description of the Sherrifdom of Ranfrow [Renfrew] holden of the Princes and Stewarts of Scotland'.
74 NKA: 'The Description of Murray'.
75 NKA: 'Ane description of most of ye Highlands & West Isles a manuscript'.
76 'Manuscripts sent to me by ye Parson of Rothiemay' [James Gordon]. Adv. MS 34.2.8, f. 359 lists 26 separate Latin items. 'Descriptions in English sent to me by ye parson of Rothiemay' – 23 works listed (f. 360).
77 NKA: 'Presbytery of Forfar'; 'Presbytery of Dundee'; 'Presbytery of Meigle'; 'Presbytery of Brechin'; 'Presbytery of Arbroth': Adv. MS. 34.2.8, ff. 294–328.

# Bibliography

## Primary sources

Manuscript sources


*Aberdeen University Library*
MS K 45 Minute Book of King's College [1761–5]
MS K 153 William Black's dictates on geography, 1692–3
MS M41 Minute Book of Marischal College [1729–90]
MS 189 William Beattie's notes on Robert Hamilton's geography teaching
MS 2092 George Skene's dictates on geography, 1701–4
MS 2131/8/V/I Thomas Reid's [geography] course outline of 1752
MS CT Box 43 Letter of J. Lorimer to R. MacLeod, 25 June 1794

*Bodleian Library*
Ashmolean MSS 1813 Papers relating to James Gregory
Ashmolean MSS 1817a Letter of Andrew Symson to Edward Lhuyd

*British Library*
Sloane MSS 4046 Letter from John MackGregory to Sloane, 12 March 1722
Sloane MSS 4060 Letter from Charles Preston to Sloane, 22 November 1698
Add MSS 22,911 Letter of John MackGregory, 23 October 1722

*City of Edinburgh District Council Archives*
Edinburgh Town Council Minutes, 10 June 1741 [Colin Maclaurin's Proposal for an Observatory]

*Edinburgh University Library (Special Collections)*
MS DK. 1.1.2 Letter of William Carstares to George Ridpath, 1708
MS DK.5.27 John Wishart's dictates on geography, 1679–80
MS Gen 422D Society of Arts lecture programme, 1836–7
MS Gen 1996/9/70–88 T. J. Torrie-Jameson correspondence
MS Gen 1999 Cairns–Jameson correspondence

Laing MS III.535 Sir Robert Sibbald's 'Memoirs of my Life'
Dc.1.60 A Prospect for a Navigation and Writing School [J. Gregory]
Dc.8.35 Sibbald–Sloane correspondence

*Edinburgh University Library (New College Library)*
MS Box 3.1.2 Dictates on Botany, by J. H. Balfour [1847–8]
MS CHA.4.10.53 Chalmers–Cleland correspondence
MSS THO 2 Letter of Thomas Pennant to David Skene, 7 January 1770

*Glasgow University Library*
MS Gen. 1355/105 Notes on James Millar's geography class [December 1821]
MS Gen. 1355/106 Notes on James Millar's geography class [undated; 1821–2]
MS GUA 26720 Senatus Minutes, 1908–

*National Library of Scotland*
Adv. MS 15.2.17 Sir James Balfour's Order of King Charles Entering Edinburgh, 1633
Adv. MS 33.2.27 Sir James Balfour's Topographical Descriptions Relating to Scotland, *c.*1632–*c.*1654
Adv. MS 15.1.1 'Account of the Scotish Atlas'
Adv. MS 33.5.1 Sir Robert Sibbald's 'Memoirs of my Life'
Adv. MS 33.1.16; Sir Robert Sibbald's papers
Adv. MS 33.3.16;
Adv. MS 33.5.15;
Adv. MS 33.5.16;
Adv. MS 33.5.19;
Adv. MS 34.2.8
Adv. MS 83.1.8 Subscription list for the Company of Scotland, 1696
Adv. MS 83.7.3 William Paterson's proposals for a colony at Darien
Crawford MS MB 277 Sibbald's instructions to those responding to his 1682 'General Queries'
MS Acc. 5000 Oliver and Boyd Archive
MS Acc. 100015 Sir James Balfour's and Sir John Lawder's 'Geographical Dictionary' [*c.*1640]
MS Acc. 11592 Playfair-Hood on the sale of Playfair's *System of Geography*
MS 1389 Martin Martin correspondence
MS 2257 'A Catholick Pile to purge out Christianity, prescribed by Dr Sibbald' [1686]
MS 2801 Student notes on James Millar's geography class, 1802–3
MS 5391 Letter of Ebenezer MacFait to A. Stuart, 24 November 1782
MS 8250 Notice of Ebenezer MacFait's death
MS 14100 Geography exercise book, 22 December 1815
MS 14138 School geography exercise book, July 1827
MS14142–54 Notebooks containing essays and notes from lectures, kept by W. S. Walker as an undergraduate at Edinburgh University, 1828–31
MS 14292 'Geometry, vol. I, Perth Academy, 1828'
MS 15048 John Rae to Edward Ellice, 4 December 1854

*Perth City Archives*

MS B 59/24– Correspondence relating to geography teaching in Perth Academy, 1779–94

Perth Town Council Minutes (1752, 1770)

*Royal College of Physicians, Edinburgh*

MS 17/125–19/132 Richard Poole Manuscripts

*Royal Scottish Geographical Society*

Uncatalogued MSS Minutes Books of Council, Vols. 1–8 Correspondence files, 1884–

Uncatalogued MSS, Correspondence, 1884–1939

*Royal Society*

EL.P1 Preston–Sloane correspondence

EL.S2.3, 6 Sibbald–Sloane correspondence

LBC 11, 12 Letter books

*Scottish Record Office (National Archives of Scotland)*

CH 2/154/2 Register minutes of the Synod of Fife, 1641–9

GD 18/50536 Notes on Thomas Blackwell's geography teaching

GD 24/1/564 Rev. Dr. John Walker's 1764 Report on the Hebrides

GD 45/1/161 Memorial on behalf of the Scots Company Trading to Africa and the Indies, 1699

GD 342/23/4 Correspondence of the Educational Institute of Scotland Primary Education Committee, 1929–31

GD 406/1/6489 Letter from the Council in Caledonia to the Company in Scotland, New Edinburgh, 28 December 1698.

PC 12/7/13A Memorial in favour of a Scottish colony, 28 February 1681

PC 13/3 Act of the General Assembly for a Fast Day, 4 March 1701

RH 1/2/511 Letter of Duke of Cumberland to King George

RH 14/203 'Act in favour of John Adair, geographer, for surveying the kingdom of Scotland, and navigating the coasts and isles thereof'

*Society of the Antiquaries of Scotland*

Letters Books, 1781–4

*Strathclyde University Library*

MS T-GED Patrick Geddes Papers

Printed primary sources

*Books and pamphlets*

Adair, J., *QUERIES in order to a true description, and an account of the natural curiousitys, and antiquities* (n.p., 1694).

Anon., *A first book of geography, for the use of schools and private teachers* (Edinburgh, 1819).

Anon., *A short account from, and description of the isthmus of Darien, where the Scots collony are settled with a particular map of the isthmus and enterance to the river of Darien. According to our late news, and Mr Dampier, and Mr Wafer* (Edinburgh, 1699).

Anon., *The people of Scotland's groans and lamentable complaints* (Edinburgh, 1700).

Bacon, F., *Reasons for an union between the kingdome of England and Scotland* (London, 1705).

Barclay, J., *A treatise on education* (Edinburgh, 1743).

Boece, H., *The history and chronicles of Scotland: written in Latin by Hector Boece, Canon of Aberdeen; and translated by John Bellenden, Archdean of Moray and canon of Ross* (Edinburgh, 1821 edition, translated by James Aikman).

Buchanan, G., *The History of Scotland* (Glasgow, 1827 edition).

Cowley, J., *A new and easy introduction to the study of geography* (Edinburgh, 1742).

Drummond, G., *A short treatise of geography, general and special* (Edinburgh, 1708).

Drummond, W., *The entertainment of the high and mighty monarch Charles king of Great Britaine, France and Ireland, into his ancient and royall city of Edinburgh, the fifteenth of June, 1633* (Edinburgh, 1633).

Eachard, L., *Compleat compendium of geography* (London, 1691).

*Edinburgh Journal of Natural and Geographical Science* (Edinburgh, London and Dublin, 1829–31).

Foggo, D., *Elements of modern geography* (Edinburgh, 1822).

Gall, J., *An account of the recent discoveries which have been made for facilitating the education of the blind* (Edinburgh, 1837).

Gordon, P., *Geography anatomiz'd: or, a compleat geographical grammar* (London, 1693).

Gordon, R., *A genealogical history of the earldom of Sutherland* (Edinburgh, 1813).

Guthrie, W., *A new geographical, historical and commercial grammar* (London, 1770).

Haig, T., *Geography of Scotland for the use of the blind with historical notes* (Edinburgh, 1851).

Harrison, T., *Arches of triumph* (London, 1604).

Heyleyn, P., *Cosmography in four books. containing the chorography and history of the whole world; and all the principal kingdoms, provinces, seas, and the isles there of* (London, 1682 edition).

Howell, J., *A perfect description of the people and country of Scotland* (London, 1648).

*Londonopolis . . . the imperial chamber, and chief emporium of Great Britain* (London, 1657).

Ingram, A., *Principles of geography* (Edinburgh, 1799).

James VI, *Basilicon doron* (Edinburgh, 1599).

MacFait, E., *A new system of general geography* (Edinburgh, 1780).

Mair, J., *A brief survey of the terraqueous globe* (Edinburgh, 1762).

Mark, G., *For publishing by subscription, an accurate map or geometrical survey of the shires, of Lothian, Tweddale, and Clydsdale* (Edinburgh, 1728).

Martin., M., *A late voyage to St Kilda* (London, 1698).

*A description of the western isles of Scotland* (London, 1703).

Melville, J., *Mr James Melville's diary, 1556–1601* (Edinburgh, 1829).

Mill, H. R., *Life interests of a geographer* (East Grinstead, 1944).

Mitchell, A. (ed.), *Geographical collections relating to Scotland made by Walter Macfarlane* (Edinburgh, 1906–8).
Murray, H., *Encyclopedia of geography* (Edinburgh, 1834).
*Mr MackGregory's advertisement to gentlemen and ladies* (London, 1713, 1715).
Nicholson, W., *Scottish historical library* (London, 1702).
Ogilby, J., *The entertainment of his most excellent majestie Charles II in his passage through the city of London to his coronation* (London, 1662).
*The king's coronation* (London, 1685).
'Philo-Caledon' [Archibald Foyer], *A defence of the Scots settlement at Darien* (Edinburgh, 1699).
Playfair, J., *System of geography, ancient and modern* (Edinburgh, 1808–14).
Pont, R., *De unione Britanniae* (Edinburgh, 1604).
*Prospectus of the Scottish Geographical Society* (Edinburgh, 1884).
Ross, T., *A compendious system of geography* (Edinburgh, 1804).
Salmon, T., *A new geographical and historical grammar* (London, 1749).
Sibbald, R., *An account of the Scotish atlas, or the description of Scotland ancient and modern* (Edinburgh, 1683).
*Nuncius Scoto-Britannus, sive admonito de atlante Scotico seu descriptione Scotiae antiquae et modernae* (Edinburgh, 1684).
*Scotia Illustrata, sive prodromus historie naturalis* (Edinburgh, 1684).
*The history, ancient and modern, of the sheriffdoms of Fife and Kinross* (Edinburgh, 1710).
*The history, ancient and modern, of the sheriffdoms of Linlithgow and Stirling* (Edinburgh, 1710).
*Vindiciae Scotiae illustratae, sive prodromi naturalis historiae Scotiae, contra prodromomastiges* (Edinburgh, 1710).
*The description of the isles of Orkney and Zetland* (Edinburgh, 1711).
Sinclair, J., *Analysis of the statistical account of Scotland* (London, 1826).
Symson, M., *Geography compendiz'd; or the world survey'd* (Edinburgh, 1702).
*Encheiridion geographicum* (Edinburgh, 1704).
Thornborough, J., *The joyful and blessed reuniting of the two mighty and famous kingdomes, England and Scotland into their ancient name of Great Brittaine* (Oxford, 1604).
*Transactions of the Society of the Antiquaries of Scotland,* I (1792).

*Newspapers and Periodicals*
*Aberdeen Journal*
*Blackwood's Magazine*
*Caledonian Mercury*
*Critical Review*
*Dumfries Weekly Journal*
*Edinburgh Advertiser*
*Edinburgh Evening Courant*
*Edinburgh Literary Journal*
*Edinburgh Monthly Review*
*Edinburgh Observer*

*Edinburgh Philosophical Journal*
*Edinburgh Review*
*Galloway Advertiser and Wigtownshire Free Press*
*Glasgow Herald*
*Monthly Review*
*Montrose Review*
*North British Daily Mail*
*North British Review*
*Stirling Journal and Advertiser*

*Parliamentary Papers*

BPP, 'Evidence, oral and documentary, taken before the Commissioners for visiting the Universities of Scotland', 1837, XXXVIII

BPP, 'Trigonometrical Survey, Great Britain', 1837, XLVII

BPP, 'Answers made by schoolmasters in Scotland to queries circulated in 1838 by order of the Select Committee on Education in Scotland', 1841, XIX

**Secondary sources**

Abbatista, G., 'Establishing "the Order of Time and Place": "Rational Geography", French erudition and the emplacement of history in Gibbon's mind', in D. Womersley (ed.), *Edward Gibbon: Bicentenary essays*, 45–72.

Adams, I. H. (ed.), *Descriptive list of plans in the Scottish record office* (4 volumes, Edinburgh, 1966–88).

Adams, I. H., Crosbie, A. J., and Gordon, G., *The making of Scottish geography: one hundred years of the Royal Scottish Geographical Society* (Edinburgh, 1984).

Adams, I. H., and Fortune, G. (eds.), *Alexander Lindsay: a rutter of the Scottish seas* (Greenwich, 1980).

Alfrey, N., and Daniels, S., (eds.), *Mapping the landscape: essays on art and cartography* (Nottingham, 1990).

Allen, D. E., *The naturalist in Britain: a social history* (Princeton, 1994 edition).

'Walking the swards: medical education and the rise and spread of the botanical field class', *Archives of Natural History*, 27 (2000), 335–67.

Alwall, E., *The religious trend in secular Scottish school-books 1850–1861 and 1873–1882* (London, 1970).

Allan, D., *Virtue, learning and the Scottish enlightenment: ideas of scholarship in early modern history* (Edinburgh, 1993).

Anderson, A., *Ben Peach's Scotland: landscape sketches by a Victorian geologist* (Inverness, 1980).

Anderson, B., *Imagined communities* (London, 1991 edition).

Anderson, J., *The rise of the modern state* (Brighton, 1986).

Anderson, K., and Gales, S. (eds.), *Inventing places* (London, 1992).

Andrews, H., *The search for the picturesque: landscape aesthetics and tourism in Britain, 1760–1800* (Aldershot, 1989).

Anon., 'Geography and the cinematograph', *Scottish Geographical Magazine*, 30 (1914), 100–1.

Anon., 'Geography without tears', *The Glasgow Herald* (6 November 1936).

Anon., *Memorabilia of the City of Perth* (Perth, 1806).
Anon., 'Scotland and geographical work', *Scottish Geographical Magazine*, 1 (1885), 17–25.
Anon., 'The Scotch Educational Department and the teaching of geography', *Scottish Geographical Magazine*, 28 (1912), 321–3.
Anon., 'The use of the cinema in the teaching of geography', *Education Times*, February 1921, 85–6.
Anon., 'The use of the cinema in the teaching of geography', *Geographical Teacher*, 10 (1920), 280–2.
Armitage, D., 'The Scottish vision of empire: intellectual origins of the Darien venture', in Robertson (ed.), *A union for empire*, 97–120.
Ash, M., *The strange death of Scottish history* (Edinburgh, 1980).
Ashworth, J. H., Gibson, K., Newbigin, M., and Smith, J. (eds.), *Edinburgh's place in scientific progress* (Edinburgh, 1921).
Ashworth, W., 'Natural history and the emblematic world view', in Lindberg and Westman (eds.), *Reappraisals of the scientific revolution*, 303–32.
Aspromourgos, T., 'Political economy and the social division of labour: the economics of William Petty', *Scottish Journal of Political Economy*, 33 (1986), 28–45.
Atkinson, D., 'Geopolitics, cartography and geographical knowledge: envisioning Africa from Fascist Italy', in Bell, Butlin and Heffernan (eds.), *Geography and imperialism*, 265–97.
Baigent, E., 'Recreating our past: geography and the rewriting of the Dictionary of National Biography', *Transactions of the Institute of British Geographers*, 19 (1994), 225–8.
Baines, P., 'Ossianic geographies: Fingalian figures on the Scottish tour, 1760–1830', *Scotlands*, 4 (1997), 44–61.
Baker, J. N. L., 'The geography of Bernhard Varenius', *Transactions of the Institute of British Geographers*, 21 (1955), 51–60.
*The history of geography* (Oxford, 1963).
Balfour, J. H., and Babbington, C., 'Account of a botanical excursion to Skye and the Outer Hebrides during the month of August 1841', *Transactions of the Botanical Society of Edinburgh*, 1 (1844), 133–44.
Banes, D., 'The Portuguese voyages of discovery and the emergence of modern science', *Journal of the Washington Academy of Sciences*, 78 (1988), 47–58.
Barber, P., 'Necessary and ornamental: map use in England under the later Stuarts, 1660–1714', *Eighteenth-Century Life*, 14 (1990), 1–28.
Barber, P., 'A friend at a distance: Mercator and the mapping of Britain 1538–1595', in Watelet (ed.), *Gerardus Mercator rupelmundanus*, 27–39.
Barclay, J., *The film in Scottish schools* (Edinburgh, 1993).
Barker-Benfield, G. J., *The culture of sensibility: sex and society in eighteenth-century Britain* (Chicago, 1992).
Barnes, B., and Shapin, S., (eds.), *Natural order – historical studies of scientific culture* (London, 1979).
Barnes, T. J., and Duncan, J. S. (eds.), *Writing worlds: discourse, text and metaphor* (London, 1992).
Barnett, C., 'Awakening the dead: who needs the history of geography?', *Transactions of the Institute of British Geographers*, 20 (1995), 417–19.

Bartholomew, J. G., *The Royal Scottish Geographical Society's Atlas of Scotland* (Edinburgh, 1895).

'A plea for a national institute of geography', *Scottish Geographical Magazine*, 18 (1902), 144–8.

Bassin, M., 'Russia between Europe and Asia: the ideological construction of geographical space', *Slavic Review*, 50 (1991), 1–17.

'Inventing Siberia: visions of the Russian East in the early nineteenth century', *American Historical Review*, 96 (1991), 763–94.

*Imperial visions: nationalism and geographical imagination in the Russian Far East, 1840–1865* (Cambridge, 1999).

Bayley, C., *Empire and information* (Cambridge, 1996).

Beddoe, J., 'A last contribution to Scottish ethnology', *Journal of the Anthropological Institute of Great Britain*, 38 (1908), 212–20.

Belford, A. J., *Centenary handbook of the Educational Institute of Scotland* (Edinburgh, 1946).

Bell, M., 'Edinburgh and empire: geographical science and citizenship for a 'New' age, ca. 1900', *Scottish Geographical Magazine*, 111 (1995), 139–49.

Bell, M., Butlin, R. A., and Heffernan, M. J., (eds.), *Geography and imperialism, 1820–1940* (Manchester, 1995).

Bender, J., 'A new history of the Enlightenment?', *Eighteenth-Century Life*, 16 (1992), 1–20.

Benjamin, M., (ed.), *Science and sensibility: gender and scientific enquiry, 1780–1945* (Oxford, 1991).

*A question of identity: women, science and literature* (New Brunswick, 1993).

Berdoulay, V., 'Stateless national identity and French-Canadian geographic discourse', in Hooson (ed.), *Geography and national identity*, 184–96.

Beveridge, C., and Turnbull, R., *The eclipse of Scottish culture: inferiorism and the intellectuals* (Edinburgh, 1989).

Beveridge, H., 'School excursions in Scotland', *The Geographical Teacher*, 1 (1911–12), 79–82.

Biagioli, M., *Galileo, courtier: the practices of science in the culture of absolutism* (Chicago, 1993).

Birch, T., *History of the Royal Society* (London, 1756–1757).

Black, J., 'Mapping early modern Europe', *European History Quarterly*, 25 (1995), 431–42.

Black, J., (ed.), *Culture and society in Britain, 1660–1800* (Manchester, 1997).

Blouet, B. W. (ed.), *The origins of academic geography in the United States* (Connecticut, 1981).

Blumenreich, J., *The missionary: his trials and triumphs. Being nine years experience in the wynds and closes of Edinburgh* (Edinburgh, 1864).

Blunt, A., *Travel, gender and imperialism: Mary Kingsley and West Africa* (New York, 1994).

Blunt, A., and Rose, G. (eds.), *Writing women and space: colonial and postcolonial geographies* (New York, 1994).

Bone, T. R., *School inspection in Scotland, 1840–1966* (London, 1968).

Boog Watson, W. N., 'The Scottish Marine Station for Scientific Research, Granton, 1884–1903', *Book of the Old Edinburgh Club*, 23 (1969), 5–18.

Borland, F., *The history of Darien* (Glasgow, 1779).
Boud, R., 'The Highland and Agricultural Society of Scotland and the Ordnance Survey of Scotland, 1837–1875', *The Cartographic Journal*, 23 (1986), 3–26.
Bourdieu, P., *In other words* (Cambridge, 1992).
Bowen, M., *Empiricism and geographical thought: from Francis Bacon to Alexander von Humboldt* (Cambridge, 1981).
Bowler, P., *The environmental sciences* (London, 1992).
Boyd, *Education in Ayrshire through seven centuries* (London, 1961).
Bravo, M., *The accuracy of ethnoscience: a study of Inuit cartography and cross-cultural commensurability* (Manchester Papers in Social Anthropology, 2. Manchester, 1996).
'The anti-anthropology of highlanders and islanders', *Studies in the History and Philosophy of Science*, 29 (1998), 369–89.
'Ethnographic navigation and the geographical gift', in Livingstone and Withers (eds.), *Geography and Enlightenment*, 199–235.
Brewer, J., *The pleasures of the imagination: English culture in the eighteenth century* (London, 1997).
Bridges, R., 'The foundation and early years of the Aberdeen centre of the Royal Scottish Geographical Society', *Scottish Geographical Magazine*, 101 (1985), 77–84.
Broadie, A., *The circle of John Mair: logic and logicians in pre-Reformation Scotland* (Oxford, 1985).
Broc, N., *La géographie de la renaissance* (Paris, 1980).
Brock, W., 'Humboldt and the British: a note on the character of British science', *Annals of Science*, 50 (1993), 365–72.
Broman, T., 'The Habermasian public sphere and "science *in* the enlightenment"', *History of Science*, 36 (1998), 124–49.
Brotton, J., *Trading territories: mapping the early modern world* (London, 1997).
Broun, D., 'The origin of Scottish identity in its European context', in Crawford (ed.), *Scotland in dark age Europe* (St Andrews, 1994), 21–31.
'Defining Scotland and the Scots before the wars of independence', in Broun, Finlay, and Lynch (eds.), *Image and identity*, 4–17.
*The Irish identity of the kingdom of Scots* (London, 1989).
'The birth of Scottish history', *Scottish Historical Review*, 76 (1997), 4–22.
Broun, D., Finlay, R. J., and Lynch, M. (eds.), *Image and identity: the making and remaking of Scotland through the ages* (Edinburgh, 1998).
Brown, K., 'The vanishing emperor: British kingship and its decline 1603–1701', in Mason (ed.), *Scots and Britons*, 58–87.
Brown, P. H., *The Union of 1707* (Glasgow, 1907).
Brown, P. H. (ed.), *Early travellers in Scotland* (Edinburgh, 1978).
Brown, S. J., *Thomas Chalmers and the Godly commonwealth* (Oxford, 1982).
Browne, J., *The secular ark: studies in the history of biogeography* (New York and London, 1983).
Bruckner, D., 'Lessons in geography: maps, spellers, and other grammars of nationalism in the early republic', *American Quarterly*, 51 (1999), 311–43.
Bryden, D., 'The Edinburgh observatory 1736–1811: a story of failure', *Annals of Science*, 47 (1990), 445–74.

Buck, P., 'Seventeenth-century political arithmetic: civil strife and vital statistics', *Isis*, 68 (1977), 67–84.

Bud, R., and Cozzens, S. (eds.), *Invisible connections:instruments, institutions and science* (Washington, 1992).

Buisseret, D., (ed.), *Monarchs, ministers and maps: the emergence of cartography as a tool of government in early modern Europe* (Chicago, 1992).

Bunyan, I., *No ordinary journey: John Rae: Arctic explorer, 1813–1893* (Edinburgh, 1993).

Burnett, G., 'The history of cartography and the history of science', *Isis*, 90 (1999), 775–80.

Burns, J., 'George Buchanan and the antimonarchomachs', in Mason (ed.), *Scots and Britons*, 138–158.

Bushnell, R., 'George Buchanan, James VI and neoclassicism', in Mason (ed.), *Scots and Britons*, 91–111.

Butlin, R. A., 'George Adam Smith and the historical geography of the Holy Land: contents, contexts and connections', *Journal of Historical Geography*, 14 (1988), 381–404.

'Regions in England and Wales', in Dodgshon and Butlin (eds.), *An historical geography of England and Wales* (Cambridge, 1991), 223–54.

*Historical geography: through the gates of space and time* (London, 1993).

'Historical geographies of the British Empire, c.1887–1925', in Bell, Butlin and Heffernan (eds.), *Geography and imperialism*, 151–89.

Buttimer, A., 'On people, paradigms and "progress" in geography', in Stoddart (ed.), *Geography, ideology and social concern*, 81–98.

Buttimer, A., and Wallin, L. (eds.), *Nature, culture and human identity in cross-cultural perspective* (Dordrecht, 1999).

Buttimer, A., Brunn, S., and U. Wardenga (eds.), *Text and image: social construction of regional knowledges* (Leipzig, 1999).

Bynum, W. F., and Porter, R. (eds.), *William Hunter and the eighteenth-century medical world* (London, 1985).

Cable, J. A., 'The early history of Scottish popular science', *Studies in Adult Education*, 4 (1972), 34–45.

Calhoun, C., (ed.), *Habermas and the public sphere* (Cambridge, 1992).

Camerini, J., 'Wallace in the field', in Kuklick and Kohler (eds.), *Science in the field*, 44–65.

'Remains of the day: early Victorians in the field', in Lightman (ed.), *Victorian science in context*, 354–77.

Cameron, E., *Land for the people? the British government and the Scottish Highlands, c.1886–1925* (East Linton, 1996).

Cameron, J., *James V the personal rule, 1528–1542* (East Linton, 1997).

Campbell, I., 'James IV and Edinburgh's first triumphal arches', in May (ed.), *The architecture of Scottish cities*, 26–33.

Campbell, J. L., and Thomson, D. S., *Edward Lhuyd in the Scottish highlands, 1699–1700* (Oxford, 1963).

Campbell, R., and Skinner, A. (eds.), *The origins and nature of the Scottish enlightenment* (Edinburgh, 1982).

Cannon, S., *Science in culture: the early Victorian period* (New York, 1978).

Capel, H., 'Institutionalization of geography and strategies of change', in Stoddart (ed.), *On geography and its history*, 37–79.

Carey, D., 'Compiling nature's history: travellers and travel narratives in the early Royal Society', *Annals of Science*, 54 (1997), 269–92.

Carrillo, J., 'From Mt Ventoux to Mt Mayasa: the rise and fall of subjectivity in early modern travel narrative', in Elsner and Rubiés (eds.), *Voyages and visions*, 57–73.

Carter, J., and Pittock, J. H. (eds.), *Aberdeen and the enlightenment* (Aberdeen, 1987).

Cavers, K., *A vision of Scotland: the nation observed by John Slezer, 1671–1717* (Edinburgh, 1993).

Chalmers, T., *The Christian and civic economy of large towns* (Glasgow, 1821–6).

*The sufficiency of a parochial system, without a poor rate, for the right management of the poor* (Glasgow, 1841).

Chartier, R., *The order of books* (Cambridge, 1994).

Chisholm, G. G., 'The meaning and scope of geography', *Scottish Geographical Magazine*, 24 (1908), 561–75.

'Geography', in Ashworth (ed.), *Edinburgh's place in scientific progress*, 168–75.

Christian, J., 'Paul Sandby and the Military Survey of Scotland', in Alfrey and Daniels (eds.), *Mapping the landscape*, 18–22.

Christie, J. R. R., 'The origins and development of the Scottish scientific community, 1680–1760', *History of Science*, 12 (1974), 122–41.

Clark, W., Golinski, J., and Schaffer, S. (eds.), *The sciences in enlightened Europe* (Chicago, 1999).

Clarke, J., *Bishop Gilbert Burnet as educationalist being his thoughts on education with notes and life of the author* (Aberdeen, 1914).

Clayton, D., *Islands of truth: the imperial fashioning of Vancouver island* (Vancouver, 2000).

Cleland, J., *Enumeration of the inhabitants of Scotland* (Glasgow, 1828).

*Enumeration of the inhabitants of Glasgow* (Glasgow, 1832).

*Historical accounts of Bills of Mortality of the probability of human life in Glasgow and other large towns* (Glasgow, 1836).

Clerk, J., *Justification of Mr Murdoch Mackenzie's nautical survey of the Orkney Islands and Hebrides in answer to the accusations of Doctor Anderson* (Edinburgh, 1785).

Clery, E. J., 'Women, publicity and the coffee-house myth', *Women: a cultural review*, 2 (1991), 168–77.

Clifford, J., *Routes: travel and translation in the late twentieth century* (Cambridge, 1997).

Clifton, G., *Directory of British scientific instrument makers 1550–1851* (London, 1995).

Cody, E. G. (ed.), *The history of Scotland, wrytten first in Latin by the most reverend and worthy Jhone Leslie* (Edinburgh, 1888).

Cohen, I. B., *Revolution in science* (Cambridge, Mass., 1985).

Cohen, M., (ed.), *Fashioning masculinity: national identity and language in the eighteenth century* (London, 1996).

Conroy, J., 'The Scottish National Antarctic Expedition', *The Scottish Naturalist*, 111 (1999), 159–82.

Cooter, R., and Pumphrey, S., 'Separate spheres and public spaces: reflections on the history of science popularization and science in popular culture', *History of Science*, 32 (1994), 237–67.

Cormack, L. B., '"Good fences make good neighbours": geography as self-definition in early modern England', *Isis*, 82 (1991), 639–61.

'Twisting the lion's tail: practice and theory at the court of Prince Henry of Wales', in Moran (ed.), *Patronage and institutions*, 67–84.

'The fashioning of an empire: geography and the state in Elizabethan England', in Godlewska and Smith (eds.), *Geography and empire*, 15–30.

*Charting an empire: geography at the English universities, 1580–1620* (Chicago, 1997).

Cosgrove, D. E. (ed.), *Nature, environment, landscape: European attitudes and discourses 1920–1970* (London, 1993).

*Mappings* (London, 1999).

Cosgrove, D. E., and Daniels, S. J. (eds.), *The iconography of landscape* (Cambridge, 1988).

Couper, W., *The Edinburgh periodical press* (2 volumes, Stirling, 1908).

Cowan, E. J., 'Myth and identity in early medieval Scotland', *Scottish Historical review*, 63 (1984), 116–22.

'Identity, freedom and the declaration of Arbroath', in Broun, Finlay and Lynch (eds.), *Image and identity*, 38–68.

Craig., W. S., *History of the Royal College of Physicians of Edinburgh* (Oxford, 1976).

Crawford, B. E. (ed.), *Scotland in dark age Europe* (St Andrews, 1994).

Crombie, A., *Styles of scientific thinking in the European tradition* (3 volumes, London, 1994).

Crosland, M., 'History of science in a national context', *British Journal for the History of Science*, 10 (1977), 95–113.

Crowhurst, A., 'Empire theatres and the empire: the popular geographical imagination in the age of empire', *Environment and Planning D: Society and Space*, 15 (1997), 155–74.

Cunningham, A., 'The culture of gardens', in Jardine, Secord and Spary (eds.), *Cultures of natural history*, 38–56.

Cunningham, I. (ed.), *The nation survey'd* (East Linton, 2001).

Currie, R. I., 'Marine science', *Proceedings of the Royal Society of Edinburgh*, Series B, 84 (1983), 231–50.

Curry, P., *Prophecy and power: astrology in early modern England* (Cambridge, 1989).

Curtin, P., 'The environment beyond Europe and the European theory of empire', *Journal of World History*, 1 (199), 131–50.

Daniels, S., 'The political iconography of woodland in later Georgian England', in Cosgrove and Daniels (eds.), *The iconography of landscape*, 43–84.

*Fields of vision: landscape imagery & national identity in England & the United States* (Cambridge, 1993).

Danziger, K., *Constructing the subject: historical origins of psychological research* (Cambridge, 1999).

Darnton, R., 'In search of the Enlightenment: recent attempts to create a social history of ideas', *Journal of Modern History*, 63 (1971), 113–32.

Daston, L., and Park, K., *Wonders and the order of nature, 1150–1750* (New York, 1999).

Davie, G., *The democratic intellect: Scotland and her universities in the nineteenth century* (Edinburgh, 1999 edition).

Deacon, M., *Scientists and the sea 1650–1900: a study of marine science* (London, 1971).

'The *Challenger* expedition and geology', *Proceedings of the Royal Society of Edinburgh*, Series B, 72 (1972), 145–53.

Deacon, R., *John Dee: scientist, geographer, astrologer and secret agent to Elizabeth I* (London, 1968).

Dear, P., '*Totius in Verba*: rhetoric and authority in the early Royal Society', *Isis*, 76 (1985), 145–61.

*The scientific enterprise in early modern Europe* (Chicago, 1997).

Destombes, M., 'André Thevet (1504–1592) et sa contribution à la cartographie et à l'oceanographié', *Proceedings of the Royal Society of Edinburgh*, Series B, 72 (1972), 123–31.

Dettelbach, M., 'Humboldtian science', in Jardine, Secord and Spary (eds.), *Cultures of natural history*, 287–304.

Dickinson, V., *Drawn from life: science and art in the portrayal of the new world* (Toronto, 1998).

Dickie, G., *Flora Abredonensis, comprehending a list of the flowering plants and ferns found in the neighbourhood of Aberdeen* (Aberdeen, 1838).

Dickinson, W. C., *Two students at St. Andrews, 1711–1716* (Edinburgh, 1952).

Dickson, D., 'Science and political hegemony in the seventeenth century', *Radical Science Journal*, 8 (1979), 7–38.

Dirks, N., (ed.), *Colonialism and culture* (Ann Arbor, 1992).

Dixon, C., and Heffernan, M. (eds.), *Colonialism and development in the contemporary world* (London, 1991).

Doig, A., Ferguson, J. P. S., Milne, I. A., and Passmore, R., (eds.), *William Cullen and the eighteenth-century medical world* (Edinburgh, 1993).

Domosh, M., 'Toward a feminist historiography of geography', *Transactions of the Institute of British Geographers*, 16 (1991), 95–104.

'Beyond the frontiers of geographical knowledge', *Transactions of the Institute of British Geographers*, 16 (1991), 488–90.

Donnachie, I., and Whatley, C., (eds.), *The manufacture of Scottish history* (Edinburgh, 1992).

Dorflinger, J., *Der Géographie in der 'Encyclopédie' ein wissenschaftsgeschichlische Studie* (Vienna, 1976).

Downes, A., 'The bibliographic dinosaurs of Georgian geography (1714–1830)', *Geographical Journal*, 137 (1971), 379–87.

Drayton, R., 'Knowledge and empire', in Marshall (ed.), *The Oxford history of the British empire: the eighteenth century*, 231–52.

*Nature's government: science, imperial Britain and the 'improvement' of the world* (New Haven, 2000).

Driver, F., 'Moral geographies: social science and the urban environment in mid-nineteenth-century England', *Transactions of the Institute of British Geographers*, 13 (1988), 275–87.

'Henry Morton Stanley and his critics: geography, exploration and empire', *Past and Present*, 133 (1991), 134–67.

'Geography's empire: histories of geographical knowledge', *Environment and Planning D: Society and Space*, 10 (1992), 23–40.

'New perspectives on the history and philosophy of geography', *Progress in Human Geography*, 18 (1994), 92–100.

'Visualising geography: a journey to the heart of the discipline', *Progress in Human Geography*, 19 (1995), 19–34.

'David Livingstone and the culture of exploration in mid-Victorian Britain', in *David Livingstone and the Victorian encounter with Africa*, 109–38.

'Histories of the present: the history and philosophy of geography', *Progress in Human Geography*, 20 (1996), 100–9.

*Geography Militant: cultures of exploration and empire* (Oxford, 2000).

Driver, F., and Gilbert, D. (eds.), *Imperial cities* (Manchester, 1999).

Driver, F., and Maddrell, A. M., 'Geographical education and citizenship: introduction', *Journal of Historical Geography*, 22 (1996), 371–2.

Driver, F., and Rose, G. (eds.), *Nature and science: essays in the history of geographical knowledge* (Cheltenham, 1992).

Driver, F., Matless, D., Rose, G., Barnett, C., and Livingstone, D., 'Geographical traditions: rethinking the history of geography', *Transactions of the Institute of British Geographers*, 20 (1995), 403–4.

Dunbar, G., 'Geographic education in early Charleston', *Journal of Geography*, 69 (1970), 348–50.

Dunbar, G. (ed.), *From traveller to scientist: the professionalization and institutionalization of geography in Europe and North America since 1870* (Dordrecht, forthcoming).

Duranti, A., 'Mediated encounters with Pacific cultures: three Samoan dinners', in Miller and Reill, *Visions of empire*, 326–34.

Durkan, J., *Bibliography of George Buchanan* (Glasgow, 1994).

Durkan, J., and Ross, A., *Early Scottish libraries* (Glasgow, 1961).

Durkan, J., and Kirk, J., *The University of Glasgow 1451–1977* (Glasgow, 1977).

Eamon, W., *Science and the secrets of nature: books of secrets in medieval and early modern culture* (Princeton, 1994).

Edney, M., 'Mathematical cosmography and the social ideology of British cartography, 1780–1820', *Imago Mundi*, 46 (1994), 101–16.

*Mapping an empire: the geographical construction of British India, 1765–1843* (Chicago, 1997).

Elder, J., 'A proposal for uniting Scotland with England', in *Miscellany of the Bannatyne Club Volume I* (Edinburgh, 1827), 1–18.

Ellis, S. G., 'Tudor state formation and the shaping of the British Isles', in Ellis and Barber, *Conquest & union*, 40–64.

Ellis, S. G., and Barber, S. (eds.), *Conquest & union: fashioning a British state, 1485–1725* (London, 1995).

Elsner, J., and Rubiés, J-P. (eds.), *Voyages and visions: towards a cultural history of travel* (London, 1999).

Ely, G., 'Nations, publics, and political cultures: placing Habermas in the nineteenth century', in Calhoun, *Habermas and the public sphere*, 189–240.

Emerson, J., *Poetical descriptions of Orkney* (Edinburgh, 1835).

Emerson, R. L., 'Natural philosophy and the problem of the Scottish Enlightenment', *Studies on Voltaire and the Eighteenth Century*, 242 (1986), 243–91.

'Sir Robert Sibbald, Kt, the Royal Society of Scotland and the origins of the Scottish enlightenment', *Annals of Science*, 45 (1988), 41–72.

*Professors, patronage and politics: the Aberdeen universities in the eighteenth century* (Aberdeen, 1992).

Emery, F. V., 'The geography of Robert Gordon, 1590–1661, and Sir Robert Sibbald, 1641–1722', *Scottish Geographical Magazine*, 74 (1958), 2–12.

'English regional studies from Aubrey to Defoe', *Geographical Journal*, 124 (1958), 315–25.

'Irish geography in the seventeenth century', *Irish Geography*, 3 (1958), 259–270.

'A "geographical description" of Scotland prior to the statistical accounts', *Scottish Studies*, 3 (1959), 1–16.

Escolar, M., 'Promotion and diffusion of geographical knowledges: Argentine editorial policies and the nation's geographical body representation (1863–1916)', in Buttimer, Brunn and Wardenga (eds.), *Text and image*, 91–9.

Feingold, M., *The mathematicians' apprenticeship: science, universities, and society in England, 1560–1640* (Cambridge, 1984).

Ferguson, W., 'Imperial crowns: a neglected fact of the background to the treaty of union', *Scottish Historical Review*, 53 (1974), 22–44.

*The identity of the Scottish nation: an historic quest* (Edinburgh, 1998).

Fernandez-Armesto, F., *Truth: a history and a guide for the perplexed* (London, 1997).

Findlen, P., 'Controlling the experiment: rhetoric, court patronage and the experimental method of Francesco Redi', *History of Science*, 31 (1993), 35–64.

*Possessing nature: museums, collecting and scientific culture in early modern Italy* (Berkeley, 1994).

'Translating the new science: women and the circulation of knowledge in enlightenment Italy', *Configurations*, 2 (1995), 167–206.

'Courting Nature', in Jardine, Secord and Spary (eds.), *Cultures of natural history*, 57–74.

Finlay, R., 'Controlling the past: Scottish historiography and Scottish identity in the 19th and 20th centuries', *Scottish Affairs*, 9 (1994), 127–43.

'The rise and fall of popular imperialism in Scotland, 1850–1950', *Scottish Geographical Magazine*, 113 (1997), 13–21.

'Imperial Scotland: the British empire and Scottish national identity, c.1850–1914', in MacKenzie (ed.), *Scotland and the British Empire* (Edinburgh, forthcoming).

*First Annual Report of the Botanical Society of Edinburgh* (Edinburgh, 1837–1840).

*First Annual Report of the Edinburgh School of Art* (Edinburgh, 1822).

Fletcher, H. R., and Brown, W. H., *The Royal Botanic Garden in Edinburgh 1670–1970* (Edinburgh, 1970).

Flower, R., 'Lawrence Nowell and the discovery of England in Tudor times', *Proceedings of the British Academy*, 21 (1935), 41–54.

Forbes, E., 'Mathematical cosmography', in Rousseau and Porter (eds.), *Ferment of knowledge*, 417–48.

Forgan, S., 'The architecture of display: museums, universities and objects in nineteenth-century Britain', *History of Science*, 32 (1994), 139–62.

'"But indifferently lodged . . .": perception and place in building for science in Victorian London', in Smith and Agar (eds.), *Making space for science*, 195–215.

Forgan, S., and Gooday, G., 'Constructing South Kensington: the buildings and politics of T. H. Huxley's working environments', *British Journal for the History of Science*, 29 (1996), 435–68.

Forsyth, D., 'Empire and union: imperial and national identity in nineteenth century Scotland', *Scottish Geographical Magazine*, 113 (1997), 6–12.

Forte, A. D. M., 'Kenning be kenning and course be course: Alexander Lindsay's rutter and the problematics of navigation in fifteenth- and sixteenth-century Scotland', *Review of Scottish Culture*, 11 (1998–99), 32–45.

Foucault, M., *The archaeology of knowledge* (London, 1972).

*The order of things: an archaeology of the human sciences* (London, 1974).

'Of other spaces', *Diacritics*, 16 (1986), 22–7.

Fox, C., Porter, R., and Wokler, R. (eds.), *Inventing human science: eighteenth-century domains* (Berkeley, 1995).

Frangsmyr, T., Heilbron, J., and Rider, R. L. (eds.), *The quantifying spirit in the eighteenth century* (Berkeley, 1990).

Fraser, W., *The Sutherland Book* (Edinburgh, 1892).

Freeman, T. W., *A hundred years of geography* (Manchester, 1971).

French, R., *Ancient natural history* (London, 1994).

Friel, B., Andrew, J., and Barry, K., 'Translations and a paper landscape: between fiction and history', *The Crane Bag*, 7 (1983), 119–127.

Galloway, B. R., and B. P. Levack, *The Jacobean union: six tracts of 1604* (Edinburgh, 1985).

Gambi, L., 'Geography and imperialism in Italy: from the unity of the nation to the "new" Roman empire', in Godlewska and Smith (eds.), *Geography and empire*, 74–91.

Garcia-Ramon, M. D., and Nogue-Font, J., 'Nationalism and geography in Catalonia', in Hooson (ed.), *Geography and national identity*, 197–211.

Gardner, L., *Bartholomew: 150 years* (Edinburgh, 1976).

Gardiner, R. A., 'William Roy, surveyor and antiquary', *Geographical Journal*, 143 (1977), 439–50.

Gascoigne, J., *Science in the service of the empire: Joseph Banks, the British state and the uses of science in the age of revolution* (Cambridge, 1998).

'The Royal Society and the emergence of science as an instrument of state policy', *British Journal for the History of Science*, 32 (1999), 171–84.

Gavine, D., 'Astronomy in Scotland, 1745–1900' (Ph.D. thesis, Open University, 1982).

'Navigation and astronomy teachers in Scotland outside the universities', *Marriner's Mirror*, 76 (1990), 5–12.

Geddes, P., 'Note on a draft plan for an institute of geography', *Scottish Geographical Magazine*, 18 (1902), 141–4.

*Course of nature study and geography in education* (Edinburgh, 1904).

*City development: a study of parks, gardens and culture-institutes* (Birmingham, 1904).

Gentry, J. R., 'English chorographers 1656–1695: artists of the shire' (Ph.D. thesis, University of Utah, 1985).

Gibson, E., *Camden's Britannia* (London, 1695).

Gibson-Wood, C., 'Classification and value in a seventeenth-century museum: William Courten's collection', *Journal of the history of collections*, 9 (1997), 61–77.

Gilbert, E. W., *British pioneers in geography* (Newton Abbot, 1972).
Gillies, J., and Vaughan, V. M. (eds.), *Playing the globe: genre and geography in English Renaissance drama* (London, 1998).
Glick, T. (ed.), *The comparative reception of Darwinism* (Austin, Tex., 1974).
Goblet, Y., *La transformation de la géographie politique de l'Irlande au XVIIe siècle* (Paris, 1930).
Godlewska, A., 'Map, text and image. The mentality of enlightened conquerors: a new look at the *Description de l'Egypte', Transactions of the Institute of British Geographers*, 20 (1995), 5–28.
*Geography unbound: French geographic science from Cassini to Humboldt* (Chicago, 1999).
Godlewska, A., and Smith, N. (eds.), *Geography and Empire* (Oxford, 1994).
Goldgar, A., *Impolite learning: conduct and community in the republic of letters, 1680–1750* (New Haven, 1996).
Goldman, L., 'The origins of British social science: political economy, natural science and statistics, 1830–1835', *Historical Journal*, 26 (1983), 587–616.
Goldstein, T., *The dawn of modern science* (Boston, 1988).
Golinski, J., 'Peter Shaw: chemistry and communication in Augustan England', *Ambix*, 30 (1983), 19–29.
'Science *in* the Enlightenment', *History of Science*, 24 (1986), 411–24.
*Science as public culture: chemistry and enlightenment in Britain* (Cambridge, 1992).
*Making natural knowledge: constructivism and the history of science* (Cambridge, 1998).
Good, G. (ed.), *Sciences of the earth: an encyclopedia of events, people and phenomena* (2 volumes, New York, 1998).
Goodare, J., *State and society in early modern Scotland* (Oxford, 1999).
Goodare, J., and Lynch, M., 'The Scottish state and its borderlands', in Goodare and Lynch (eds.), *The reign of James VI*, 186–207.
Goodare, J., and Lynch, M. (eds.), *The reign of James VI* (East Linton, 2000).
Gooday, G., 'Precision measurement and the genesis of physics teaching laboratories in late Victorian Britain', *British Journal for the History of Science*, 23 (1990), 25–51.
'"Nature in the laboratory": domestication and discipline with the microscope in Victorian life science', *British Journal for the History of Science*, 24 (1991), 307–71.
'Teaching telegraphy and electrotechnics in the physics laboratory: William Ayrton and the creation of an academic space for electrical engineering in Britain 1873–1884', *History of Technology*, 13 (1991), 73–111.
'Instrumentation and interpretation: managing and representing the working environments of Victorian experimental science', in Lightman (ed.), *Victorian science in context*, 409–37.
'The premisses of premises: spatial issues in the historical construction of laboratory credibility', in Smith and Agar (eds.), *Making space for science*, 216–45.
Goodman, D., 'Public sphere and private life: towards a synthesis of current historiographical approaches to the old regime', *History and Theory*, 31 (1992), 1–20.
'Introduction: the public and the nation', *Eighteenth-Century Studies*, 29 (1995), 1–2.
Goodman, D., and Russell, C. (eds.), *The rise of scientific Europe 1500–1800* (Milton Keynes, 1991).

Goodwin, J., *Athanasius Kircher: a renaissance man and the quest for lost knowledge* (London, 1979).
Gordon, [Mrs], *The home life of Sir David Brewster* (Edinburgh, 1870).
Grant, A., *The story of the University of Edinburgh* (London, 1884).
Grant, E., *Memoirs of a highland lady* (Edinburgh, 1988 edition).
Gray, J., 'Memoir on the pigmentation survey of Scotland', *Journal of the Royal Anthropological Institute*, 37 (1907), 375–401.
Gray, J., and Tocher, J., 'The physical characteristics of adults and school children in East Aberdeenshire', *Journal of the Royal Anthropological Institute*, 30 (1900), 104–24.
Gray, T., and Birley, E. (eds.), 'Bishop Nicholson's diary (1703–4): Part II', *Transactions of the Cumberland and Westmorland Antiquarian and Archaeological Society*, 50 (1951), 110–34.
Greenblatt, S., *Renaissance self-fashioning from More to Shakespeare* (Chicago, 1980).
*Marvellous possessions: the wonder of the new world* (Oxford, 1991).
Gregory, D., *Geographical imaginations* (Oxford, 1993).
Gregory, D., 'Between the book and the lamp: imaginative geographies of Egypt, 1849–1850', *Transactions of the Institute of British Geographers*, 20 (1995), 29–57.
'Discourse', in Johnston, Gregory, Pratt and Watts (eds.), *The dictionary of human geography*, 180–1.
'Post-colonialism', in Johnston, Gregory, Pratt and Watts (eds.), *The dictionary of human geography*, 612–5.
Greville, R. K., *Flora Edinensis: or a description of plants growing near Edinburgh* (Edinburgh, 1824).
Grove, R., 'Cressey Dymock and the draining of the fens: an early agricultural model', *Geographical Journal*, 147 (1981), 27–37.
Gruffudd, P., 'Back to the land: historiography, rurality and the nation in interwar Wales', *Transactions, Institute of British Geographers*, 19 (1994), 61–77.
'"A crusade against consumption": environment, health and social reform in Wales, 1900–1950', *Journal of Historical Geography*, 21 (1995), 39–54.
Gunther, R. T., *Early science in Oxford: volume XIV: life and letters of Edward Lhuyd* (Oxford, 1945).
Habermas, J., *The structural transformation of the public sphere: an inquiry into a category of bourgeois society* (Cambridge, 1989).
'Further reflections on the public sphere', in Calhoun (ed.), *Habermas and the public sphere*, 421–61.
Haggett, P., and Chorley, R., 'Models, paradigms and the new geography', in Haggett and Chorley (eds.), *Models in geography*, 19–42.
Haggett, P., and Chorley, R. J. (eds.), *Models in geography* (London, 1967).
Hanham, H. J., 'Mid-nineteenth-century Scottish nationalism: romantic and radical', in Robson, *Ideas and institutions of Victorian Britain*, 143–79.
Hanna, W., *Memoirs of the life and writings of Thomas Chalmers* (Edinburgh, 1852).
Hardy, M., 'Botanical geography and the biological utilisation of the soil', *Scottish Geographical Magazine*, 18 (1902), 225–37.
'A note upon the methods of botanical geography', *Scottish Gographical Magazine*, 18 (1902), 406–13.

Harley, J. B., 'Maps, knowledge and power', in Cosgrove and Daniels (eds.), *The iconography of landscape*, 277–312.
'Secrecy and silences: the hidden agenda of state cartography in early modern Europe', *Imago Mundi*, 11 (1988), 111–30.
*Maps and the Columbian encounter* (Milwaukee, 1990).
'Deconstructing the map', in Barnes and Duncan (eds.), *Writing worlds*, 231–47.
Hart, V., *Art and magic in the court of the Stuarts* (London, 1994).
Harvey, M., and Holly, B., 'Paradigms, philosophy and geographic thought', in Harvey and Holly (eds.), *Themes in geographic thought*, 11–32.
Harvey, M., and Holly, B. (eds.), *Themes in geographic thought* (Beckenham, 1981).
Harwood, J., *Styles of scientific thought* (Chicago, 1993).
Hastings, A., *The construction of nationhood: ethnicity, religion and nationalism* (Cambridge, 1997).
Hearn, J., *Claiming Scotland: national identity and liberal culture* (Edinburgh, 2000).
Heffernan, M., 'The limits of utopia: Henri Duveyrier and the exploration of the Sahara in the nineteenth century', *Geographical Journal*, 155 (1989), 342–52.
The ambiguity of the Sahara: French images of the North African desert in the nineteenth century', *Maghreb Review*, 17 (1992), 178–88.
'A state scholarship: the political geography of French international science during the nineteenth century', *Transactions of the Institute of British Geographers*, 19 (1994), 21–45.
'On geography and progress: Turgot's *Plan d'un Ouvrage sur la géographie politique* (1751) and the origins of modern progressive thought', *Political Geography*, 13 (1994), 328–43.
'The science of empire: the French geographical movement and the forms of French imperialism, 1870–1920', in Godlewska and Smith (eds.), *Geography and empire* (Oxford, 1994), 92–114.
'The spoils of war: the Société de Géographie de Paris and the French empire, 1914–1919', in Bell, Butlin and Heffernan (eds.), *Geography and imperialism*, 221–64.
Heffernan, M., and Sutton, K., 'The landscape of colonialism: the impact of French colonial rule on the Algerian rural settlement pattern, 1830–1987', in Dixon and Heffernan (eds.), *Colonialism and development in the contemporary world*, 121–51.
Heilbron, J. A., *Electricity in the 17th and 18th centuries: a study of early modern physics* (Berkeley, 1979).
Helgerson, R., 'The land speaks: cartography, chorography and subversion in renaissance England', *Representations*, 16 (1986), 51–85.
*Forms of nationhood: the Elizabethan writing of England* (Chicago, 1992).
Henderson, D. M., and Dickson, J. H., *A naturalist in the Highlands. James Robertson, his life and travels, 1767–1771* (Edinburgh, 1994).
Henry, J., 'Magic and science in the sixteenth and seventeenth centuries', in Olby, Cantor, Christie and Hodge (eds.), *Companion to the history of modern sciences*, 583–96.
*The scientific revolution and the origins of modern science* (London and New York, 1997).

Herbertson, A. J., 'The parlous plight of geography in Scottish education', *Scottish Geographical Magazine*, 14 (1899), 81–8.
Herdman, W. A., 'Oceanography', in Ashworth (ed.), *Edinburgh's place in scientific progress*, 126–44.
Heron, R. [John Pinkerton], *Letters of literature* (London, 1785).
Hett, F. P., *The memoirs of Sir Robert Sibbald (1641–1722)* (London, 1932).
Heyck, T. W., *The transformation of intellectual life in Victorian England* (London, 1982).
Hill, C., 'Science and magic in seventeenth-century England', in Samuel and Stedman Jones (eds.), *Culture, ideology and politics*, 176–93.
Hobsbawm, E., *The age of empire* (London, 1987).
Höem, I., 'The scientific endeavor and the natives', in Miller and Reill (eds.), *Visions of empire*, 305–25.
Hohendahl, P., 'The public sphere: models and boundaries', in Calhoun (ed.), *Habermas and the public sphere*, 99–108.
Holloway, J., and Errington, L., *The discovery of Scotland: the appreciation of Scottish scenery through two centuries of painting* (Edinburgh, 1978).
Home, R., 'Humboldtian science revisited: an Australian case study', *History of Science*, 33 (1995), 1–22.
Hook, A and Sher, R. B. (eds.), *The Glasgow Enlightenment* (East Linton, 1997).
Hooker, W. J., *Flora Scotica, or a description of Scottish plants* (Edinburgh, 1821).
Hooson, D. M. (ed.), *Geography and national identity* (Oxford, 1994).
Hooykaas, R. J., *Humanism and the voyages of discovery in 16th century Portuguese science and letters* (Amsterdam, 1979).
'The rise of modern science: when and why', *British Journal for the History of Science*, 20 (1987), 453–73.
Hoppen, K. T., *The common scientist in the seventeenth century: a study of the Dublin Philosophical Society* (London, 1970).
Horsman, F., 'Plant distribution patterns: the first British map', *Archives of Natural History*, 26 (1999), 279–86.
Houston, R., and Knox, W. (eds.), *Penguin history of Scotland* (London, forthcoming).
Howard, P., *Landscapes – the artists' vision* (London, 1991).
Howarth, D., *Art and patronage in the Caroline courts* (Cambridge, 1993).
Hudson, B., 'The new geography and the new imperialism: 1870–1918', *Antipode*, 9 (1977), 12–19.
Hull, C. (ed.), *The economic writings of Sir William Petty* (New York, 1963).
Humes, W., and H. Paterson (eds.), *Scottish culture and Scottish education 1800–1900* (Edinburgh, 1983)
Hunter, M., *Science and society in Restoration England* (Cambridge, 1981).
*Science and the shape of orthodoxy: intellectual change in late seventeenth-century Britain* (Woodbridge, 1995).
Hutchinson, A. F., *History of the High School of Stirling* (Striling, 1934).
Iliffe, R., 'Material doubts: Hooke, artisan culture and the exchange of information in 1670s London', *British Journal for the History of Science*, 28 (1995), 285–318.
Inkster, I., 'The public lecture as an instrument of science education for adults – the case of Great Britain c.1750–1850', *Paedogogica Historica: International Journal for the History of Education*, 20 (1980), 80–107.

Inkster, I., and Morrell, J. B. (eds.), *Metropolis and province: science in British culture, 1780–1850* (London, 1983).
Insch, G. P. (ed.), *Papers relative to the ships and voyages of the Company of Scotland trading to Africa and the Indies, 1696–1707* (Edinburgh, 1924).
*International Exhibition of Industry, Science and Art: Official Catalogue* (Edinburgh, 1886).
Jackson, P., *Race and racism: essays in social geography* (London, 1987).
Jacob, M., 'The mental landscape of the public sphere: a European perspective', *Eighteenth-Century Studies*, 28 (1994), 95–113.
Jankovic, V., 'The place of nature and the nature of place: the chorographic challenge to the history of British provincial science', *History of Science*, 38 (2000), 79–113.
Jardine, L., *Ingenious pursuits: building the scientific revolution* (London, 1999).
Jardine, N., Secord, J. A., and Spary, E. (eds.), *Cultures of natural history* (Cambridge, 1996).
Jessop, J. C., *Education in Angus* (London, 1931).
Johnston, R. J., 'Paradigms and revolutions or evolution?: observations on human geography since the Second World War', *Progress in Human Geography*, 2 (1978), 189–206.
Johnston, R. J., Gregory, D. J., Pratt, G., and Watts, M. (eds.), *The dictionary of human geography*, fourth edition (Oxford, 2000).
Johnston, S., 'Mathematical practitioners and instruments in Elizabethan England', *Annals of Science*, 48 (1991), 319–44.
Johnston, T. B. (ed.), *In memoriam of the late A. Keith Johnston L. L. D., Geographer to the Queen for Scotland* (Edinburgh, 1873).
Jolly, W., 'Opening address: the scientific materials of the north of Scotland', *Transactions of the Inverness Scientific Society and Field Club*, 1 (1875–1880), 3–19.
Jones, J., 'James Hutton's agricultural research and his life as a farmer', *Annals of Science*, 42 (1985), 573–601.
Jones, P. (ed.), *Philosophy and science in the Scottish Enlightenment* (Edinburgh, 1989).
Kafker, F., and Kafker, S., *The encyclopedists as individuals: a biographic dictionary of the authors of the Encyclopédie. Studies on Voltaire and the Eighteenth Century* (Oxford, 1988).
Kearns, G., 'The imperial subject: geography and travel in the work of Mary Kingsley and Halford Mackinder', *Transactions of the Institute of British Geographers*, 22 (1997), 450–72.
Keltie, J. S., 'Geographical education', *Scottish Geographical Magazine*, 1 (1885), 497–504.
'Geographical education – Report to the Council of the Royal Geographical Society', *Supplementary Paper of the Royal Geographical Society*, 1 (1886), 439–594.
Kermack, W. R., 'The making of Scotland: an essay in historical geography', *Scottish Geographical Magazine*, 29 (1913), 295–305.
*Historical geography of Scotland* (London, 1926 edition).
*Human environment and progress* (London and Edinburgh, 1927).
Kidd, C., *Subverting Scotland's past: Scottish Whig historians and the creation of an anglo-British identity, 1689–c.1830* (Cambridge, 1993).

Kinchin, P., and Kinchin, J., *Glasgow's great exhibitions 1888, 1901, 1911, 1938, 1988* (Bicester, 1990).
Klein, L., 'Gender, conversation and the public sphere in early eighteenth-century England', in Still and Worton, *Textuality and sexuality*, 100–15.
'Gender and the public/private distinction in the eighteenth century: some questions about evidence and analytical procedure', *Eighteenth-Century Studies*, 29 (1995), 97–109.
Knorr-Cetina, K., and Mulkay, M. (eds.), *Science observed: perspectives on the social study of science* (Beverly Hills and London, 1983).
Knowles, R. (ed.), *The entertainment of his most excellent majestie Charles II in his passage through the city of London to his coronation* (London, 1988 facsimile edition).
Koerner, L., 'Linnaeus' floral transplants', *Representations*, 47 (1994), 144–69.
*Linnaeus: nature and nation* (Harvard, 1999).
Konvitz, J., *Cartography in France 1660–1848: science, engineering and statecraft* (Chicago, 1987).
Kuklick, H., and Kohler, L. (eds.), *Science in the field* (Chicago, 1996).
Lacoste, Y., *La géographie, ça sert, d'abord, à faire la guerre* (Paris, 1976).
Langlands, R., 'Britishness or Englishness? the historical problem of national identity in Britain', *Nations and Nationalism*, 5 (1999), 53–69.
Larson, J., 'Not without a plan: geography and natural history in the late eighteenth century', *Journal of the History of Biology*, 19 (1986), 447–88.
Latour, B., 'Give me a laboratory and I will raise the world', in Knorr-Cetina and Mulkay, *Science observed*, 141–70.
*Science in action: how to follow scientists and engineers through society* (Milton Keynes, 1987).
Laurie, S. S., *Occasional addresses on education subjects* (Cambridge, 1988).
*The training of teachers and methods of instruction* (Edinburgh, 1901).
Law, A., *Teachers in Edinburgh in the eighteenth century* (Edinburgh, 1965)
*Education in Edinburgh in the eighteenth century* (Edinburgh, 1965).
*Schoolbooks and textbooks in Scotland in the eighteenth century* (Edinburgh, 1989).
*Edinburgh schools of the nineteenth century* (Edinburgh, 1996).
Lawrence, C., 'The nervous sytem and society in the Scottish enlightenment', in Barnes and Shapin (eds.), *Natural order – historical studies of scientific culture*, 19–40.
'Ornate physicians and learned artisans: Edinburgh medical men, 1726–1776', in Bynum and Porter (eds.), *William Hunter and the eighteenth-century medical world*, 153–76.
Lee, M., *Great Britain's Solomon: James VI and I and his three kingdoms* (Urbana and Chicago, 1990).
Levack, B. P., *The formation of the British state: England, Scotland and the union 1603–1707* (Oxford, 1987).
Lightfoot, J., *Flora Scotica: or, a systematic arrangement, in the Linnaean method of the native plants of Scotland and the hebrides* (London, 1777).
Lightman, B., *Victorian science in context* (Chicago, 1997).
Lindberg, D. C., and Westman, R. S. (eds.), *Reappraisals of the scientific revolution* (Cambridge, 1990).
Lindley, D., *Court masques* (Oxford, 1995).

Lindley, D. (ed.), *The court masque* (Manchester, 1984).

Livingstone, D. N., 'The history of science and the history of geography: interpretations and implications', *History of Science*, 22 (1984), 271–301.

'Natural theology and neo-Lamarckism: the changing context of nineteenth-century geography in the United States and Great Britain', *Annals of the Asssociation of American Geographers*, 74 (1984), 9–28.

*Darwin's forgotten defenders: the encounter between evangelical theology and evolutionary thought* (Edinburgh, 1987).

'Science, magic and religion: a contextual reassessment of geography in the sixteenth and seventeenth centuries', *History of Science*, 26 (1988), 269–94.

'Geography, tradition, and the scientific revolution: an interpretative essay', *Transactions of the Institute of British Geographers*, 15 (1990), 359–73.

'The moral discourse of climate: historical considerations on race, place and virtue', *Journal of Historical Geography*, 17 (1991), 413–34.

*The geographical tradition: episodes in the history of a contested enterprise* (Oxford, 1992).

'Darwinism and Calvinism: the Belfast–Princeton connection', *Isis*, 83 (1992), 408–28.

'"Never shall ye make the crab walk straight": an inquiry into the scientific sources of racial geography', in Driver and Rose (eds.), *Nature and science*, 31–43.

'Climate's moral economy: science, race and place in post-Darwinian British and American geography', in Godlweska and Smith (eds.), *Geography and empire*, 132–54.

'Science and Religion: foreword to the historical geography of an encounter', *Journal of Historical Geography*, 20 (1994), 367–83.

'Geographical traditions', *Transactions of the Institute of British Geographers*, 20 (1995), 420–2.

'The spaces of knowledge: contributions towards a historical geography of science', *Environment and Planning D: Society and Space*, 13 (1995), 5–34.

'A chapter in the historical geography of Darwinism: a Belfast-Edinburgh case study', *Scottish Geographical Magazine*, 113 (1997), 51–7.

'Tropical climate and moral hygiene: the anatomy of a Victorian debate', *British Journal for the History of Science*, 32 (1999), 93–110.

*Spaces for science* (Chicago, forthcoming).

Livingstone, D. N., Hart, D. G., and Noll, M. (eds.), *Evangelicals and science in historical perspective* (New York and Oxford, 1999).

Livingstone, D., and Withers, C. W. J. (eds.), *Geography and Enlightenment* (Chicago, 1999).

Lochhead, E. N., 'The emergence of academic geography in Britain in its historical context' (Ph.D. thesis, University of California at Berkeley, 1980).

'Scotland as the cradle of modern academic geography in Britain', *Scottish Geographical Magazine*, 97 (1981), 98–109.

'The Royal Scottish Geographical Society: the setting and sources of its success', *Scottish Geographical Magazine*, 100 (1984), 69–80.

Lorimer, H., '"Happy hostelling in the Highlands": nationhood, citizenship and the inter-war youth movement', *Scottish Geographical Magazine*, 113 (1997), 42–50.

Love, J., *The schools and schoolmasters of Falkirk from the earliest times* (Falkirk, 1898).
Lowenthal, D., 'British national identity and the English landscape', *Rural History*, 2 (1991), 205–30.
Lux, D. S., and Cook, H. J., 'Closed circles or open networks?: communicating at a distance during the scientific revolution', *History of Science*, 36 (1998), 179–211.
Lynch, M., 'A nation born again? Scottish identity in the sixteenth and seventeenth centuries', in Broun, Finlay and Lynch (eds.), *Image and identity*, 82–104.
'James VI and the Highland problem', in Goodare and Lynch (eds.), *The origin of James VI*, 208–27.
Lynch, M. (ed.), *Oxford companion to Scottish history* (Oxford, forthcoming).
MacAlister, D., 'Geography at the University of Glasgow', *Scottish Geographical Magazine*, 37 (1921), 53–7.
Macdonald, G., 'General William Roy and his Military Antiquities of the Romans in Britain', *Archaeologia*, 68 (1917), 161–228.
Macdougall, N. (ed.), *Church, politics and society: Scotland 1408–1929* (Edinburgh, 1983).
MacInnes, A., *Charles I and the making of the covenanting movement, 1625–1641* (Edinburgh, 1991).
MacKenzie, J. M., 'The provincial geographical societies in Britain, 1884–1914', in Bell, Butlin and Heffernan (eds.), *Geography and imperialism*, 93–124.
'David Livingstone and the worldly after-life: imperialism and nationalism in Africa', in *David Livingstone and the Victorian encounter with Africa* (National Portrait Gallery, exhibition catalogue, London, 1996), 201–219.
'"The second city of the empire": Glasgow – imperial municipality', in Driver and Gilbert (eds.), *Imperial cities*, 215–37.
MacKenzie, J. M. (ed.), 'Essay and reflection: Scotland and the British empire', *International History Review*, 4 (1993), 714–39.
*Scotland and the British Empire: studies in imperialism* (Manchester, forthcoming).
Maclean, K., 'Scottish geographical roots 3: Rev. Alexander Mackay', *Scottish Association of Geography Teachers' Journal*, 27 (1998), 71–84.
Macleod, F. (ed.), *Togail Tir marking time: the mapping of the western isles* (Stornoway, 1989).
McCorkle, B., 'The maps of Patrick Gordon's *Geography Anatomiz'd:* an eighteenth-century success story', *The Map Collector*, 66 (1994), 11–15.
McCrie, T., *Life of Andrew Melville* (Edinburgh, 1856).
McCulloch, J. R., *The highlands and islands of Scotland* (London, 1824).
*Memoirs to His Majesty's Treasury respecting the geological map of Scotland* (London, 1836).
McEwan, C., '"The mother of all the peoples": geographical knowledge and the empowering of Mary Slessor', in Bell, Butlin and Heffernan (eds.), *Geography and imperialism*, 125–50.
'Gender, science and physical geography in nineteenth-century Britain', *Area*, 30 (1998), 215–23.
'Cutting power lines within the palace?: countering paternity and eurocentrism in the geographical tradition', *Transactions of the Institute of British Geographers*, 23 (1998), 371–84.

McGrane, B., *Beyond anthropology: society and the other* (New York, 1989).
Mackay, A., *A rhyming geography for little boys and girls* (Edinburgh, 1873).
Macmillan, D., *Painting in Scotland: the golden age* (Oxford, 1986).
MacPherson, D., *Geographical illustrations of Scottish history* (London, 1796).
Maddrell, A. M. C., 'Empire, emigration and school geography: changing discourses of imperial citizenship, 1880–1925', *Journal of Historical Geography*, 22 (1996), 373–87.
'Scientific discourse and the geographical work of Marion Newbigin', *Scottish Geographical Magazine*, 113 (1997), 33–41.
'Discourses of race and gender and the comparative method in geography school texts 1830–1918', *Environment and Planning D: Society and Space*, 16 (1998), 81–103.
Mah, H., 'Phantasies of the public sphere: rethinking the Habermas of historians', *Journal of Modern History*, 72 (2000), 153–82.
Maidment, J., *Remains of Sir Robert Sibbald, KNT, MD, containing his autobiography, memoirs of the Royal College of Physicians, portions of his literary correspondence, and an account of his MSS* (Edinburgh, 1837).
Mair, A., 'Thomas Kuhn and understanding geography', *Progress in Human Geography*, 10 (1986), 345–69.
Maisenschein, J., 'Epistemic styles in German and American embryology', *Science in Context*, 4 (1991), 407–28.
Maitland, P., 'Freshwater science', *Proceedings of the Royal Society of Edinburgh*, Series B, 84 (1983), 171–210.
Malament, B. C. (ed.), *After the reformation: essays in honour of J. H. Hexter* (Philadelphia, 1980).
Mander, A. M., 'Geography, gender and the state: a critical evaluation of the development of geography 1830–1918' (D.Phil. thesis, University of Oxford, 1995).
Marsden, W. E., 'Sir Archibald Geikie (1835–1924) as geographical educationist', in Marsden (ed.), *Historical perspectives on geographical education*, 54–65.
'"All in a good cause": geography, history and the politicization of the curriculum in nineteenth and twentieth-century England', *Journal of Curriculum Studies*, 21 (1989), 506–26.
Marsden, W. E., (ed.), *Historical perspectives on geographical education* (London, 1980).
Marshall, P. J. (ed.), *The Oxford history of the British empire: the eighteenth century* (Oxford, 1998).
Marshall, T., *Theatre and empire: Great Britain on the London stages under James VI and I* (London, 2000).
Martin, B., *The young gentleman and lady's philosophy, in a continual survey of the works of nature and art; by way of dialogue* (London, 1772).
Martin, J., 'A study of the entertainments for the state entry of Charles I, Edinburgh, 15 June 1633' (MA thesis, University of Edinburgh, 1990).
Marwick, W., 'Early adult education in Edinburgh', *Journal of Adult Education*, 5 (1930–2), 389–404.
'Mechanics' Institutes in Scotland', *Journal of Adult Education*, 6 (1932–43), 292–309.

Mason, R. A., 'Covenant and commonweal: the language of politics in reformation Scotland', in Macdougall, *Church, politics and society*, 97–126.

'Scotching the brut: politics, history and national myth in sixteenth-century Britain', in Mason, *Scotland and England 1286–1815*, 8–84.

'Kingship, nobility and anglo-Scottish union: John Mair's *History of Greater Britain* (1520)', *Innes review*, 41 (1990), 182–222.

'Chivalry and citizenship: aspects of national identity in renaissance Scotland', in Mason and Macdougall, *People and power in Scotland*, 50–73.

'Imagining Scotland: Scottish political thought and the problem of Britain 1560–1660', in Mason, *Scots and Britons*, 3–13.

'George Buchan, James VI and the presbyterians', in Mason, *Scots and Britons*, 112–37.

*Kingship and the commonweal: political thought in renaissance and reformation Scotland* (East Linton, 1998).

Mason, R. A. (ed.), *Scotland and England 1286–1815* (Edinburgh, 1987).

*Scots and Britons: Scottish political thought and the union of 1603* (Cambridge, 1994).

Mason, R A., and Macdougall, N. (eds.), *People and power in Scotland* (Edinburgh, 1992).

Mather, A. S., 'Geddes, geography and ecology: the golden age of vegetation mapping in Scotland', *Scottish Geographical Journal*, 115 (1999), 35–52.

Matless, D., 'Regional surveys and local knowledges: the geographical imagination in Britain, 1918–1939', *Transactions of the Institute of British Geographers*, 17 (1992), 464–80.

'The art of right living: landscape and citizenship, 1918–1939', in Pile and Thrift (eds.), *Mapping the subject*, 93–122.

'Effects of history', *Transactions of the Institute of British Geographers*, 20 (1995), 405–9.

*Landscape and Englishness* (London, 1999).

Mayhew, R., 'Landscape, religion and knowledge in eighteenth-century England', *Ecumene*, 3 (1996), 454–71.

Mayhew, R., *Geography and literature in historical context: Samuel Johnson and eighteenth-century conceptions of geography* (University of Oxford School of Geography, Research Papers, 154, 1997).

'Was William Shakespeare an eighteenth-century geographer?: constructing histories of geographical knowledge', *Transactions of the Institute of British Geographers*, 23 (1998), 21–38.

'The character of English geography, *c.*1660–1800', *Journal of Historical Geography*, 24 (1998), 385–412.

'Geography in eighteenth-century British education', *Paedagogica Historica*, 34 (1998), 731–69.

'William Guthrie's *Geographical grammar*, the Scottish Enlightenment and the politics of British geography', *Scottish Geographical Journal*, 115 (1999), 19–34.

*Enlightenment geography: the political languages of British geography, 1650–1850* (London, 2000).

Mays, D. (ed.), *The architecture of Scottish cities: essays in honour of David Walker* (East Linton, 1997).

Mechie, S., *The Church and Scottish social development, 1780–1870* (London, 1960).

Meller, H., *Patrick Geddes: social evolutionist and city planner* (London, 1993).

*Memorial of the Council of the Royal Scottish Geographical Society to the Commissioners appointed by Parliament in the Universities (Scotland) Act, 1889.*

Mendyk, S., 'Scottish regional historians and the *Britannia* project', *Scottish Geographical Magazine*, 101 (1985), 165–73.

'Early British chorography', *Sixteenth-Century Journal*, 17 (1986), 459–81.

*Speculum Britanniae: regional study, antiquarianism, and science in Britain to 1700* (Toronto, 1989).

Merton, R. K., 'Science, technology and society in seventeenth-century England', *Osiris*, 4 (1938), 360–632.

Merz, J., *A history of European thought in the nineteenth century* (4 volumes, Edinburgh, 1896–1914).

Mill, A., *Mediaeval plays in Scotland* (Edinburgh, 1927).

Mill, H., 'Recollections of the Society's early years', *Scottish Geographical Magazine*, 50 (1934), 257–69.

Mill, H., *Life interests of a geographer, 1861–1944: an experiment in autobiography* (East Grinstead, 1944).

Millburn, J. R., *Benjamin Martin: author, instrument maker and 'country showman'* (Leyden, 1976).

'The London evening courses of Benjamin Martin and James Ferguson, eighteenth-century lecturers on experimental philosophy', *Annals of Science*, 40 (1983), 437–55.

Miller, D., 'Joseph Banks, empire, and "centers of calculation" in late Hanoverian London', in Miller and Reill (eds.), *Visions of empire*, 21–37.

'The usefulness of natural philosophy: the Royal Society and the culture of practical utility in the later eighteenth century', *British Journal for the History of Science*, 32 (1999), 185–201.

Miller, D., and Reill, P. H. (eds.), *Visions of empire: voyages, botany, and representations of nature* (Cambridge, 1996).

Miller, D. P., 'The usefulness of natural philosophy: the Royal Society of London and the culture of practical utility in the later eighteenth century', *British Journal for the History of Science*, 32 (1999), 185–202.

Miller, J., *History of Dunbar* (Dunbar, 1859).

Mitchison, R., 'Nineteenth-century Scottish nationalism: the cultural background', in Mitchison (ed.), *The roots of nationalism*, 131–42.

Mitchison, R. (ed.), *The roots of nationalism: studies in northern Europe* (Edinburgh, 1980).

Moir, D. G. (ed.), *The early maps of Scotland to 1850: volumes 1 and 2* (Edinburgh, 1973).

Money, J., 'From Leviathan's air-pump to Brittania's voltaic pile: science, public life and the forging of Britain, 1660–1820', *Canadian Journal of History / Annals Canadiennes d'Histoire*, 28 (1993), 521–44.

Monro, P. A. G., 'Introduction, and the Professor's daughter', *Proceedings of the Royal College of Physicians of Edinburgh*, 26 (1996), 1–189.

Moore, J. N., 'Scottish cartography in the later Stuart era, 1660–1714', *Scottish Tradition*, 14 (1986–87), 28–44.

*The historical cartography of Scotland* (Aberdeen, 1991).

Moran, B. (ed.), *Patronage and institutions: science, technology, and medicine at the European court* (Woodbridge, 1981).

Moravia, S., 'Philosophie et géographie à la fin du XVIIIe siècle', *Studies in Voltaire and the Eighteenth Century*, 57 (1967), 937–1071.

'The enlightenment and the sciences of man', *History of Science*, 18 (1980), 247–68.

Morgan, A. (ed.), *University of Edinburgh: charters, statutes, and acts of the town council and the senatus 1583–1858* (Edinburgh, 1937).

Morrell, J. B., 'The University of Edinburgh in the late eighteenth century: its scientific eminence and academic structure', *Isis*, 62 (1971), 158–71.

Morrell, J. B., and Thackray, A. (eds.), *Gentlemen of science: early years of the British Association for the Advancement of Science* (Oxford, 1981).

*Gentlemen of science: early correspondence of the British Association for the Advancement of Science* (Oxford, 1984).

Morris, R. J., and Morton, G., 'Where was nineteenth-century Scotland?', *Scottish Historical Review*, 73 (1994), 89–99.

Morton, A. (ed.), 'Science lecturing in the eighteenth century', *British Journal for the History of Science*, 28 (1995), 1–100.

Morton, A., and Wess, J., *Public and private science: the King George III Collection* (Oxford, 1993).

Morton, G., 'Scottish rights and "centralisation" in the mid-nineteenth century', *Nations and Nationalism*, 2 (1996), 257–79.

'What if . . . ? the significance of Scotland's missing nationalism in the nineteenth century', in Broun, Finlay and Lynch (eds.), *Image and identity*, 157–76.

*Unionist-nationalism: the governing of urban Scotland, 1830–1860* (East Linton, 1998).

'Civil society, municipal government and the state: enshrinement, empowerment and legitimacy – Scotland, 1800–1929', *Urban History*, 25 (1998), 348–367.

'Historical clubs', in Lynch (ed.), *Oxford companion to Scottish history* (Oxford, forthcoming).

Morton, G., and Morris, R. J., 'Civil society, governance and nation: Scotland 1830–1914', in Houston and Knox (eds.), *Penguin history of Scotland* (London, forthcoming).

Morus, I., 'Currents from the underworld: electricity and the technology of display in early Victorian England', *Isis*, 84 (1993), 50–69.

Muir, J., *Stories of my boyhood and youth* (Edinburgh, 1985).

Mundy, B., *The mapping of new Spain: indigenous cartography and maps of the relaçiones geograficas* (Chicago, 1996).

Murray, A., *The northern flora: or a description of the wild plants belonging to the north and east of Scotland* (Edinburgh, 1836).

Murray, H., 'On the ancient geography of central and eastern Asia, with illustrations derived from recent discoveries in the north of India [1816]', *Transactions of the Royal Society of Edinburgh*, VIII (1818), 171–204.

'Observations on the information collected by the Ashantee Mission, respecting the course of the River Niger and the interior of Africa', *Edinburgh Philosophical Journal*, 1 (1819), 163–170.

'Observations on the results of the late expedition of Captain Parry, including a view

of previous discoveries made in the same direction', *Edinburgh Philosophical Journal*, 11 (1824), 225–49.
Murray, J., 'Scottish geography in school textbooks', *Scottish Geographical Magazine*, 29 (1913), 198–200.
Murray, J., and Pullar, J., *Bathymetrical survey of the freshwater lochs of Scotland* (Edinburgh, 1910).
Murray, J. A. H., *The dialect of the southern counties of Scotland* (London, 1873).
Mykkanen, J., '"To methodize and regulate them": William Petty's governmental science of statistics', *History of the Human Sciences*, 7 (1994), 65–88.
Naiden, J. R., *The Sphera of George Buchanan (1506–1582)* (Washington, 1969).
Newbigin, M., 'Editorial: the value of geography', *Scottish Geographical Magazine*, 21 (1905), 1–4.
'The study of weather as a branch of nature knowledge', *Scottish Geographical Magazine*, 23 (1907), 627–48.
'Geography in Scotland since 1889', *Scottish Geographical Magazine*, 29 (1913), 471–9.
'Race and nationality', *Geographical Journal*, 50 (1917), 313–35.
*Man and his conquest of nature* (fourth edition, London, 1926).
*A new regional geography of the world* (London, 1929).
'The Royal Scottish Geographical Society: the first fifty years', *Scottish Geographical Magazine*, 50 (1934), 257–69.
Nichols, J., *The progresses, processions and magnificent festivities of King James the first* (London, 1828).
*Illustrations of the literary history of the eighteenth century* (London, 1840).
Nicholson, M., 'National styles, divergent classifications: a comparative case study from the history of French and American plant ecology', *Knowledge and Society*, 8 (1989), 139–86.
Numbers, R. L., *Darwinism comes to America* (Cambridge, Mass., 1998).
Nye, M., '"National styles"?: French and English chemistry in the nineteenth and early twentieth centuries', *Osiris*, 8 (1993), 30–52.
O'Donoghue, Y., *William Roy 1726–1790: pioneer of the Ordnance Survey* (London, 1977).
Ogborn, M., 'The capacities of the state: Charles Davenant and the management of the excise', *Journal of Historical Geography*, 24 (1998), 289–312.
*Spaces of modernity: London's geographies, 1680–1780* (New York, 1998).
'Georgian geographies', *Journal of Historical Geography*, 24 (1998), 218–24.
Olby, R. C., Cantor, G. N., Christie, J. R. R., and Hodge, M. J. S. (eds.), *Companion to the history of modern sciences* (London, 1990).
Oldroyd, D., *The Highlands controversy: constructing geological knowledge through fieldwork in nineteenth-century Britain* (Chicago, 1990).
Ophir, A., and Shapin, S., 'The place of knowledge: a methodological survey', *Science in Context*, 4 (1991), 3–21.
Orgel, S., *The illusion of power* (Berkeley, 1975).
Ouston, H., 'York in Edinburgh: James VII and the patronage of learning in Scotland, 1679–1688', in Dwyer, Mason and Murdoch (eds.), *New perspectives on the politics and culture of early modern Scotland*, 133–55.

Outram, D., *The enlightenment* (Cambridge, 1995).

'New spaces in natural history', in Jardine, Secord and Spary (eds.), *Cultures of natural history*, 249–65.

Pallot, J., 'Imagining the rational landscape in late imperial Russia', *Journal of Historical Geography*, 26 (2000), 273–91.

Pagden, A., *European encounters with the New World* (New Haven and London, 1993).

*Lords of all the world: ideologies of empire in Spain, Britain and France, c.1500–1800* (Yale, 1995).

Paradis, J. G., 'The natural historian as antiquary of the world: Hugh Miller and the rise of literary natural history', in Shortland (ed.), *Hugh Miller*, 122–50.

Parker, J., *Books to build an empire* (Amsterdam, 1965).

Parry, G., *The golden age restor'd: the culture of the Stuart court 1603–1642* (Manchester, 1981).

*The trophies of time* (Oxford, 1995).

Parry, J. H., *The age of reconnaisance: discovery, exploration and settlement, 1450–1650* (Berkeley, 1981).

Passmore, R. (ed.), *Proceedings of the Royal College of Physicians: ter-centenary congress* (Edinburgh, 1982).

Peck, L. (ed.), *The mental world of the Jacobean court* (Cambridge, 1991).

Peet, R., 'The social origins of environmental determinism', *Annals of the Association of American Geographers*, 75 (1985), 309–33.

Phillips, P., *The scientific lady: a social history of women's scientific interest, 1520–1918* (London, 1990).

Phillips, R., *Mapping men and empire: a geography of adventure* (London, 1996).

Phillipson, N., 'Culture and society in the eighteenth-century province: the case of Edinburgh and the Scottish Enlightenment', in Stone (ed.), *The university in society*, Vol. II, 407–20.

'The Scottish Enlightenment', in Porter and Teich (eds.), *The enlightenment in national context*, 19–40.

Phillipson, N. J., and Skinner, Q. (eds.), *Political discourse in early modern Britain* (Cambridge, 1993).

Pile, S., and Thrift, N. (eds.), *Mapping the subject: geographies of cultural transformation* (London, 1995).

Pinkerton, J., *Dissertation on the origin and progress of the Scythians or Goths, being an introduction to the ancient and modern history of Europe* (London, 1786).

*A new and complete system of universal geography* (London, 1796).

*Modern geography* (London, 1802).

Ploszajska, T., 'Constructing the subject: geographical models in English schools, 1870–1944', *Journal of Historical Geography*, 22 (1996), 388–98.

*Geographical education, empire and citizenship: geographical teaching and learning in English schools, 1870–1944* (Historical Geography Research Group Publication Series, 35, Cambridge, 1999).

Porter, R., 'Science, provincial culture and public opinion in Enlightenment England', *British Journal for Eighteenth-Century Studies*, 3 (1980), 22–46.

*The enlightenment* (London, 1990).

'The terraqueous globe', in Rousseau and Porter (eds.), *The ferment of knowledge*, 285–324.

'Measure of ideas, role of language: mathematics and language in the eighteenth century', in Frangsmyr, Heilbron and Rider (eds.), *The quantifying spirit*, 113–140.
'Medical lecturing in Georgian London', *British Journal for the History of Science*, 28 (1995), 91–100.
*Enlightenment Britain and the creation of the modern world* (London, 2000).
Porter, R., and Teich, M. (eds.), *The enlightenment in national context* (Cambridge, 1981).
*The scientific revolution in national context* (Cambridge, 1992).
*Proceedings of the Perthshire Society of Natural Science* (Perth, 1886).
Proudfoot, W., *Biographical memoir of James Dinwiddie* (Liverpool, 1868).
Pumfrey, S., 'Who did the work?: experimental philosophers in Augustan England', *The British Journal for the History of Science*, 28 (1995), 131–56.
Pyenson, L., and Sheets-Pyenson, S., *Servants of nature: a history of scientific institutions, enterprises and sensibilities* (London, 1999).
Raven, C., *English naturalists from Neckham to Ray: a study of the making of the modern world* (Cambridge, 1947).
Ravenstein, E. G., 'On the Celtic languages in the British Isles: a statistical survey', *Journal of the Royal Statistical Society* (1879), 579–643.
Reid, J. S., 'Late eighteenth-century adult education in the sciences at Aberdeen: the natural philosophy classes of Professor Patrick Copland', in Carter and Pittock (eds.), *Aberdeen and the enlightenment*, 168–79.
Reingold, N., 'The peculiarities of the Americans, or are there national styles in the sciences?', *Science in Context*, 4 (1991), 347–66.
Renwick, J. (ed.), *L'Invitation au voyage: studies in honour of Peter France* (Oxford, 2000).
*Report of the Education Department of the Glasgow Authority: film in the classroom* (Glasgow, 1933).
Revel, J., 'Knowledge of the territory', *Science in Context*, 4 (1991), 133–61.
Richards, P., 'Kant's geography and mental maps', *Transactions of the Institute of British Geographers*, 61 (1974), 1–16.
Richards, R. L., *Dr John Rae* (Whitby, 1985).
Robbins, K., *Great Britain: identities, institutions and the idea of Britishness* (London, 1998).
Robertson, C. J., 'Scottish geographers: the first hundred years', *Scottish Geographical Magazine*, 89 (1973), 5–18.
Robertson, J. A., 'Empire and union: two concepts of the early modern European political order', in Robertson (ed.), *A union for empire*, 3–36.
'An elusive sovereignty: the course of the union debate in Scotland, 1698–1707', in Robertson (ed.), *A union for empire*, 198–227.
'The Enlightenment above national context: political economy in eighteenth-century Scotland and Naples', *The Historical Journal*, 40 (1997), 667–97.
Robertson, J. A. (ed.), *A union for empire: political thought and the British union of 1707* (Cambridge, 1995).
Robic, M-C., 'National identity in Vidal's *Tableau de la géographie de la France*: from political geography to human geography', in Hooson (ed.), *Geography and national identity*, 58–70.
Robson, R. (ed.), *Ideas and institutions of Victorian Britain* (London, 1967).

Rose, G., 'Tradition and paternity: same difference', *Transactions of the Institute of British Geographers*, 20 (1995), 414–6.

Ross, K., *Church and creed in Scotland: the Free Church case 1900–1904 and its origins* (Edinburgh, 1988).

Rousseau, G., and Porter, R. (eds.), *The ferment of knowledge* (Cambridge, 1980).

Roy, G. R., (ed.), *The letters of Robert Burns* (Oxford, 1985).

Roy, W., 'An account of the measurement of a base on Hounslow Heath', *Philosophical Transactions*, 75 (1785), 379–97.

Rudmose Brown, R. N., 'Scotland and some trends in geography: John Murray, Patrick Geddes and Andrew Herbertson', *Geography* 33 (1948), 107–120.

Rupke, N., 'A geography of enlightenment: the critical reception of Alexander von Humboldt's Mexico work', in Livingstone and Withers (eds.), *Geography and enlightenment*, 319–39.

Rupp, J., 'The new science and the public sphere in the premodern era', *Science in Context*, 8 (1995), 487–507.

Russell, J. L., 'Cosmological teaching in the seventeenth-century Scottish universities', *Journal of the History of Astronomy*, 5 (1974), 122–32; 145–54.

Ryan, J. R., 'Imperial landscapes: photography, geography and British overseas exploration, 1858–1872', in Bell, Butlin and Heffernan (eds.), *Geography and imperialism*, 53–80.

*Picturing empire: photography and the visualization of the British empire* (London, 1997).

Sacks, D. H., and Kelley, D. R. (eds.), *The historical imagination of early modern Britain: history, rhetoric, and fiction, 1500–1800* (Cambridge, 1999).

Said, E., *Orientalism* (London, 1978).

*Culture and imperialism* (London, 1991).

Samuel, R., and Stedman Jones, G. (eds.), *Culture, ideology and politics* (London, 1982).

Sander, G., 'In search of identity: German nationalism and geography, 1871–1910', in Hooson (ed.), *Geography and national identity*, 71–91.

Sander, G., and Rossler, M., 'Geography and empire in Germany, 1871–1945', in Godlewska and Smith (eds.), *Geography and empire*, 115–29.

Sanders, L., (ed.), *Celebrities of the century* (London, 1890).

Scargill, D. I., 'The R. G. S. and the foundation of geography at Oxford', *Geographical Journal*, 142 (1976), 438–61.

Schaffer, S., 'Natural philosophy and public spectacle in the eighteenth century', *History of Science*, 21 (1983), 1–43.

'Late Victorian metrology and its instrumentation: a manufactory of ohms', in Bud and Cozzens (eds.), *Invisible connections*, 23–56.

'Machine philosophy: demonstration devices in Georgian mechanics', *Osiris*, 9 (1994), 157–82.

'Physics laboratories and the Victorian country house', in Smith and Agar (eds.), *Making space for science*, 149–80.

Schuster, J. A., 'The scientific revolution', in Olby, Cantor, Christie and Hodge (eds.), *Companion to the history of modern sciences*, 217–42.

Schwartz, J., '*The Geography Lesson*: photographs and the construction of imaginative geographies', *Journal of Historical Geography*, 22 (1996), 16–45.

Scotch Education Department, *Memorandum on the teaching of geography in Scottish primary schools* (London, 1912).
Scottish Geographical Society, *Scheme for the encouragement of geography in Scottish schools* (Edinburgh, 1886).
Scowen, J., 'A study in the historical geography of an idea: Darwinism in Edinburgh, 1859–1875', *Scottish Geographical Magazine*, 114 (1998), 148–56.
Secord, A., 'Scientists in the pub: artisan botanists in early nineteenth-century Lancashire', *History of Science*, 32 (1994), 269–315.
'Artisan botany', in Jardine, Secord and Spary (eds.), *Cultures of natural history*, 378–93.
Secord, J., 'Newton in the nursery: Tom Telescope and the philosophy of tops and balls, 1761–1838', *History of Science*, 23 (1985), 127–51.
Shapin, S., 'The audience for science in eighteenth-century Edinburgh', *History of Science*, 12 (1974), 95–121.
'Property, patronage and the politics of science: the founding of the Royal Society of Edinburgh', *British Journal for the History of Science*, 7 (1974), 1–41.
'"Nibbling at the teats of science": Edinburgh and the diffusion of science in the 1830s', in Inkster and Morrell (eds.), *Metropolis and province*, 151–78.
'The house of experiment in seventeenth-century England', *Isis*, 79 (1988), 373–404.
'Science and the public', in Olby, Cantor, Christie and Hodge (eds.), *Companion to the history of modern sciences*, 990–1007.
'"A scholar and a gentleman:" the problematic identity of the scientific practitioner in early modern England', *History of Science*, 29 (1991), 279–327.
*A social history of truth: civility and science in seventeenth-century England* (Chicago, 1994).
*The scientific revolution* (Chicago, 1996).
'Placing the view from nowhere: historical and sociological problems in the location of science', *Transactions of the Institute of British Geographers*, 23 (1998), 5–12.
Shapin, S., and Schaffer, S., *Leviathan and the air-pump: Hobbes, Boyle and the experimental life* (Princeton, 1995).
Sharp, L. (ed.), *Early letters of Robert Wodrow 1698–1709* (Edinburgh, 1937).
Shaw, D., 'Geographical practice and its significance in Peter the Great's Russia', *Journal of Historical Geography*, 22 (1996), 160–76.
Shepherd, C., 'Philosophy and science in the arts curriculum of the Scottish universities in the seventeenth century' (Ph.D. thesis, University of Edinburgh, 1975).
'Newtonianism in Scottish universities in the seventeenth century', in Campbell and Skinner (eds.), *The origins and nature of the Scottish enlightenment*, 65–85.
Shinn, T., and Whitley, R. (eds.), *Expository science: forms and functions of popularisation* (Dordrecht, 1985).
Shire, H., *Song, dance and poetry of the court of Scotland under James VI* (Edinburgh, 1969).
Shortland, M. (ed.), *Hugh Miller and the controversies of Victorian science* (Oxford, 1996).
Shteir, A., *Cultivating women, cultivating science: flora's daughters and botany in England 1760–1860* (Baltimore, 1996).
Sidaway, J., 'The production of British geography', *Transactions of the Institute of British Geographers*, 22 (1997), 488–504.

Simpson, A. D. C., 'Sir Robert Sibbald – founder of the College', in R. Passmore (ed.), *Proceedings of the Royal College of Edinburgh: ter-centenary congress* (Edinburgh, 1982), 59–91.

'Globe production in the period 1770–1830', *Der Globusfreund*, 35–7 (1987), 21–36.

Sitwell, O. F. G., 'John Pinkerton: an armchair geographer of the early nineteenth century', *Geographical Journal*, 138 (1972), 470–9.

*Four centuries of special geography* (Vancouver, 1993).

Skelton, R. A., 'Bishop Leslie's Map of Scotland, 1578', *Imago Mundi*, 7 (1950), 103–6.

'The Military Survey of Scotland 1747–1755', *Scottish Geographical Magazine*, 83 (1977), 5–16.

Skinner, Q., 'The origins of the Calvinist theory of revolution', in Malament (ed.), *After the reformation*, 309–30.

Sloan, P., 'Natural history, 1670–1802', in Olby, Cantor, Christie and Hodge (eds.), *Companion to the history of modern sciences*, 295–313.

Sloan, P., 'The gaze of natural history', in Fox, Porter and Wokler, *Inventing human science*, 112–51.

Smart, E., *History of Perth Academy* (Perth, 1932).

Smellie, W., *Account of the institution and progress of the society of the antiquaries of Scotland* (Edinburgh, 1782).

Smith, A., *A summer in Skye* (Edinburgh, 1880).

Smith, A. H., 'The Dundee centre of the Royal Scottish Geographical Society 1884–1985', *Scottish Geographical Magazine*, 101 (1985), 184–6.

Smith, C., and Agar, J., 'Introduction: making space for science', in Smith and Agar (eds.), *Making space for science*, 1–24.

Smith, C., and Agar, J. (eds.), *Making space for science: territorial themes in the shaping of knowledge* (London, 1998).

Smith, D., *A history of the modern British Isles 1601–1707: the divided crown* (Oxford, 1998).

Smith, D. C. F., 'The progress of the *Orcades* survey, with biographical notes on Murdoch Mackenzie senior (1712–1797)', *Annals of Science*, 44 (1987), 277–88.

Smith, J., *History of Kelso Grammar school* (Kelso, 1909).

Smith, J. V., 'Manners, morals and mentalities: reflections on the popular enlightenment of early nineteenth-century Scotland', in Humes and Paterson (eds.), *Scottish culture and Scottish education*, 25–54.

Smith, N., and Godlewska, A., 'Introduction: critical histories of geography', in Godlewska and Smith, *Geography and empire*, 1–8.

Smith, R., 'Botanical Survey of Scotland I. Edinburgh District', *Scottish Geographical Magazine*, 16 (1900), 385–416.

'Botanical Survey of Scotland II. North Perthshire District', *Scottish Geographical Magazine*, 16 (1900), 441–67.

Smith, S. J., 'Soundscape', *Area*, 26 (1994), 232–40.

'Beyond geography's visible worlds: a cultural politics of music', *Progress in Human Geography*, 21 (1997), 502–29.

Smout, T. C., 'The Highlands and the roots of green consciousness, 1750–1990', *Proceedings of the British Academy*, 76 (1991), 237–63.

Solkin, D., *Visual art and the public sphere in eighteenth-century England* (New Haven, 1992).

Somerville, M., *Personal recollections, from early life to old age of Mary Somerville with selections from her correspondence* (London, 1873).
Sorrenson, R., 'The ship as a scientific instrument in the eighteenth century', *Osiris*, 11 (1996), 221–36.
'Towards a history of the Royal Society in the eighteenth century', *Notes and Records of the Royal Society in London*, 50 (1996), 29–46.
Spate. O., 'Geography and national identity in Australia', in Hooson, *Geography and national identity*, 277–82.
Speak, P., 'William Speirs Bruce 1867–1921', *Geographers' Biobibliographical Studies*, 17 (1997), 17–25.
Speak, P. (ed.), *The log of the* Scotia *expedition* (Edinburgh, 1992).
Stafford, B., *Artful science: enlightenment entertainment and the eclipse of visual education* (Cambridge, 1994).
Stanley, H. M., 'Inaugural address delivered before the Scottish Geographical Society at Edinburgh, 3rd December 1885', *Scottish Geographical Magazine*, 1 (1895), 1–17.
Staum, M., 'The Enlightenment transformed: the Institute Prize contests', *Eighteenth-Century Studies*, 19 (1985), 153–79.
'Human geography in the French institute: new discipline or missed opportunity?', *Journal of the History of the Behavioural Sciences*, 23 (1987), 332–40.
'The Paris Geographical Society constructs the Other, 1821–1850', *Journal of Historical Geography*, 26 (2000), 222–38.
Steel, R. W., *The Institute of British Geographers: the first fifty years* (London, 1984).
Stefansson, V., 'Arctic controversy: the letters of John Rae', *Geographical Journal*, 120 (1954), 486–93.
Stephan, N., *The idea of race in science: Great Britain 1800–1960* (London, 1982).
Steuart, J., *Enquiry into the principles of oeconomy* (Edinburgh, 1767).
Stevenson, D., 'Cartography and the kirk: aspects of the making of the first atlas of Scotland', *Scottish Studies*, 26 (1982), 1–12.
Stewart, L., 'Public lectures and private patronage in Newtonian England', *Isis*, 75 (1986), 47–58.
*The rise of public science: rhetoric, technology, and natural philosophy in Newtonian Britain, 1660–1750* (Cambridge, 1992).
'Other centres of calculation, or, where the Royal Society didn't count: commerce, coffee-houses and natural philosophy in early modern London', *British Journal for the History of Science*, 32 (1999), 133–54.
Still, J., and Worton, M. (eds.), *Textuality and sexuality: reading theories and practices* (Manchester, 1993).
Stocking, G., *Functionalism historicised: essays on British social anthropology* (Wisconsin, 1984).
Stoddart, D. R., 'The paradigm concept and the history of geography', in Stoddart (ed.), *Geography, ideology and social concern*, 70–80.
*On geography and its history* (Oxford, 1986).
'Do we need a feminist historiography of geography – and if we do, what should it be?', *Transactions of the Institute of British Geographers*, 16 (1991), 484–7.
Stoddart, D. R. (ed.), *Geography, ideology and social concern* (Oxford, 1981).

Stone, J. C., *The Pont manuscript maps of Scotland: sixteenth century origins of a Blaeu Atlas* (Tring, 1978).

'Robert Gordon of Straloch: cartographer or chorographer?', *Northern Scotland*, 4 (1981), 7–22.

'Robert Gordon and the making of the first atlas of Scotland', *Northern Scotland*, 18 (1998), 15–29.

Stone, L. (ed.), *The university in society* (2 volumes, Princeton, 1975).

Stott, P., 'History of biogeography' in Taylor, *Themes in biogeography*, 1–24.

Strauss, G., *Sixteenth-century Germany: its topography and topographers* (Madison, 1959).

Strawhorn, J., *750 years of a Scottish school: Ayr Academy 1233–1983* (Ayr, 1983).

Strong, R., *Art and power: renaissance festivals 1450–1650* (Woodbridge, 1984).

Sutton, G., *Science for a polite society: gender, culture, and the demonstration of enlightenment* (Boulder, 1996).

Sweet, J. M., 'Robert Jameson and the explorers: the search for the north-west passage part I', *Annals of Science*, 31 (1974), 21–47.

Tait, J. B., 'Oceanography in Scotland during the XIXth and early XXth centuries', *Bulletin de Institute du Oceanographie de Monaco*, 1 (1968), 281–92.

Takeuchi, K., 'Nationalism and geography in modern Japan, 1880s to 1920s', in Hooson (ed.), *Geography and national identity*, 104–11.

Taylor, E. G. R., *Tudor geography, 1485–1583* (London, 1930).

*Late Tudor and early Stuart geography, 1583–1650* (London, 1934).

'Robert Hooke and the cartographical projects of the late seventeenth century (1666–1696)', *Geographical Journal*, 90 (1937), 529–40.

*The mathematical practitioners of Tudor and Stuart England* (London, 1954).

Taylor, J. A. (ed.), *Themes in biogeography* (London, 1984).

*Third Report of the Dunbar Mechanics' Institute* (Haddington, 1828).

Thompson, D. Wentworth, 'The history of science in Scotland', in Wright and Snodgrass (eds.), *Scotland and its people*, 241–72.

Thomson, J., *Atlas of Scotland* (Edinburgh, 1820).

Three of the Staff, *Voyage of the Scotia* (Edinburgh and London, 1906).

Thrift, N., Driver, F., and Livingstone, D., 'The geography of truth', *Environment and Planning D: Society and Space*, 13 (1995), 1–3.

Thrower, N. J. W., (ed.), *The compleat platt-maker: essays on chart, map and globe-making in England in the seventeenth and eighteenth centuries* (Berkeley, 1987).

Tilly, C. A., (ed.), *The formation of nations in western Europe* (Princeton, 1975).

Tocher, J., 'The necessity for a national eugenic survey', *Eugenics Review* (July 1910), 124–41.

Tomaselli, S., 'The Enlightenment debate on women', *History Workshop Journal*, 20 (1985), 101–24.

*Transactions of the Glasgow and Clydesdale Statistical Society.*

Turnbull, D., 'Cartography and science in early modern Europe: mapping the construction of knowledge spaces', *Imago Mundi*, 46 (1996), 5–24.

Turner, F., 'Public science in Britain, 1880–1919', *Isis*, 71 (1980), 589–608.

Tyacke, S., *London mapsellers 1660–1720* (Tring, 1978).

Unwin, T., *The place of geography* (London, 1992).

Urry, J., 'Englishmen, Celts and Iberians: the ethnographic survey of the United Kingdom, 1892–1899', in Stocking (ed.), *Functionalism historicised*, 83–105.
Van der Velde, P., 'The Royal Dutch Geographical Society and the Dutch East Indies, 1873–1914: from colonial lobby to colonial hobby', in Bell, Butlin and Heffernan (eds.), *Geography and imperialism*, 80–92.
Van Eerde, K. S., *John Ogilby and the taste of his times* (London, 1978).
Vickers, B., 'Francis Bacon and the progress of knowledge', *Journal of the History of Ideas*, 53 (1992), 495–518.
Wahrman, D., 'National society, communal culture: an argument about the recent historiography of eighteenth-century Britain', *Social History*, 17 (1992), 43–72.
Wallis, H., 'Geographie is better than divinitie: maps, globes, and geography in the days of Samuel Pepys', in Thrower, *The compleat platt-maker*, 1–43.
Walters, A. N., 'Conversation pieces: science and politeness in eighteenth-century England', *History of Science*, 35 (1997), 121–54.
Wardenga, U., 'Constructing regional knowledge in German geography: the Central Commission on the Regional Geography of Germany, 1882–1941', in Buttimer, Brunn and Wardenga (eds.), *Text and image*, 77–84.
Warntz, W., '*Geographia Generalis* and the earliest development of American geography', in Blouet (ed.), *The origins of academic geography in the United States*, 245–63.
Warntz, W., 'Newton, the Newtonians, and the *Geographia Generalis Varenii*', *Annals of the Association of American Geographers*, 79 (1989), 165–91.
Watelet, M. (ed.), *Gerardus Mercator rupelmundanus* (Antwerp, 1994).
Watson, F., 'The enigmatic lion: Scotland, kingship and national identity in the wars of independence', in Broun, Finlay and Lynch, *Image and identity*, 18–37.
Watson, J. A., and Amery, G. D., 'Early Scottish agricultural writers (1697–1790)', *Transactions of the Highland and Agricultural Society of Scotland*, 43 (1931), 60–85.
Watson, J. T. S., *Pathmakers in the isles* (Edinburgh, 1956).
Webster, C., *From Paracelsus to Newton: magic and the making of modern science* (Cambridge, 1982).
Webster, D. C. F., 'A cartographic controversy: in defence of Murdoch Mackenzie', in MacLeod, *Togail tir marking time*, 33–42.
Webster, H. A., 'What has been done for the geography of Scotland and what remains to be done', *Scottish Geographical Magazine*, 1 (1885), 487–96.
Werritty, A., and Reid, L., 'Debating the geographical tradition', *Scottish Geographical Magazine*, 111 (1995), 196–8.
Westfall, R., 'Scientific patronage: Galileo and the telescope', *Isis*, 76 (1985), 11–30.
Whitaker, I., 'The reports on the parishes of Scotland, 1627', *Scottish Studies*, 3 (1959), 229–32.
Whitley, 'Knowledge producers and knowledge acquirers: popularisation as a relation between scientific fields and their publics', in Shinn and Whitley (eds.), *Expository science*, 3–28.
Whittington, G., and Gibson, A., *The Military Survey of Scotland 1747–1755: a critique* (Lancaster, 1986).
Whyte, I. D., *Agriculture and society in seventeenth-century Scotland* (Edinburgh, 1979).

Wilczynski, W., 'Gatekeeping and regional knowledge in Poland', in Buttimer, Brunn and Wardenga (eds.), *Text and image*, 111–21.

Williams, J., *Great Britains Salomon. A sermon preached at the magnificent funerall, of the most high and mighty king James, the late king of Great Britaine* (London, 1625).

Williamson, A., *Scottish national consciousness in the age of James VI* (Edinburgh, 1979).

'Number and national consciousness: the Edinburgh mathematicians and Scottish political culture at the union of the crowns', in Mason, *Scots and Britons*, 187–212.

Williamson, A. H., 'George Buchan, civic virtue and commerce: European imperialism and its sixteenth-century critics', *Scottish Historical Review*, 75 (1997), 20–37.

'Scots, indians and empires: the Scottish politics of civilization, 1519–1609', *Past and Present*, 150 (1996), 46–83.

Wilson, K., *The sense of the people: politics, culture and imperialism in England, 1715–85* (Cambridge, 1995).

'Citizenship, empire and modernity in the English provinces, *c.*1720–1790', *Eighteenth-century Studies*, 29 (1995), 69–96.

Wilson, N., *Presenting Scotland: a film survey* (Edinburgh, 1945).

Winlow, H., 'Anthropometric cartography: constructing Scottish racial identity in the early twentieth century', *Journal of Historical Geography* (forthcoming).

Wintle, M., 'Renaissance maps and the construction of the idea of Europe', *Journal of Historical Geography*, 25 (1999), 137–65.

Wise, M., 'A university teacher of geography', *Transactions of the Institute of British Geographers*, 66 (1975), 1–16.

Withers, C. W. J., *Gaelic in Scotland 1698–1981: the geographical history of a language* (Edinburgh, 1984).

'Improvement and enlightenment: agriculture and natural history in the work of the Rev. Dr. John Walker (1731–1803)', in Jones, *Philosophy and science in the Scottish enlightenment*, 102–16.

'William Cullen's agricultural lectures and writings and the development of agricultural science in eighteenth-century Scotland', *Agricultural History Review*, 37 (1989), 144–56.

'Natural knowledge as cultural property: disputes over the "ownership" of natural history in late eighteenth-century Edinburgh', *Archives of Natural History*, 19 (1992), 289–303.

'The historical creation of the Scottish Highlands', in Donnachie and Whatley, *The manufacture of Scottish history*, 143–56.

'Geography in its time: geography and historical geography in Diderot and d'Alembert's *Encyclopédie*', *Journal of Historical Geography*, 19 (1993), 255–64.

'On georgics and geology: James Hutton's '*Elements of Agriculture*' and agricultural science in eighteenth-century Scotland', *Agricultural History Review*, 42 (1994), 138–49.

'Picturing Highland landscapes: George Washington Wilson and the photography of the Scottish Highlands', *Landscape Research*, 19 (1994), 68–79.

'How Scotland came to know itself: geography, national identity and the making of a nation, 1680–1790', *Journal of Historical Geography*, 21 (1995), 371–97.

'Geography, natural history and the eighteenth-century enlightenment: putting the world in place', *History Workshop Journal*, 39 (1995), 136–63.

'Notes towards a historical geography of geography in early modern Scotland', *Scotlands*, 3 (1996), 111–24.

'Encyclopaedism, modernism and the classification of geographical knowledge', *Transactions of the Institute of British Geographers*, 21 (1996), 363–98.

'Geography, science and national identity in early modern Britain: the case of Scotland and the work of Sir Robert Sibbald (1641–1722)', *Annals of Science*, 53 (1996), 29–73.

'Geography, royalty and empire: Scotland and the making of Great Britain, 1603–1661', *Scottish Geographical Magazine*, 113 (1997), 22–32.

'Sir Robert Sibbald (1641–1722)', *Geographers' Biobibliographical Studies*, 19 (1997), 12–21.

'Towards a history of geography in the public sphere', *History of Science*, 37 (1999), 45–78.

'Reporting, mapping, trusting: practices of geographical knowledge in the late seventeenth century', *Isis*, 90 (1999), 497–521.

'Contested visions: nature, culture and the morality of landscape in the Scottish Highlands', in Buttimer and Wallin, *Nature, culture and human identity*, 271–86.

'John Adair (1660–1718)', *Geographers' Biobibliographical Studies*, 20 (2000), 1–8.

'John Ogilby (1600–1676), *Geographers' Biobibliographical Studies*, 20 (2000), 77–84.

'Travel and credibility: towards a geography of trust', *Géographie et Culture*, 33 (2000), 3–17.

'Travel and trust in the eighteenth century', in Renwick (ed.), *L'Invitation au voyage*, 47–54.

'Authorising landscape: "authority", naming and the Ordnance Survey's mapping of the Scottish Highlands in the nineteenth century', *Journal of Historical Geography*, 26 (2000), 532–54.

'Toward a historical geography of enlightenment in Scotland', in Wood, *The Scottish enlightenment*, 63–97.

'Pont in context: chorography, map-making and national identity in the late sixteenth century', in Cunningham ed., *The nation survey'd*, 139–54.

Withers, C. W. J., and Livingstone, D. N., 'Introduction: on geography and enlightenment', in Livingstone and Withers (eds.), *Geography and Enlightenment*, 1–28.

Withrington, D., '"Scotland a half-educated nation" in 1834? reliable critique or persuasive polemic?', in Humes and Paterson (eds.), *Scottish culture and Scottish education*, 55–74.

'What was distinctive about the Scottish enlightenment?', in Carter and Pittock (eds.), *Aberdeen and the enlightenment*, 10–25.

Withrington, D., and Grant, I. R. (eds.), *The statistical account of Scotland* (Edinburgh, 1983).

Womersley, D. (ed.), *Edward Gibbon: bicentenary essays. Studies in Voltaire and the Eighteenth Century*, *355* (Oxford, 1997).

Wonders, K., *Habitat dioramas: illusions of wilderness in museums of natural history* (Uppsala, 1993).

Wood, M. (ed.), *Extracts from the records of the burgh of Edinburgh* (London, 1940).
*Extracts from the records of the burgh of Edinburgh, 1665–1680* (Edinburgh and London, 1950).
Wood, P., *The Aberdeen enlightenment: the arts curriculum in the eighteenth century* (Aberdeen 1993).
'Science, the universities and the public sphere in eighteenth-century Scotland', *History of the Universities*, 13 (1994), 99–135.
Wood, P. (ed.), *The Scottish enlightenment: essays in reinterpretation* (Rochester, 2000).
Wormald, J., 'James VI and I: two kings or one?', *History*, 68 (1983), 187–190.
'James VI and I, *Basilicon Doron* and *The Trew Law of Free Monarchies*: the Scottish context and the English translation', in Peck (ed.), *The mental world of the Jacobean court*, 36–54.
Wormald, L., 'The creation of Britain: multiple kingdome or core and colonies', *Transactions of the Royal Historical Society*, 6 (1992), 175–94.
Wright, J. N., and Snodgrass, N. S. (eds.), *Scotland and its people* (Edinburgh and London, 1942).
Yanni, C., *Nature's museum: Victorian science and the architecture of display* (London, 1999).
Yeoman, L., 'Maps, Melvilles and mischief: Pont in context' (Conference paper to Project Pont Seminar, National Library of Scotland, Edinburgh, 2 October 1996).
Young, J., 'The Scottish parliament and national identity from the union of the crowns to the union of the parliaments, 1603–1707', in Broun, Finlay and Lynch (eds.), *Image and identity*, 105–43.

# Index

*Italic* page numbers (e.g., *121*) refer to figures in the text

## *Cambridge Studies in Historical Geography*

1 Period and place: research methods in historical geography. *Edited by* ALAN R. H. BAKER and M. BILLINGE

2 The historical geography of Scotland since 1707: geographical aspects of modernisation. DAVID TURNOCK

3 Historical understanding in geography: an idealist approach. LEONARD GUELKE

4 English industrial cities of the nineteenth century: a social geography. R. J. DENNIS*

5 Explorations in historical geography: interpretative essays. *Edited by* ALAN R. H. BAKER and DEREK GREGORY

6 The tithe surveys of England and Wales. R. J. P. KAIN and H. C. PRINCE

7 Human territoriality: its theory and history. ROBERT DAVID SACK

8 The West Indies: patterns of development, culture and environmental change since 1492. DAVID WATTS*

9 The iconography of landscape: essays in the symbolic representation, design and use of past environments. *Edited by* DENIS COSGROVE and STEPHEN DANIELS*

10 Urban historical geography: recent progress in Britain and Germany. *Edited by* DIETRICH DENECKE *and* GARETH SHAW

11 An historical geography of modern Australia: the restive fringe. J. M. POWELL*

12 The sugar-cane industry: an historical geography from its origins to 1914. J. H. GALLOWAY

13 Poverty, ethnicity and the American city, 1840–1925: changing conceptions of the slum and ghetto. DAVID WARD*

14 Peasants, politicians and producers: the organisation of agriculture in France since 1918. M. C. CLEARY

15 The underdraining of farmland in England during the nineteenth century. A. D. M. PHILLIPS

16 Migration in Colonial Spanish America. *Edited by* DAVID ROBINSON

17 Urbanising Britain: essays on class and community in the nineteenth century. *Edited by* GERRY KEARNS and CHARLES W. J. WITHERS

18 Ideology and landscape in historical perspective: essays on the meanings of some places in the past. *Edited by* ALAN R. H. BAKER and GIDEON BIGER

19 Power and pauperism: the workhouse system, 1834–1884. FELIX DRIVER

20 Trade and urban development in Poland: an economic geography of Cracow from its origins to 1795. F. W. CARTER

21 An historical geography of France. XAVIER DE PLANHOL

22 Peasantry to capitalism: Western Östergötland in the nineteenth century. GÖRAN HOPPE and JOHN LANGTON

23 Agricultural revolution in England: the transformation of the agrarian economy, 1500–1850. MARK OVERTON*

24 Marc Bloch, sociology and geography: encountering changing disciplines. SUSAN W. FRIEDMAN

25 Land and society in Edwardian Britain. BRIAN SHORT

26 Deciphering global epidemics: analytical approaches to the disease records of world cities, 1888–1912. ANDREW CLIFF, PETER HAGGETT and MATTHEW SMALLMAN-RAYNOR*

27 Society in time and space: a geographical perspective on change. ROBERT A. DODGSHON*

28 Fraternity among the French peasantry: sociability and voluntary associations in the Loire valley, 1815–1914. ALAN R. H. BAKER

29 Imperial Visions: nationalist imagination and geographical expansion in the Russian Far East, 1840–1865. MARK BASSIN

30 Hong Kong as a global metropolis. DAVID M. MEYER

31 English seigniorial agriculture, 1250–1450. BRUCE M. S. CAMPBELL

32 Governmentality and the mastery of territory in nineteenth-century America. MATTHEW G. HANNAH*

*Titles marked with an asterisk * are available in paperback.*